Springer Series on Atomic, Optical, and Plasma Physics

Volume 75

For further volumes:
http://www.springer.com/series/411

The Springer Series on Atomic, Optical, and Plasma Physics covers in a comprehensive manner theory and experiment in the entire field of atoms and molecules and their interaction with electromagnetic radiation. Books in the series provide a rich source of new ideas and techniques with wide applications in fields such as chemistry, materials science, astrophysics, surface science, plasma technology, advanced optics, aeronomy, and engineering. Laser physics is a particular connecting theme that has provided much of the continuing impetus for new developments in the field, such as quantum computation and Bose-Einstein condensation. The purpose of the series is to cover the gap between standard undergraduate textbooks and the research literature with emphasis on the fundamental ideas, methods, techniques, and results in the field.

Hans Kleinpoppen · Bernd Lohmann
Alexei N. Grum-Grzhimailo

Perfect/Complete Scattering Experiments

Probing Quantum Mechanics on Atomic and Molecular Collisions and Coincidences

Springer

Hans Kleinpoppen
Berlin
Germany

Bernd Lohmann
Institut für Theoretische Physik
Westfälische Wilhelms-Universität Münster
Münster
Germany

Alexei N. Grum-Grzhimailo
Skobeltsyn Institute of Nuclear Physics
Lomonosov Moscow State University
Moscow
Russian Federation

ISSN 1615-5653 ISSN 2197-6791 (electronic)
ISBN 978-3-662-50079-8 ISBN 978-3-642-40514-3 (eBook)
DOI 10.1007/978-3-642-40514-3
Springer Heidelberg New York Dordrecht London

Printed on acid-free paper

Springer is part of Springer Science+Business Media (www.springer.com)

Herbert Walther

Foreword

The experimental and theoretical study of atomic and molecular collision processes has progressed rapidly in recent years. This is partly because a detailed knowledge of these processes is required in many applications and partly because a study of collision processes provides an ideal means of understanding the dynamics of many-particle systems at a quantum level. On the experimental side, advances include the absolute measurement of cross sections, the development of coincidence techniques using very high-resolution electron beams, and the application of light sources including synchrotron radiation and intense laser sources. On the theoretical side, developments include methods which allow highly accurate excitation and ionization cross sections to be calculated at low and intermediate energies. Of particular importance in this work is the close collaboration between experiment and theory where the accuracy of new theory and calculations can be verified by detailed experiments. The development of experimental methods and the close coordination between experiment and theory is clearly expressed in this book. In particular, the emphasis on experimental methods which provide data on perfect/complete scattering experiments enables a deep insight into the collision mechanism to be obtained which is used to provide a stringent test of theory. This has formed a basis for the development of theoretical methods, which have been used world-wide, to predict with confidence data for collision processes of importance in applications which cannot themselves be easily studied in the laboratory. Three areas of research are reviewed in this book: (i) the analysis of atomic collisions including electron, atom, and ion collisions with atoms and ions, (ii) Auger emission and inner-shell hole experiments, and (iii) complete experiments in atomic photoionization. Together, they provide an up-to-date reference for research in these fundamental areas of atomic and molecular collision processes which will have wide applicability.

Belfast Philip G. Burke

Preface

This book on *Perfect and-or Complete Scattering Experiments as a means for Probing Quantum Mechanics on Atomic and Molecular Collisions and Coincidences* is an approach to the theme of its title and also represents a Dedication to Herbert Walther who died on July 22, 2007 in Garching, Germany.

Herbert Walther's achievements in physics will best be described by the many events which have taken place and referred to him in the past. We particularly like to refer to Wolfgang Ketterle's and Gerhard Rempe's Memorial Article in Physics Today 79, June 2007; it describes well Herbert Walther's passion for his scientific work, his world-wide recognition, his strength and stimulating teaching ability, and also his way of organizing scientific meetings and conferences. It is clear that Herbert Walther will be remembered by a very large scientific community and, of course, by his immediate family. We, the Authors of this book, are most pleased and feel honored to dedicate this book to Herbert Walther.

Combining the random of data, atomic and molecular scattering physics, using and applying the up-to-date knowledge of the theoretical applications and the experimental methods reveals a wealth of information which even encounters strange and new areas of collision and coincidence physics. As a hint, we mention the field of quantum information which might be seen as the successor of polarization analysis.

We, the Authors of this book—Alexei Grum-Grzhimailo, Bernd Lohmann, and Hans Kleinpoppen—like to thank in the highest way many of our colleagues and friends for their continual supports and advices on the completion of this comprehensive book. We are particularly impressed by the effectiveness and completion of the research involved in the many sections and subsections. Many insights and knowledges on detailed atomic and molecular collision physics will be evident and apprehensive.

We are thankful to Prof. Uwe Becker, and Dr. Burkhard Langer and the GPTA GmbH company, for providing many of the experimental data, and for generous financial support to one of us, BL, which eventually enables for the finalization of this book.

Bernd Lohmann is particularly dearful to Claudia for the very most support while editing this book, and to ongrowing little Tanya watching the miracles happening around her. Both of them for standing tall, and, giving and providing all the necessities needed to complete and finalize this book.

Hans Kleinpoppen is particularly thankful to Mrs. Helga von Kosing for her constant concerns and helpfulness in dealings with editing this book.

During the final procedures of this book, we received the message of the passing away of Prof. Vsevolod V. Balashov, who has been a close collaborator, a colleague, a friend, and a supervisor over the years to all of us. We all, in the scientific community, will certainly miss him. This simply shows the speed of life and science.

Stirling, Berlin — Hans Kleinpoppen
Münster — Bernd Lohmann
Moscow — Alexei N. Grum-Grzhimailo

Contents

Chapter 1
Introductory Remarks

Structural analysis of atoms and molecules in their ground and excited states belong to topics of *atomic and molecular spectroscopy* including actions of static or dynamical fields on free atoms and molecules. Atoms, molecules, ions, electrons, photons can also collide with each other which causes a large series of phenomena that we call *atomic and molecular collision processes*.

Physically important properties of such collision processes are their development as a function of time, the various types of interactions between the colliding partners, the intensities of all these processes as a function of initial conditions, such as the relative energies of the colliding atomic particles, their mutual potential energies and their quantum numbers before and after the collision. The total entity of all these physical processes of atomic collisions is determined both by atomic structure as well as by parameters of the collision processes. While atomic and molecular structure is an important parameter in collision processes, atomic collision dynamics is the central problem of the physics of atomic collision processes.

A huge number of possible atomic and molecular collision processes exists; any arbitrary atom or ion can collide with another atom or ion or with an electron or photon. The energies of the colliding atomic particles can be chosen arbitrarily; after the collision, the particles involved propagate in all possible directions. The manifold character of collisions requires a classification and ordering system based upon physical processes; these will be described and developed up to a certain point. A more introductory description of atomic and molecular collision processes can be found in the specialized literature, e.g. Johnson (1982).

Since about the 1960s, the field of research on atomic and molecular collisions has advanced through a variety of new development of experimental and theoretical methods; these have led to a more detailed understanding of quantum-mechanical collision dynamics. In analogy with nuclear and elementary-particle physics, especially the *technology of coincidence and spin experiments* have contributed to the advancement of new knowledge of atomic and molecular collision physics. Applications of atomic and molecular collision processes are found in astrophysics, atmospheric physics, plasma physics, nuclear fusion physics and chemical reactions. We want to

H. Kleinpoppen et al., *Perfect/Complete Scattering Experiments*,
Springer Series on Atomic, Optical, and Plasma Physics 75,
DOI: 10.1007/978-3-642-40514-3_1, © Springer-Verlag Berlin Heidelberg 2013

emphasize that the present book also refers to the article in honor of Ben Bederson (Kleinpoppen et al. 2005) but does not claim to represent an approach to review on *Perfect and-or complete Scattering Experiments*; it only encloses a selection of topics of our personal choices. More detailed reviews are available and are partly listed in the references, e.g. Andersen and Bartschat (2000), Becker and Crowe (2001), Hanne et al. (2003).

The existing data of complete/perfect scattering experiments provide a deep insight into physical mechanisms of atomic and molecular scattering processes. They reveal what types of processes and interactions occur or compete with each other in the collisional process.

Scattering amplitudes and their phase differences and also atomic target parameters extracted from these type experiments have successfully been applied *as most sensitive tests of modern collision theories*. Coincidence and spin experiments do not, in selected cases, average or sum over atomic cross sections for various sub-processes or interactions of the collisional processes. Such coincidence and spin experiments resulting in collision amplitudes, phases and target parameters have been classified as *third generation* type of experiments going well beyond the more limited kind of information obtained from differential (*second generation* type of experiments) or total (*first generation* type of experiments) cross section measurements.

In a detailed analysis of complete/perfect atomic and molecular collisions we are dividing the task into the various sub-parts, namely on electron-photon coincidence experiments, on atomic and electron spin experiments and comparisons between electron and positron scattering.

Only recently a start has been made to combine spin and coincidence experiments; we will briefly refer to this newest development.

While Chap. 2 is devoted to the general analysis methods and the detection of the scattering residuals, i.e., electrons, ions, atoms, or molecules which might, if still excited, decay further by photon emission or another scattering process, we will focus on the discussion of nowadays angle and spin resolved Auger emission experiments and related research in Chap. 3. As has been pointed out in Chap. 2 such kind of experiments yield more refined information about the electron emission process, particularly the scattering phase. This can be seen as a step towards a complete experiment, i.e., to determine all elements of the describing density matrix, as has been first requested by Bederson (1969a, b). The basics of the Auger effect and the early experiments will be shortly reviewed. Ongoing, open shell atoms will be discussed, focusing on the angular distribution of the excited sodium KLL Auger transitions. The investigation of open shell atoms yields surprising results concerning the cross section, as well as the angular distribution, due to the open shell character. Eventually, the angle and spin resolved Auger emission analysis will be investigated, experimentally and numerically, stressing the example of the angle and spin resolved resonantly excited $Ar^*(2p_{1/2}^{-1}4s_{1/2})_{J=1}$ and $Ar^*(2p_{1/2}^{-1}3d_{3/2})_{J=1}$ states and their subsequent $L_2M_{2,3}M_{2,3}$ Auger emission in detail. For the generation of the intermediate excited Rydberg state two different mechanisms will be considered.

As a prerequisite to angle and spin polarized Auger emission a deep inner shell hole must be generated which should either be aligned and/or oriented with respect to its magnetic sublevels. Particularly, the process of photoexcitation reveals several advantages with respect to its theoretical description and, due to the availability of 3rd generation synchrotron beam sources, allows for a large variety of experimental approaches. Complementary, we will investigate the generation of these oriented and/or aligned hole states via photoionization, where particularly emphasis will be given to closed (sub-)shell atoms, e.g. the rare gases or the earth alkalis. The available numerical results will be compared to the experimental data. In addition, alignment and orientation of open-shell atoms are investigated, and, as a new field, orientation and alignment of radioactive elements are explored.

As an alternative process electron impact excitation of Auger states will be considered. Only a sparse number of experiments have been carried out applying this method. Though, it yields information about the Auger process which cannot be accessed in photoionization/excitation experiments.

A challenging task of performing quantum mechanically complete experiments is molecular photoionization. This is because the anisotropic molecular potential seen by the photoelectron causes an admixture of an unlimited number of outgoing partial waves. This process may be visualized for localized core electrons as partial wave mixing due to the intramolecular scattering of the photoelectron on its way out. This will be demonstrated by investigating the angle resolved molecular Auger emission from CO molecules.

The angle and spin resolved studies of the Auger decay in a combination with detection of the polarization state of the residual ion leads to complete experiments for the Auger decay; i.e. all the complex decay amplitudes can be determined. We will discuss such experiments and present their first results for both resonant and normal Auger processes. Having the validity of a two-step model of the reaction as a necessary prerequisite, the experiments for the Auger decay provide a showcase for complete experiments for so-called *half collision* processes.

Increasing brightness and wavelength tunability of synchrotron radiation sources, including those with variable polarization of the photon beam, allowed to extend considerably the photoionization studies. In combination with angle and spin resolved electron spectroscopy and laser techniques these studies developed to the level of complete photoionization experiment. Chapter 4 is concerned with phenomena related to complete experiments in photoionization. Atomic Auger decay and atomic photoionization have in common the fact that in both cases the process can be completely characterized by a small number of angle-independent partial wave amplitudes. The latter can be extracted from different kind of measurements: observation of angular distributions of photoelectrons from isotropic and polarized targets, spin polarization of photoelectrons, polarization of residual photoions, coincidence between photoelectron and secondary products of the photoionization. The discussion of the above phenomena is preceded by general theoretical introduction on complete experiment for photoionization. Special sections deal with nonresonant two-photon ionization, features of photoionization in the region of resonances and non-dipole effects in atomic photoionization.

Chapter 2
Analysis of Atomic Collisions

2.1 Classification of Atomic Collision Processes

An atomic collision process can be illustrated and classified by its geometry. An atomic particle A may collide with an atomic particle B. Particle A propagates in a collimated aligned beam of atoms, ions, electrons or photons. It may then hit particle B of an atomic target. Often the target B is also produced by a collimated beam of particles (*crossed-beam technique*).

Further classifications can be obtained as follows: In *elastic collision processes* the energy states of the collision partners are unchanged, e.g. both collision partners remain in their ground states. According to the laws of mechanics, however, kinetic energy can be transferred between the colliding particles. It is common to include electron spin exchange in elastic processes. In *inelastic processes* the energy state of one or both colliding partners can be changed; for example, the kinetic energy of the projectile can be transferred into excitation energy of the target atom. During such processes, kinetic energy can also be used to excite both atoms (A^*, B^*), i.e.,

$$\begin{aligned} A + B &\longrightarrow A^* + B \qquad \text{or} \\ &\longrightarrow A + B^* \qquad \text{or} \\ &\longrightarrow A^* + B^* \end{aligned} \tag{2.1}$$

In *reactive processes* between primary collision partners, other forms of matter such as molecules, ionized particles or photons may be produced, e.g.,

$$A + B \longrightarrow AB + h\nu \tag{2.2}$$

or

$$A + B \longrightarrow A^{n+} + B^{m+} + (m+n)e^-, \tag{2.3}$$

or even exotic atoms, in a special reaction

H. Kleinpoppen et al., *Perfect/Complete Scattering Experiments*,
Springer Series on Atomic, Optical, and Plasma Physics 75,
DOI: 10.1007/978-3-642-40514-3_2, © Springer-Verlag Berlin Heidelberg 2013

$$e^+ + H \longrightarrow p + Ps. \tag{2.4}$$

Examples of decay reactions will be discussed in the following sections in various ways.

Collision processes between electrons and atoms can be studied experimentally with primary electron currents from about 10^{-5} to 10^{-8} A; investigations with high energy resolution ($\Delta E \leq 0.1$ eV) necessarily restrict current intensities. Since the middle of the 1970s, collision processes have also been carried out with positive electrons e^+, the positrons. Such experiments were first carried out with about a primary intensity of one positron per second and have now reached about 10^3–10^5 e^+/s. In a few years, positron currents of about 10^9 e^+/s may be available. In spite of the low current intensities, noteworthy results have already been obtained in the physics of positron scattering by atoms. Here, we report physically interesting similarities and differences between electron and positron scattering.

Historically the physics of electron scattering began around the start of the last century when Lenard (1903) fired cathode rays of a cathode tube through a thin aluminium foil of 1/100 mm thickness; behind the foil Lenard measured the absorption of the cathode rays on passing through various gases. The mean-free-path length of the cathode rays behind the foil under atmospheric conditions was determined to be about 0.5 cm which is about 10^4 to 10^5 larger than that for atoms in atomic gases. Lenard concluded that atoms must be very empty since electrons are absorbed by atoms to a very small extent. This early insight into the properties of atoms was confirmed about ten years later by the Bohr-Rutherford model of the atom.

Another fundamentally important measurement is based upon the historically famous *Franck-Hertz experiment* (Franck and Hertz 1914) in which the energy relation $h\nu = eV$ for the excitation of an atomic state by electrons of energy eV and the subsequent emission of photons of energy $h\nu$ has been confirmed. We do not discuss further details of the Franck-Hertz experiment but refer to the specialized literature (Hanne 1988), in which the above energy relation is treated in connection with modern electron-photon coincidence experiments.

Before we describe various electron collision processes in the following sections, we discuss some qualitative physical facts. We begin with low-energy, elastic electron-atom collisions in which the target atom remains in the ground state. In spite of the obvious simplicity of this collision, a series of rather complicated processes may appear in elastic low-energy scattering, since the electron shell of the atom may be deformed by the passing electron. On the other hand, this distortion of the electron shell reacts on the projectile electron, whereby a polarization interaction between the electron and the atom occurs, which affects the structure of the cross section (for example in the *Ramsauer-Townsend effect*). When the energy of the incoming electron increases so that its velocity becomes comparable to that of the outer electrons of the atom, another phenomenon may arise, namely the formation of negative ions. The existence of such, sometimes very short lived, negative ions manifests itself in sharp *resonance structures* of the electron-atom cross sections. Further complications in electron-atom collisions are due to the existence of the spins of the projectile and atomic electrons. As a consequence, *spin polarization effects* are observable in electron-atom scattering. Inelastic electron-atom scattering takes place with excitation and ionization of atoms in which metastable atoms, photons, ions and secondary electrons may arise.

Analyzing these "collisional products" by means of coincidence and spin experiments provides information on quantum mechanical amplitudes which may completely describe the collision process, i.e. maximum information on the electron collision may be extracted from such experiments. We also refer to recently extended investigations on elastic and inelastic electron-ion collisions. In addition, a projectile electron can be captured by an ion in such a collision. This process is called *recombination* which plays an important role, for example, in the outer atmosphere of the sun, in the solar corona and in dense plasmas for the production of nuclear fusions.

In the above classifications we have assumed that the magnetic components m_J of the quantum number J of the states of the particles A and B are equally distributed statistically. In accordance with this distribution, the collision processes are characterized by averaging over all quantum states of the components m_J. However, because of recent progress in experimental techniques, studies of atomic collision processes can be carried out and analyzed in which the colliding partners A and B are initially in pure quantum mechanical states. Initial proposals for such types of experiments, called *complete* or *perfect*, can be traced back to the pioneering papers by Fano (1957), Bederson (1969a, b), and Kleinpoppen (1967, 1971). It follows from these papers, for example, that particles A and B may be in the quantum states $|n_A J_A m_A\rangle$ and $|n_B J_B m_B\rangle$ with the corresponding quantum numbers n, J, and m for the two particles. The interaction process of the collision with the particles in pure quantum states can, at least in principle, be described by a quantum mechanical Hamilton operator H^{int}, which is determined by the interaction potential between the colliding partners. As a consequence of the linearity of the Schrödinger equation, the total system of the particles after the collision will also be in pure quantum states. In other words, we can represent atomic collisions between atomic particles in pure quantum states as follows:

$$\boxed{|\psi_{in}\rangle = |A\rangle|B\rangle} \xrightarrow[\text{linear operator } H^{int}]{} \boxed{|\psi_{out}\rangle = |C\rangle|D\rangle\cdots|K\rangle\cdots}. \tag{2.5}$$

Before the collision, the colliding particles are in the joint quantum state $|\psi_{in}\rangle$, after the collision the *collisional products* are in the state $|\psi_{out}\rangle$.

If the state vector $|\psi_{out}\rangle$ after the collision has been extracted from an appropriate experiment, it may be described by applying the quantum mechanical superposition principle in the form

$$|\psi_{out}\rangle = \sum_m f_m |\psi_m\rangle, \tag{2.6}$$

where ψ_m are wave functions of possible substates of the state vector $|\psi_{out}\rangle$ and f_m are complex amplitudes associated with the collision process. The extraction of the state vector $|\psi_{out}\rangle$ represents the maximum of information and knowledge that can be extracted from the experimental analysis of the collision process. Measurements of all the amplitudes associated with the collision process is equivalent to performing a *complete experiment*.

On one hand, complete experiments on atomic and molecular collision processes require a high degree of experimental effort and special methods, which have only recently been successfully applied. On the other hand, they also require advanced theoretical models to describe the dynamics in a most detailed way.

2.2 Approaches to Complete/Perfect Scattering Experiments

Correlation and coincidence experiments in electronic, atomic and molecular collisions (including photoionization and spin effects) require to determine fundamental *quantum scattering amplitudes and phases* or alternatively, *irreducible tensor operators, state multipoles* or *statistical tensors* and *expectation values of angular momenta*. The state of the art was comprehensively and efficiently reviewed by Andersen and Bartschat (2000) and by Williams (2000) relating the experimental data to atomic and molecular collision processes and theories.

A typical density matrix ρ (Fig. 2.1) illustrates the state of the art for the specific case of the $n = 2$ excitation of hydrogen atoms from the ground state (Blum and Kleinpoppen 1979):

$$\Gamma_0 = |n_0 M_{S_0}, p_0 m_0\rangle \longrightarrow \Gamma_1 = |n_1 L M_L M_{S_1}, p_1 m_1\rangle; \tag{2.7}$$

the corresponding scattering amplitude is $f(\Gamma_1, \Gamma_0)$ while n_0 and n_1 denote the initial and final principal quantum numbers with the usual meaning of the quantum numbers; $\sigma(LM_L) = <|f_{LM_L}|^2>$ is the differential cross section of levels with the quantum numbers LM_L averaged over all spins (the dependence on n_1, p_1, n_0, and p_0 is suppressed). The quantities below the matrix are a selection of the

$$\rho = \left(\begin{array}{c|ccc} \sigma(00) & \langle f_{00} f_{11}^* \rangle & \langle f_{00} f_{10}^* \rangle & \langle f_{00} f_{1-1}^* \rangle \\ \hline \langle f_{11} f_{00}^* \rangle & \sigma(11) & \langle f_{11} f_{10}^* \rangle & \langle f_{11} f_{1-1}^* \rangle \\ \langle f_{10} f_{00}^* \rangle & \langle f_{10} f_{11}^* \rangle & \sigma(10) & \langle f_{10} f_{1-1}^* \rangle \\ \langle f_{1-1} f_{00}^* \rangle & \langle f_{1-1} f_{11}^* \rangle & \langle f_{1-1} f_{10}^* \rangle & \sigma(1-1) \end{array}\right)$$

$$\boxed{\lambda, \chi, P, \mu, \beta, O^{Col}, A^{Col}, T_{KQ}}$$

Fig. 2.1 Density matrix for excitation of atomic hydrogen from the ground state to the $n = 2$ state. The quantities below the matrix are data extracted from a *complete/perfect collision experiment* (Blum and Kleinpoppen 1979; Kleinpoppen and Williams 1980, see text)

typical physical quantities determined by *complete experiments* which are described in further sub-sections.

The density matrix is divided into four sub-matrices. The upper one consists only of one element, the differential cross section for the 2s state excitation. The quadratic sub-matrix contains all elements characterizing the 2p excitation with the definite angular momentum $L = 1$ (2p sub-matrix). Its diagonal elements are the differential cross sections for exciting the various magnetic sub-levels, its off-diagonal elements characterize the coherence between the different magnetic substates. The remaining elements of the matrix characterize the interference between the 2s- and 2p states.

This fundamental matrix has been applied particularly in connection with beam-foil and atomic collision experiments.

Before we describe selected examples of *complete/perfect* scattering experiments for atomic and molecular processes we like to consider some basic scattering events for reference and comparison.

2.3 Basic Atomic and Molecular Scattering Processes

As mentioned previously collision experiments of the *first* and *second types* are primarily related to measurements and relevant theories on total and differential cross sections of which Fig. 2.2 illustrates the general geometric schemes.

The present section picks out examples of Basic Atomic and Molecular Scattering Processes which are particularly appropriate to demonstrate transitions from Basic Scattering to processes of the analysis by Perfect/Complete Collision Experiments.

In this book important experimental and theoretical technologies will generally be described in short sub-sections with references to more comprehensive literature.

In recent years applications of so-called synchrotron radiation arising from orbiting electrons of an accelerator played important roles in atomic spectroscopy and collision physics. The energy of photons of synchrotron radiation ranges from about 10^{-10} eV to 100 keV.

2.3.1 Total and Differential Cross Section

As its name indicates, the cross section is an effective area which is one of the associated physical quantities that describes the collision process. The **total cross section** can be defined and measured as follows. A beam of particles A (see Fig. 2.2a) with (ideally!) equal velocities and direction of motion passes through the target volume of particles B. Normally, the target particles B consist of free atoms at low or room temperature, so that their velocities are small in comparison to those of the projectile particles. In order to carry out a quantitative analysis of the collision process, we first introduce the intensity I of the particle current as the number of particles A per area and time. We will also assume that due to the collision along

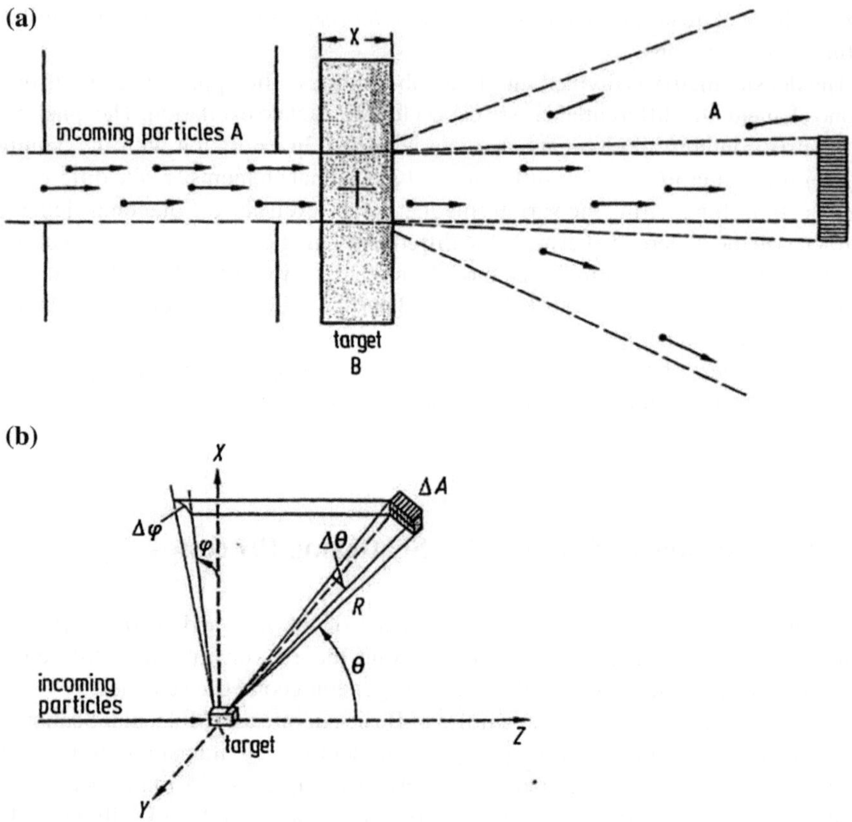

Fig. 2.2 **a** Scattering of particles A by the target particles B; **b** experimental geometry for measuring the differential cross section with the detector area $\Delta A = (R\Delta\theta)(R\sin\theta\Delta\phi)$

the target of length x the primary intensity has been reduced to the quantity $I(x)$ after the target. By varying the target thickness by a small amount Δx, the intensity behind the target will be changed by

$$\Delta I = -Qn_B I(x)\Delta x, \tag{2.8}$$

with n_B as atomic density of the target and the proportionality factor Q as total *cross section* with the dimensions of area. For the limits $\Delta I \to 0$ and $\Delta x \to 0$ we then obtain the characteristic differential equation for an exponential decay with the solution

$$I(x) = I(0)\mathrm{e}^{-Qnx}. \tag{2.9}$$

Knowing the target density n_B (which we will denote by n in order to simplify the equations involved) and the length multiplied by the total cross section Q can be determined experimentally by measuring the intensity of the incoming particle beam in front of and behind the target. If N_0 particles A cross the target, the number Z of collision processes can be calculated from the difference of the particle number before and after the target as

$$Z = N_0 - N(x) = N_0(1 - \mathrm{e}^{-nQx}), \tag{2.10}$$

and by considering the approximation $nQx \ll 1$ we obtain

$$Z = N_0 nQx. \tag{2.11}$$

The ratio $Z/N_0 = nQx$ is approximately the probability that a particle A experiences a collision process characterized by the total cross section Q. Total cross sections of atomic collisions cover a range from about 10^{-14} to $10^{-20}\,\mathrm{cm}^2$ depending, of course, on the type and conditions of the processes.

If the power of the exponential term in (2.9) becomes unity we obtain the definition of the *mean-free-path length* λ of atoms with its density n and cross section Q, i.e. $\lambda = 1/nQ$. It is normally required in experiments with "free atoms" that the dimensions of the experimental apparatus are smaller than the mean-free-path length. Therefore such high-vacuum conditions of the apparatus are required (i.e., the residual gas pressure small enough!) that the atoms of colliding atomic beams do not collide with the residual gas atoms and molecules. The mean-free-path length for a typical total cross section of $3 \times 10^{-15}\,\mathrm{cm}^2$ and residual gas pressure of 10^{-2} Pa is about 1 m. A residual gas pressure of lower than 10^{-2} Pa is therefore required in typical atomic beam experiments.

Particles A can be scattered out of the primary beam due to the collisional interaction with the target atoms B. The angular dependent intensity of the scattered atoms is characterized by the differential cross section σ which we define according to Fig. 2.2b as follows. We differentiate the total cross section Q with respect to the solid angle $\mathrm{d}\Omega$, i.e. $\mathrm{d}Q/\mathrm{d}\Omega = \sigma$; we measure the differential cross section σ by the relation

$$\frac{\text{number of particles scattered into } \mathrm{d}\Omega}{\text{number of incoming particles}} = \frac{\mathrm{d}Q}{\mathrm{d}\Omega}\mathrm{d}\Omega = \sigma \mathrm{d}\Omega, \tag{2.12}$$

related to the same time interval Δt for measuring the number of the incoming and scattered particles. We assume that the total cross section depends only on the energy, i.e., $Q = Q(E)$. On the other hand the differential cross section will also depend on the solid angle Ω, i.e., in general on the polar angle θ and the azimuthal angle ϕ. The functional representation of the differential cross section is given by $\sigma(E, \theta, \phi)$. Integration of the differential cross section over the total solid angle range results in the total cross section

$$Q(E) = \int \sigma(E, \theta, \phi)\mathrm{d}\Omega. \tag{2.13}$$

Assuming that the scattering intensity is symmetric around the direction of the incoming particles the differential cross section becomes independent of the azimuthal angle ϕ and the integral can be split up to

$$Q(E) = \int_0^{2\pi} d\phi \int_0^{\pi} \sigma(E, \theta) \sin\theta d\theta = 2\pi \int_0^{\pi} \sigma(E, \theta) \sin\theta d\theta. \tag{2.14}$$

We are going to use these definitions for the total and differential cross sections for elastic, inelastic and reactive collision processes as introduced above. Examples will be given in the following sections.

2.3.2 Analysis of Atomic Photoionization and Scattering Processes

Photoionization of atoms is usually included as part of atomic collision physics; obviously the photons as primary "colliding" particles vanish after completion of the process, i.e.,

$$h\nu + A \longrightarrow A^+ + e^-. \tag{2.15}$$

In high-energy physics, photons are normally represented by the symbol γ whereas *energy quanta* $h\nu$ are more often used in atomic physics. Occasionally, photoionization processes with atoms are called *half collisions*; the following consideration explains this: A collision between two particles may be divided into a possible dynamical approximation process with the formation of a (excited, short-lived) collision complex and a dynamical decay of this complex that asymptotically goes over into outgoing particles, which can be exemplified by the collision of an electron with an ion A^+, i.e.

$$e^- + A^+ \longrightarrow A^* \longrightarrow A^+ + e^-. \tag{2.16}$$

The photoexcitation of an atom then only provides the second part of the above process

$$h\nu + A \longrightarrow A^* \longrightarrow A^+ + e^-. \tag{2.17}$$

Development of research on atomic photoionization spans the period from the discovery of the photoelectric effect by Hertz (1887) about a hundred years ago to the measurements of total cross sections, to angular dependencies of photoelectrons (differential cross sections) and to "complete" experiments with spin-polarized atoms and spin-polarized photoelectrons. A further new development is due to the availability of intense laser radiation, by which *multiphoton ionization processes* have been discovered; these can be described as follows,

$$nh\nu + A \longrightarrow A^+ + e^-. \tag{2.18}$$

In this reaction, we assume that the energy $h\nu$ of a single photon of the laser beam is smaller than the required ionization energy, i.e. $h\nu < E_{ion}$. Ionization may take place if $nh\nu > E_{ion}$; it implies that the process of ionization is made possible through the common action of n photons. Furthermore, pulsed and very intense laser radiation have been applied to ionize atoms several times by multiple ionization, i.e.,

$$nh\nu + A \longrightarrow A^{m+} + me^{-}. \tag{2.19}$$

By applying energetic synchrotron or laser radiation *multiple ionization processes* can also be induced by single photons,

$$h\nu + A \longrightarrow A^{m+} + me^{-}. \tag{2.20}$$

We will now describe measuring methods and report on a selection of experimental photoionization data; comparisons with theoretical calculations are also given.

2.3.3 Experimental Methods for Measuring Photoionization Cross Sections Including Applications of Synchrotron Radiation

Photoabsorption Technique

A parallel beam of light of frequency ν and intensity I_0 may be directed onto an atomic target of length x. The reduction of the intensity behind the target is given by (2.9), $I(x) = I_0 e^{-\alpha x}$ with the absorption coefficient $\alpha = nQ$, the density of the target n and the total cross section $Q = Q(h\nu)$. This technique is simple, but it cannot be used to distinguish unambiguously between single and multiple ionization processes, as discussed above. A photograph (Fig. 2.3a) of the photoionization spectrum of helium and an absolute calibration for α would enable one to determine the cross section Q. Application of photoionization spectroscopy over photon energies from about 10^{-10} eV to 100 keV revealed interesting atomic structure and collision effects; Fig. 2.3 demonstrates such collision effects in the photon energy range from 19 to 21 nm of the incoming photoionization radiation. The competing processes can occur as a direct process

$$h\nu + \mathrm{He} \longrightarrow \mathrm{He}^{+} + e^{-}, \tag{2.21}$$

and as double excitation process

$$h\nu + \mathrm{He} \longrightarrow \mathrm{He}^{**}(n_1 j_1, n_2 j_2) \longrightarrow \mathrm{He}^{+} + e^{-}. \tag{2.22}$$

These doubly excited states He** have sharp, discrete energy values in the photoabsorption. The eigenfunction of these "resonance states" can be written by a wavefunction

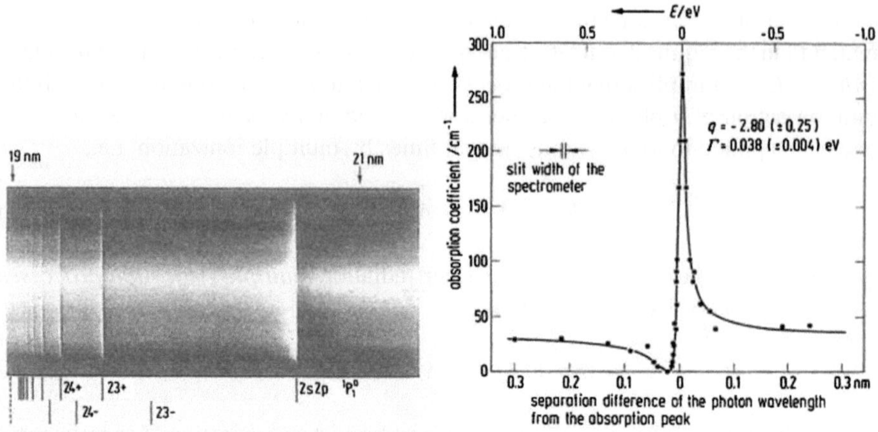

Fig. 2.3 (*left*) Absorption spectrum of helium in the spectral range from about 19 to 21 nm. The pairs of numbers below the resonances refer to the associated electron configurations 2snp and 2pns with the signs + and − for the mixed states. (*right*) Absorption profile of the resonance curve from the double excitation 2s2p $^1P_1$ of helium at about 20.7 nm. The *full curve* represents a fitted Fano profile; q is the Fano parameter; Γ is the half value width of the resonance curve (after Madden and Codling 1965)

$$\psi(\mathrm{sp}, 2n\pm) = \frac{1}{\sqrt{2}}\big(u(2\mathrm{s}n\mathrm{p}) \pm u(2\mathrm{p}n\mathrm{s})\big). \tag{2.23}$$

The pairs of numbers at the bottom of Fig. 2.3a refer to the associated electron configurations 2snp and 2pns with the + and − sign, respectively (Madden and Codling 1965). It can be proved theoretically that transitions to excited states with + signs are more probable than those with − signs. The most intense first resonance line with the lowest photon energy is the absorption spectrum which is associated with the electron configuration 2s2p $^1P_1$ in connection with the electron transition $1s^2\ ^1S_0 \rightarrow 2s2p\ ^1P_1$. Resonance lines with shorter wave lengths are linked to *spectral series* with the electron configurations $\psi(\mathrm{sp}, 2n\pm)$ as indicated above. Figure 2.3b shows the absorption profile from the double excitation 2s2p $^1P_1$ of helium at about 20.7 nm. The shape of this line is a coherent superposition of the direct photoionization process $h\nu + \mathrm{He} \rightarrow \mathrm{He}^+ + e^-$ and the double excitation process $h\nu + \mathrm{He} \rightarrow \mathrm{He}^{**}$. This coherence process is only noticeable at the sharp energy values of the resonances embedded in the continuum of the photoabsorption of the helium atoms; this phenomenon has similarity to the shape of the absorption lines (called **Beutler lines**) observed by Beutler (1934, 1935) in argon, krypton and xenon. The typical *resonance profile* of these lines has been described quantitatively by Fano (1961) as an interference spectrum between the background process (i.e. the direct photoionization) and the resonance absorption. The *Fano resonance profile of the Beutler lines* deviates from the Lorentzian curves of electric and magnetic dipole transitions. The Fano profile of resonance processes also plays an important role in atomic collision processes; e.g. see Figs. 2.26 and 2.27 later on.

Ion Collection and Charge Analysis

Figure 2.4 illustrates the method of ion collection and charge analysis by means of a time-of-flight spectrometer. The ions produced by the photo process leave the target, being accelerated by negatively charged electrodes; the ions enter the ion separator in which ions of different charges are separated from each other. A time-of-flight spectrometer analyses the time differences in the arrivals of ions with different charges after being accelerated to the energy $E = n_e E_0$ (n_e is the charge number, E_0 the energy of the singly charged ion) and passing through a drift area; a time analysis of the arriving ions at the ion detector (a channel electron multiplier, for example) marks the ion states Q^+, Q^{++}, $Q^{3+}, \ldots, Q^{m+}$ of single and multiple photoionization as a function of the energy $h\nu$ of the impinging photons.

As examples, we show applications of the single and multiple photoionization of xenon atoms induced by synchrotron and laser radiation.

In reactions $h\nu + \text{Xe} \to \text{Xe}^{n+} + ne^-$ with synchrotron photons of energies in the keV range, xenon atoms have been ionized, so far, up to the charge state ne with $n = 14$ (see Fig. 2.5). It is interesting to observe such a high degree of multiply charged ions of xenon atoms; other atoms, of course, can also be ionized into highly charged states in such photoionization processes. However, we do not discuss this here in detail.

A considerable contribution of photoelectrons may result from multiple ionization of deeper electron shells such as the M and L or even the K shells. Xenon has eight electrons in the outer O shell and 18 electrons in the next, deeper N shell. It follows from the analysis of the charge spectra of xenon, that a mean charge $\langle q \rangle = \sum q F_q$ can be extracted from the data in which F_q is the fraction of the observed ions with charge q. Figure 2.6 shows the mean charge as a function of the photon energy. If the photon energy is below the ionization threshold, energy for the L subshell, the mean charge has approximately the value of $\langle q \rangle = 6$ charge units. If the photon energy passes this threshold $\langle q \rangle$ increases in a jump by 1.2 charge units. Furthermore, $\langle q \rangle$ increases by about 0.2 charge units when the photon energies pass through the

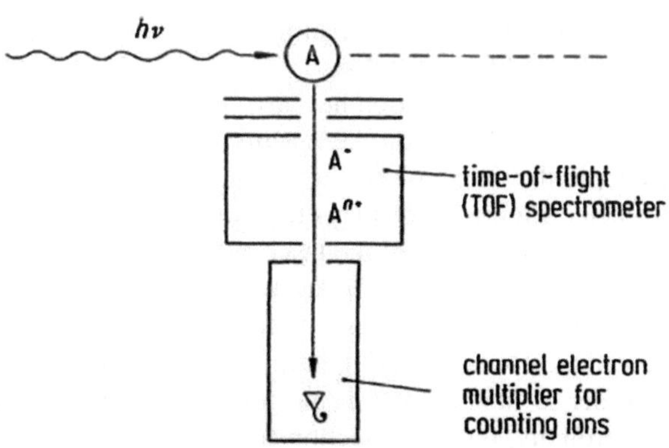

Fig. 2.4 Ion charge analysis by means of a time-of-flight (TOF) spectrometer and ion detection by a channel electron multiplier

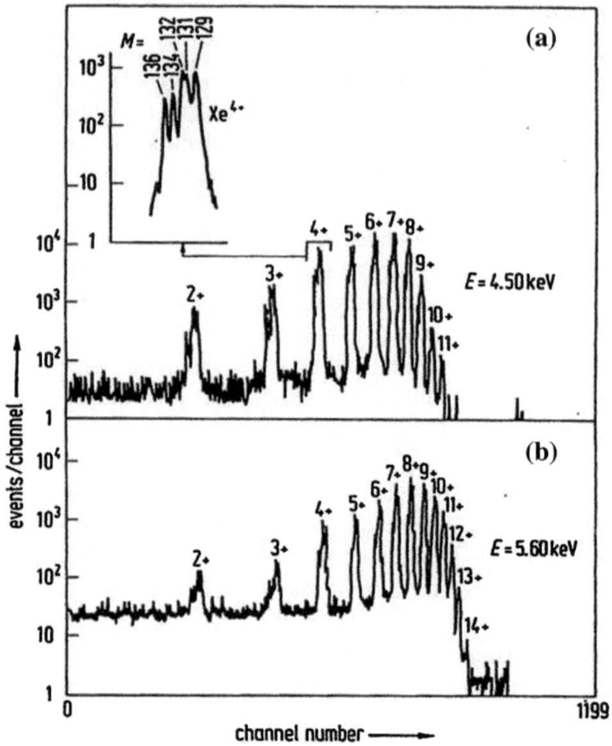

Fig. 2.5 **a** Time-of-flight charge spectrum of Xe ions with positive charges from Xe^{2+} to Xe^{11+} produced by synchrotron radiation of photon energy $h\nu = 4.50\,\text{keV}$ and a neutral Xe target; the enlarged spectrum above at the top left-hand corner shows the mass numbers of various isotopes for Xe^{4+}; **b** charge spectrum of Xe^{2+} up to Xe^{14+} at a photon energy of $h\nu = 5.60\,\text{keV}$ (after Tonuma et al. 1987)

binding energies of the L_2 and L_1 subshells. These facts, concerning the mean charge, indicate that the photoionization cross section of the L_3-subshell is larger than those of the L_2 and L_1 subshells.

Intense laser radiation has also been applied for production of multiply-charged xenon ions based upon the reaction $nh\nu + \text{Xe} \rightarrow \text{Xe}^{m+} + me^-$. As in the above example the detection of the xenon ions was carried out by a time-of-flight spectrometer. Figure 2.7 displays the yields of ions produced as a function of the laser intensity; the energy of the laser radiation was $h\nu = 2.33\,\text{eV}$ with a wavelength of $\lambda = 532.2\,\text{nm}$. As can be seen in the double logarithmic format of Fig. 2.7a, the ion yield for Xe^+ increases with the sixth power of the laser intensity; this is in accordance with the fact that at least six photons of energy 2.33 eV are necessary in order to liberate one electron of the outer shell of the xenon atom (Fig. 2.7b). Accordingly, the cross section for this multiple photoionization is proportional to the sixth power of the laser intensity, i.e. $Q(\text{Xe}^+, 6h\nu) \sim I_0(h\nu)$. Above the

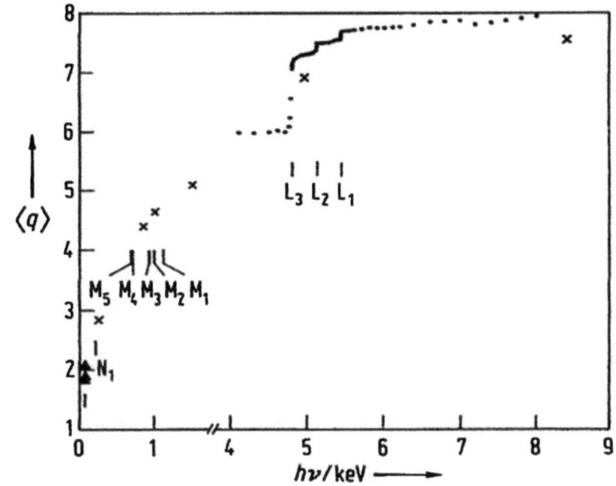

Fig. 2.6 Average charge $\langle q \rangle$ of Xe ions as a function of the photon energy of the synchrotron radiation inducing ionization of neutral xenon. The *short vertical lines* indicate the positions of the threshold energies of the L, M and N shells of xenon (after Tonuma et al. 1987)

intensity I_s, called saturation intensity, as indicated by the vertical line in Fig. 2.7a, the measured intensity deviates from that of the sixth power law of the laser intensity. This is due to the fact that for very high intensities, the density of the atoms is considerably reduced by the high degree of ionization. Two processes can lead to double ionization Xe^{++}, namely, (1) a direct process with 15 photons, i.e. $15h\nu + \mathrm{Xe} \rightarrow Xe^{2+} + 2e^-$ and (2) a two-step process according to the scheme $6h\nu + \mathrm{Xe} \rightarrow Xe^+ + e^-$ and $10h\nu + \mathrm{Xe}^+ \rightarrow Xe^{2+} + e^-$ for which 16 photons are required.

The saturation intensity for Xe^{2+} ions occurs almost precisely at the same energy as for Xe^+ ions which is in accordance with the production mechanism (1). A sudden increase in the number of Xe^{2+} ions takes place at $I = 1.5 \times 10^{12}\,\mathrm{W/cm^2}$; the measured shape of this part of Xe^{2+} yield results in 11 ± 1 photons, in accordance with the second step of process (2). When the laser intensity exceeds $1.5 \times 10^{12}\,\mathrm{W/cm^2}$, the two-step process appears to be saturated since a considerable fraction of the Xe^+ ion density in the target has been de-populated by the second ionization step. No reliable statements can be made from the data extracted from Fig. 2.7a for the production mechanisms of Xe^{3+} and Xe^{4+}. At least $15 + 33 = 48$ photons are required for the direct process of production of Xe^{4+} ions.

With the advent of free-electron lasers (FEL) operating in self-amplified spontaneous emission (SASE) mode, the reactions (2.19) have been observed in the extreme ultraviolet and X-ray wavelength range. The charge states with $m = 21$ in xenon atoms (Sorokin et al. 2007) and the $m = 10$ states in neon atoms (Young et al. 2010), i.e. the bare neon nuclei, have been observed by the method of ion collection using photons of energy 93 eV and 2 keV, respectively.

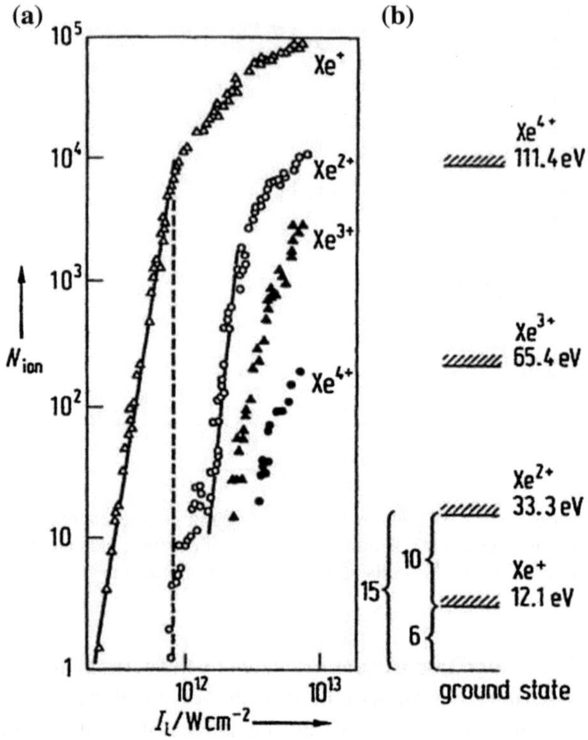

Fig. 2.7 **a** Doubly logarithmic representation of the number of Xe ions (in arbitrary units) as a function of the laser intensity I_L. The *vertical dashed line* marks the saturation intensity for the Xe^+ and Xe^{2+} ions (see text). **b** Schematic representation of the 15-photon-one-step and (6–10)-photon-two-step process for the production of Xe^{2+} ions (after L'Huiller et al. 1983)

Electron Collection and Photoelectron Spectroscopy

By applying this technique, photoelectrons of a photoionization process can be studied with regard to their energy and direction of the electron ejection (Fig. 2.8). Measuring the electron energy gives information concerning the electron shell or subshell from which the photoelectron originates. The angular distribution of the photoelectrons is determined by the angular momentum balance of the electron for the transition from the bound atomic state with the quantum numbers n_i, ℓ_i into the unbound, free state $\varepsilon\ell_f$ (ε energy and ℓ_f angular momentum quantum number of the photoelectron); according to the selection rule $\Delta\ell = \pm 1$ for electric dipole transitions induced by single photons, ℓ_f can have the two values $\ell_f = \ell_i \pm 1$. The angular distribution of photoelectrons is usually described by the expression

$$\sigma(E_{h\nu}, \theta) = \frac{Q(E_{h\nu})}{4\pi}\Big(1 + \beta P_2(\cos\theta)\Big). \tag{2.24}$$

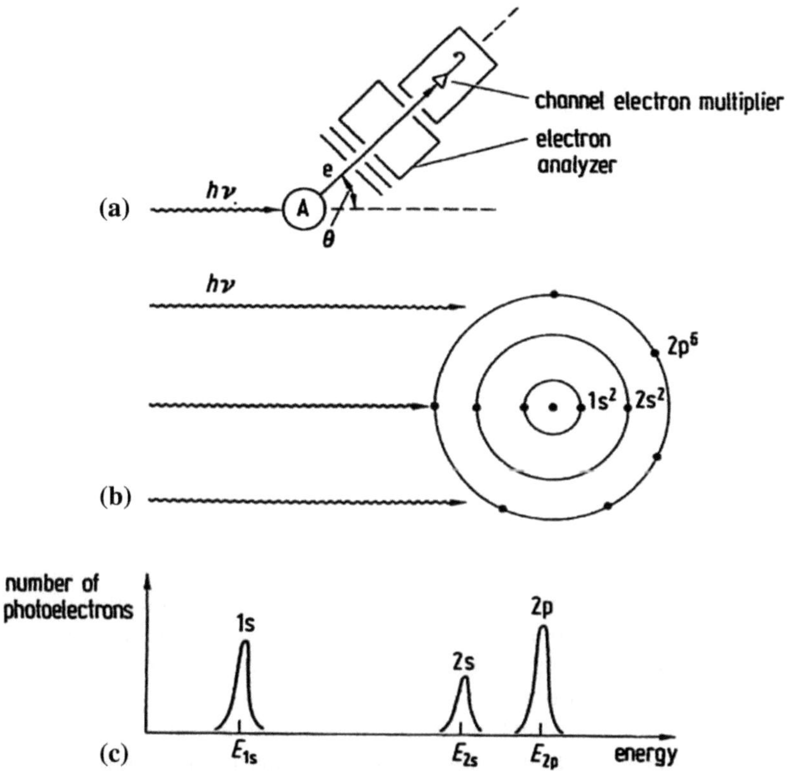

Fig. 2.8 **a** Detection of photoelectrons as a function of their energy by means of an electron energy analyzer positioned at an observation angle θ. **b** Pictorial representation of photoionization of atoms from inner shells. **c** Energy spectrum of photoelectrons (photoelectron spectroscopy) from the 1s, 2s and 2p shells of an atom

Here, $\sigma(E_{h\nu}, \theta)$ is the differential cross section for the emission of the photoelectron, under the angle θ with reference to the direction of the electric field of the linearly polarized incoming photons, $Q(E_{h\nu})$ the total photoionization cross section, $E_{h\nu}$ the energy of the incoming photons, the second Legendre polynomial $P_2(\cos\theta) = \frac{1}{2}(3\cos^2\theta - 1)$ and β is the asymmetry parameter which depends on the above angular momentum balance, the partial cross sections Q_{ℓ_i+1} and Q_{ℓ_i-1} for the photoelectrons in the $\ell_i \pm 1$ angular momenta states, and the phase differences between relevant amplitudes of the photoionization process. We restrict our discussion here to a clear-cut example, see Fig. 2.9, namely the angular dependence of the photoelectrons of the photoionization of helium in the ground state $\mathrm{He}(1s^2\,{}^1S_0)$, i.e., the process $1s^2 \to 1s\varepsilon p$ according to the above way of describing photoionization processes with a free photoelectron in the εp state which lies in the ionization continuum of energy ε and the orbital angular momentum quantum number $\ell = 1$ (p state). In accordance with (2.24) the angular dependence of photoelectrons for $\beta = 2$ is described by

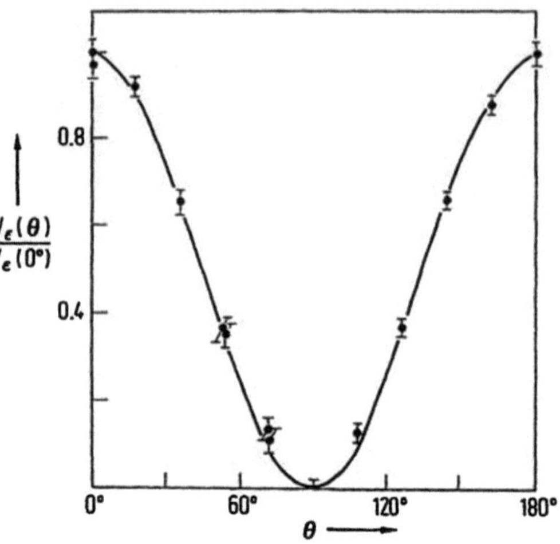

Fig. 2.9 Angular distribution of photoelectrons of the photoprocess $1s^2 \rightarrow 1s\varepsilon p$ in the helium atom; $\varepsilon = 0.5$ eV is the energy of the photoelectron, i.e. the photon energy of the linearly polarized radiation of the Glasgow 340 MeV electron synchrotron was 25.08 eV (ionization energy of helium 24.58 eV). Experimental data with error bars after Watson and Stuart (1974); the *full curve* represents the $\beta = 2$, $\cos^2\theta$ relation of (2.24)

a $\cos^2\theta$ function, the maximum of which is in the direction of the electric field strength of the linearly polarized radiation inducing the photoionization (this case is obviously fulfilled for helium in Fig. 2.9). Values of $\beta \neq 2$ are associated with more complicated atoms; the β parameter of such atoms provides some information on many-electron effects, for example electron correlations, spin-orbit effects etc.; we refer to more advanced literature for detailed descriptions as well as to Chap. 4 where complete experiments in atomic photoionization are discussed. Instead, we mention angular dependent effects of photoelectrons from multiple photon ionization of atoms irradiated by intense lasers. As an example, we consider the process $nh\nu + \text{Xe} \rightarrow \text{Xe}^+ + e^-$. Figures 2.10 and 2.11 describe schematically this process by applying the spectral lines $\lambda = 532$ nm ($h\nu = 2.33$ eV) and $\lambda = 1064$ nm ($h\nu = 1.65$ eV) of a Nd: YAG laser. The xenon ion produced by the intense laser radiation results from two channels of its ground state, namely the $^2P_{1/2}$ and $^2P_{3/2}$ states, which are energetically separated from each other by 1.31 eV. The special characteristics of this multiphoton process consist in the observation of sharp maxima in the cross section of the photoelectron spectra (Fig. 2.11); these maxima are also called ATI peaks as abbreviation for *above threshold ionization*. In other words, the photoelectron as a free unbound particle in the ionization continuum, absorbs energy in integer multiples of the photon energy $h\nu$ from the laser radiation. The angular distributions of photoelectrons in these ATI peaks in the continuum of xenon (Fig. 2.11) have also been measured; they can be fitted by the sum of Legendre polynomials of the order $2k$ where k is linked to the number L of the absorbed photons by the inequality relation $0 \leq k \leq L$. Similar experiments have been carried out with hydrogen atoms which absorb 26 photons from the 1064 nm laser radiation, 12 photons below and 14 photons above the ionization limit. As an example for atomic hydrogen, the scheme of multi-photon ionization by the 532 nm laser radiation is shown in Fig. 2.12, it also displays the angular distribution of the photoelectrons in the ATI peaks just above

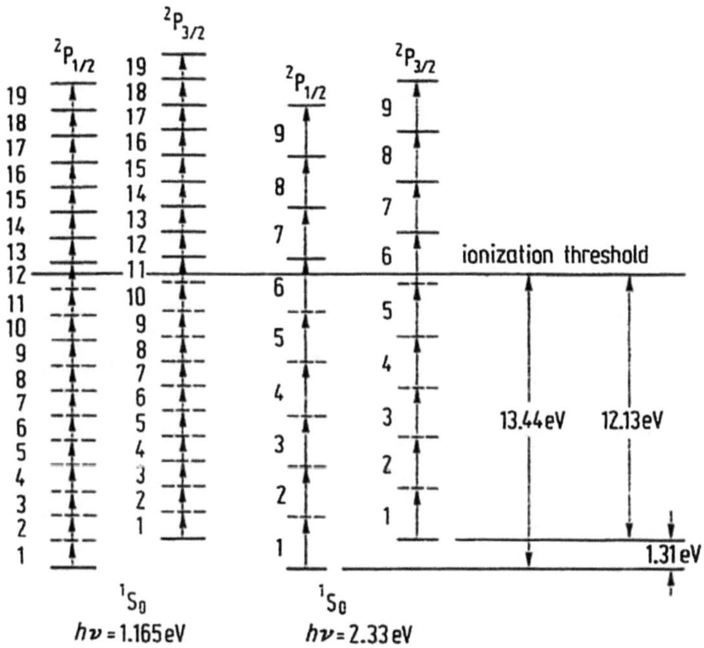

Fig. 2.10 Multiphoton ionization process of the Xe atom after Humpert et al. (1985). The *numbered rows* represent the photon process with energies $nh\nu$ ($n = 1-19$ and $n = 1-9$); the *right-hand* part of the diagram indicates the two lowest ground states of xenon and their ionization threshold

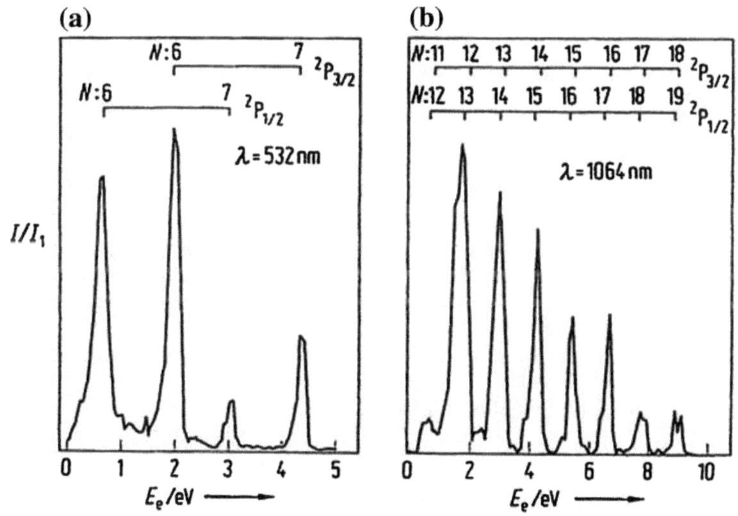

Fig. 2.11 Photoelectron spectra, including ATI peaks in the multiphoton ionization of the Xe atom. **a** Laser wavelength of the ionizing radiation $\lambda = 532$ nm; **b** laser wavelength $\lambda = 1064$ nm. The *numbered rows N* give the number of photons absorbed by the atom. E_e is the photoelectron energy, I the signal intensity of the photoelectrons in arbitrary units I_1 (after Humpert et al. 1985)

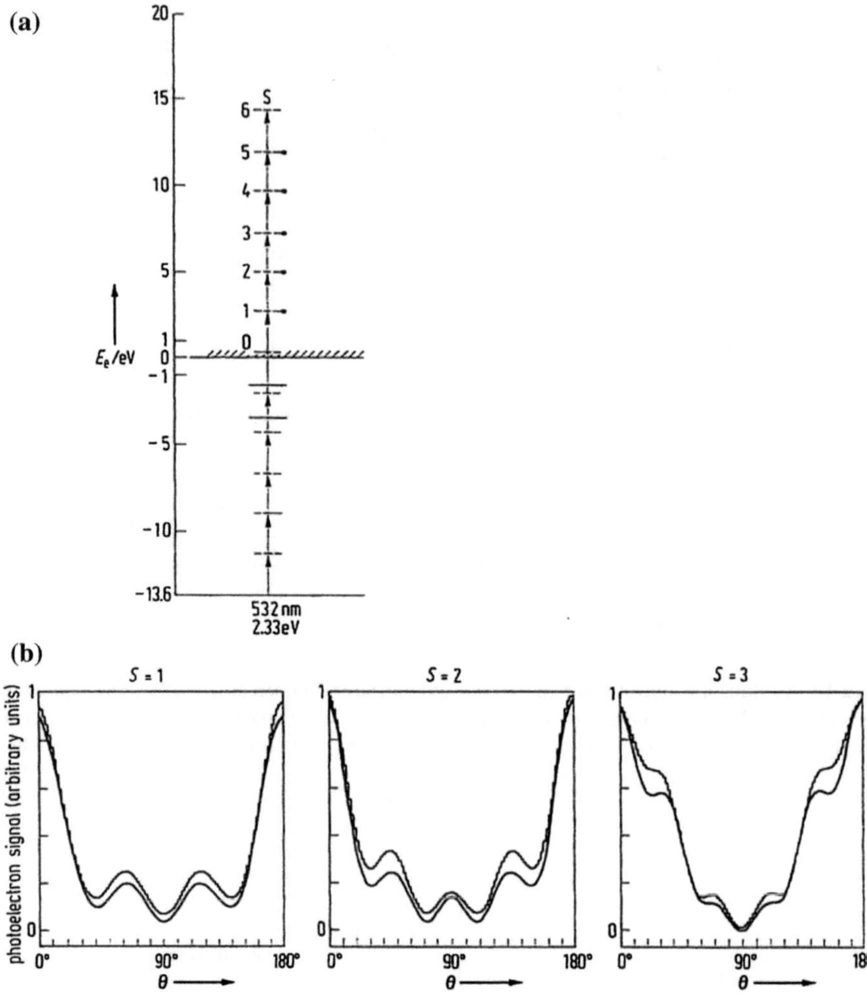

Fig. 2.12 **a** Multiphoton ionization of atomic hydrogen induced by laser radiation of wavelength $\lambda = 532$ nm; E_e photoelectron energy; the *horizontal full lines* represent hydrogen states, the *dashed lines* "virtual" in-between states. **b** Angular dependence of photoelectrons from the ATI peaks with the values $S = 1, 2,$ and 3 (see part **a**) for the number of photons with energies $h\nu$, $2h\nu$, and $3h\nu$ which are absorbed above the ionization threshold. The *full curve* corresponds to an experiment with a laser intensity of $I = 1.3 \times 10^{13}$ W cm^{-2}, the step-like curve is for a laser intensity of $I = 1.6 \times 10^{13}$ W cm^{-2} (after Wolff et al. 1988)

the ionization threshold; again the angular distribution can well be fitted by Legendre polynomials. It should be noted, however, that the form of the angular distributions in the ATI peaks depends on the intensity of the laser radiation; this influence cannot yet be described completely by theory.

Due to intense femtosecond XUV pulses generated by free-electron lasers, such non-linear key processes are now studied by means of electron spectroscopy as two-photon double ionization (Braune et al. 2007; Kurka et al. 2009), three-photon double ionization (Fukuzawa et al. 2010; Hikosaka et al. 2010; Miyauchi et al. 2011), or three-photon triple ionization (Rouźee et al. 2011). To distinguish these processes from other ionization mechanisms, it is crucial to know the energy of the outgoing electrons.

2.3.4 Results for the Total Cross Sections

In this sub-section we present some typical examples for total cross sections of atomic photoionization. Of course, atomic hydrogen again is of special importance due to its simple structure, and accordingly it plays a dominant role as a test for theoretical calculations. The Hamilton operator H for the description of photoionization of atomic hydrogen is composed of the Hamilton operator H_0 of the free atom and an interaction operator H_{int} which describes the interaction of the electron specified by its linear momentum and the electromagnetic wave of the incoming light described by a vector potential. The relevant matrix element with this interaction operator becomes $\langle\psi_\varepsilon|H_{int}|\psi_0\rangle$, including the eigenfunction ψ_0 of the ground state of atomic hydrogen and the eigenfunction ψ_ε of the continuum state of the free photoelectron with the energy ε. The eigenfunction ψ_ε of the electron in the continuum can be described by a plane wave to a first approximation. However, this eigenfunction will be modified by the Coulomb field of the ion produced (the proton for hydrogen) in the next higher approximation. Instead of a plane wave a so-called Coulomb wave will be a more satisfactory representation for the continuum eigenfunction. The total cross section, which is proportional to the square of the matrix element can then be calculated to

$$Q(h\nu) = \frac{4\pi^2\alpha}{3}\left|\langle\psi_\varepsilon|\mathbf{r}e^{i\mathbf{kr}}|\psi_0\rangle\right|^2, \tag{2.25}$$

with Sommerfeld's fine structure constant α, $\mathbf{r}$ for the coordinate of the electron in the atom, and $\mathbf{k}$ being the photon linear momentum. By approximating $e^{i\mathbf{kr}} = 1 + \mathbf{k}\,\mathbf{r} + \cdots$ and neglecting $\mathbf{kr}$ we obtain

$$Q(h\nu) = \frac{4\pi^2\alpha}{3}\left|\langle\psi_\varepsilon|\mathbf{r}|\psi_0\rangle\right|^2, \tag{2.26}$$

in the dipole approximation. Taking $\alpha \simeq 1/137$ and a typical value for the matrix element $\left|\langle\psi_\varepsilon|\mathbf{r}|\psi_0\rangle\right|^2 \approx 10^{-16}\,\text{cm}^2$ we obtain

$$Q(h\nu) \approx 10 \times 10^{-2} \times 10^{-16}\,\text{cm}^2 \approx 10^{-17}\,\text{cm}^2. \tag{2.27}$$

The ratio of the first (dipole) approximation to the approximation of second order (electric quadrupole approximation) is determined by the factor $\gamma = (2\pi r/\lambda)^2$ which becomes $\gamma \approx 10^{-6}$ for typical values of the atomic coordinate $r \approx 0.1$ nm and a photon wavelength of $\lambda = 10^2$ nm. However, for $\lambda = 0.1$ nm ($\simeq$12 keV X-rays) this ratio becomes $\gamma \approx 1$. In other words, the dipole approximation should represent a sufficiently accurate description of the photoionization cross section as long as the wavelengths of the photons are not shorter than about 0.1 nm (energetic X-rays). The situation changes for the angular distribution of photoelectrons, where interference between first and second order terms contributes. In this case, a non-dipole effect can show up at much smaller photon energies. The non-dipole effects are discussed in more detail in Sects. 4.1.2 and 4.9.

By inserting exact non-relativistic eigenfunctions for ψ_0 and ψ_ε into the matrix element of (2.26), a quantitative relation results for the total cross section of atomic hydrogen and hydrogen-like ions with nuclear charge Z, of the mass m_A of the atom and of the threshold energy $h\nu_0$ for the photoionization:

$$Q(h\nu)/\mathrm{cm}^2 = 1.05 \times 10^7 (\nu_0/\nu)^{8/3} \frac{1}{Z^2} [m_A/g]. \qquad (2.28)$$

In Born's approximation, with plane waves for the outgoing photoelectrons, the cross section can be calculated as

$$Q(h\nu) = 8.64 \times 10^7 (\nu_0/\nu)^{7/2} \frac{1}{Z^2} [m_A], \qquad (2.29)$$

which gives sufficiently accurate theoretical data for $\nu \gg \nu_0$.

It is, of course, disappointing that there is only a limited range of data available for the photoionization cross section of atomic hydrogen as basic test case (see Fig. 2.13) although it is in good agreement with the theoretical prediction of the exact dipole approximation and the $(\nu_0/\nu)^{8/3}$ dependence.

In the photoionization experiments of Beynon and Cairns (1965) atomic hydrogen was produced by dissociating the molecular gas in an intense radio-frequency discharge. The atomic gas streamed through an absorption tube and the ionizing monochromatic radiation passed through this tube for an absorption measurement. To produce atomic hydrogen Palenius et al. (1976) applied an aerodynamically heated shock-wave tube in which the hydrogen is completely dissociated. Flash lamps after Garton (1959) were used to produce intense vacuum-ultraviolet (VUV) radiation where either sharp spectral lines (Wolff et al. 1988; Beynon and Cairns 1965) or a VUV background (Palenius et al. 1976) were applied for photoabsorption measurements.

Similar to the behavior of atomic hydrogen, the total photoionization cross section of helium decreases monotonically with increasing photon energy. However, a resonance structure appears at about $\lambda = 20$ nm (Fig. 2.14), which is observable at higher energy resolution as shown in Fig. 2.3. The resonance structure is due to double excitation of both helium electrons, which may lead to autoionizing processes and is the

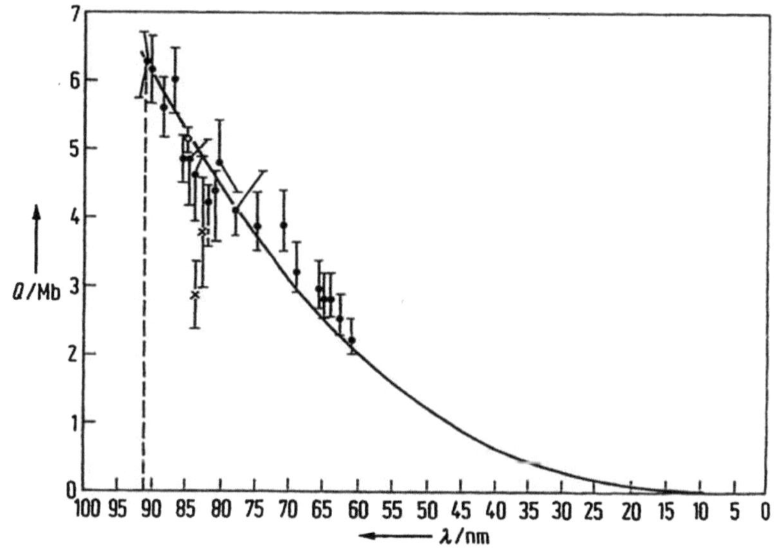

Fig. 2.13 Cross section Q of the photoionization of atomic hydrogen as a function of the wavelength λ. (——) theory based upon dipole approximation; (∘) experimental data point with the value $Q(\lambda = 85.06\,\text{nm}) = (5.15 \pm 0.18)\,\text{Mb}$ (Mb = megabarn = $10^{-18}\,\text{cm}^2$) after Beynon and Cairns (1965); (×) two experimental data points after Beynon (1966); (•) experimental data after Palenius et al. (1976)

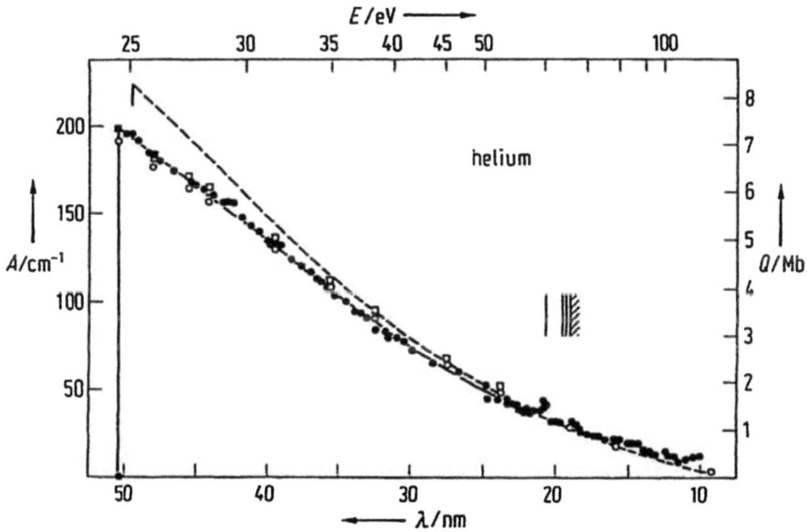

Fig. 2.14 Photoionization cross section Q and absorption coefficient A of helium as a function of the energy of the incident photons of energy E or wavelength λ. (∘) experimental data of Samson (1964) and (•) Lowry et al. (1965) in comparison to various dipole approximations; (□) Stewart and Webb (1963); (– – –) Cooper (1962). The *short vertical lines* at 20 nm indicate the positions of unresolved resonances (see Fig. 2.3)

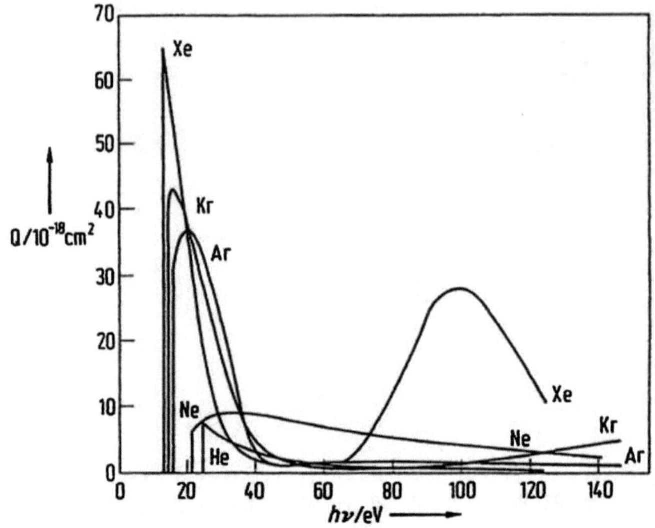

Fig. 2.15 Experimental cross sections Q of the photoionization of rare gas atoms as a function of the photon energy $h\nu$ (after Samson 1982)

origin of Fano-Beutler resonances in the cross section. Such resonance structures in photoionization cross sections are observed in many multi-electron atoms.

Photoionization cross sections of rare gas and alkali metal atoms have been extensively investigated. As seen in the Figs. 2.15 and 2.16 the structure of their cross sections deviates considerably from the monotonic shapes for atomic hydrogen and helium (apart from the resonance structures in helium, which may also appear in heavy atoms). Deep minima in the photoionization cross sections of alkali atoms are further discussed in Sect. 4.2.1.

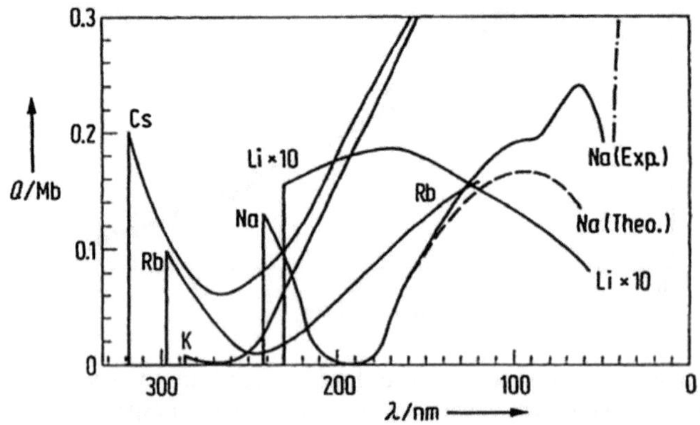

Fig. 2.16 Experimental photoionization cross sections Q of alkali atoms as a function of the wavelength λ (after Samson 1982); the *dashed line* represents calculations based on the many-body perturbation theory by Chang (1972)

2.3.5 *Differential Cross Sections, Partial Wave Analysis and Ramsauer-Townsend Effect*

In contrast to the Rutherford scattering of charged particles at atomic nuclei, scattering of electrons by atoms cannot satisfactorily be described classically. On the other hand, quantum mechanical methods only provide theoretical results approximately even for the simplest case, electron-hydrogen scattering. The reason for this is due to the dynamically, not rigorously-solvable three-particle problem (or many-particle problem for more complicated atoms). However, in this section we only sketch the general physical aspects of quantum mechanical approximations for electron-atom scattering processes, which in detail require considerably higher mathematical efforts than needed for theories of the static atomic structure previously described.

Figure 2.17 illustrates the process of electron scattering by atoms. An incoming plane wave of electrons of intensity I_0 propagates parallel to the z-direction and interacts with an atomic target of thickness d; the electron wave traversing the target experiences a reduction of its intensity to I_t, in the forward direction. With reference to the solid angle $\Delta\Omega$, the intensity scattered into the angles (θ, ϕ) is equal to $|f(\theta, \phi)|^2 I_0/r^2$ with r as (averaged) distance between the detector for the scattered electrons and the scattering center and with the quantity $f(\theta, \phi)$ which we denote as the *scattering or collision amplitude*; $f(\theta, \phi)$ is a function of the polar angle θ and azimuthal angle ϕ and also of various quantities such as the energy of the incoming electrons, the dynamics of the collision process (i.e. elastic, inelastic) and the properties of the target atoms. For $r \gg d$ we can approach asymptotically a wave

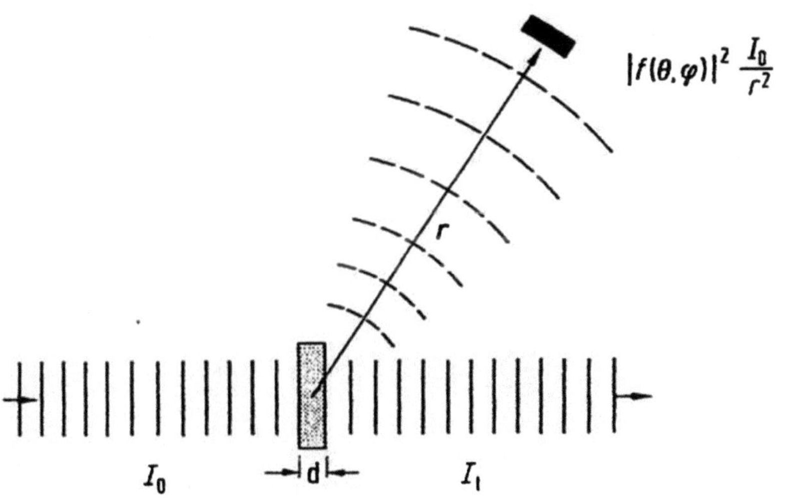

Fig. 2.17 Scheme of scattering experiments for measuring total and differential cross sections: An incident wave of intensity I_0 traverses the atomic target of thickness d; the outgoing spherical wave propagates in the direction of **r** and the transmitted wave in the forward direction with intensity I_t

function for the electrons behind the target by a superposition of the transmitted plane wave and the scattered spherical wave, i.e.,

$$\psi(r) \underset{r \gg d}{\longrightarrow} \sqrt{I_t} \mathrm{e}^{\mathrm{i}(\mathbf{p}'/\hbar)\mathbf{r}} + \sqrt{I_0} f(\theta, \phi) \frac{1}{r} \mathrm{e}^{\mathrm{i}(\mathbf{p}'/\hbar)\mathbf{r}}, \tag{2.30}$$

with $\mathbf{p}'$ as the linear momentum of the scattered electrons. The connection between the scattering amplitude and the differential cross section can be obtained as follows: $I_0 \left| f(\theta, \phi) \right|^2 \Delta\Omega$ is the probability per unit of time that an electron is scattered into the solid angle element $\Delta\Omega$. With reference $\Delta\Omega$ and the total cross section Q we obtain

$$\Delta Q = \sigma(\theta, \phi) \Delta\Omega = \left| f(\theta, \phi) \right|^2 \Delta\Omega, \tag{2.31}$$

according to (2.9)–(2.14); the differential cross section then becomes

$$\frac{\mathrm{d}Q}{\mathrm{d}\Omega} = \sigma(\theta, \phi) = \left| f(\theta, \phi) \right|^2. \tag{2.32}$$

We now describe the scattering process quantum mechanically as follows: We assume that both the atom and its potential that acts on the electron are spherically symmetric; as a consequence, the orbital angular momentum ℓ of the incoming electron and its z-component in the forward direction is conserved during the scattering process.

By taking into account the de Broglie relation $\mathbf{p} = \hbar\mathbf{k}$, the incoming plane wave $\mathrm{e}^{ikr\cos\theta}$ of the primary electron and the outgoing spherical wave $\frac{1}{r}\mathrm{e}^{\mathrm{i}\mathbf{kr}}$ of the scattered electron can be expressed by appropriate superpositions of waves with orbital angular momentum ℓ. Scattering of electrons by atoms takes place when phase differences η_ℓ, between the outgoing spherical waves of angular momenta $\ell = 0, 1, 2, \ldots$ are formed with and without the scattering center. The mathematical description of such a physical scattering process for the partial waves with angular momenta ℓ results in the following equation for the scattering amplitude, which becomes independent of the azimuthal angle ϕ:

$$\begin{aligned} f(\theta) &= \frac{1}{2k\mathrm{i}} \sum_{\ell=0}^{\infty} (2\ell + 1)(\mathrm{e}^{2\mathrm{i}\eta_\ell} - 1) P_\ell(\cos\theta) \\ &= \frac{1}{k} \sum_{\ell=0}^{\infty} (2\ell + 1) \mathrm{e}^{\mathrm{i}\eta_\ell} \sin\eta_\ell P_\ell(\cos\theta), \end{aligned} \tag{2.33a}$$

where $P_\ell(\cos\theta)$ are the Legendre polynomials. We obtain expressions for differential and total cross sections (again under the assumption of cylindrical symmetry around the z-direction, see (2.14), (2.32)) as follows:

$$\sigma(\theta, \phi) = \left|f(\theta, \phi)\right|^2 = \frac{1}{k^2}\left|\sum_{\ell=0}^{\infty}(2\ell+1)\mathrm{e}^{\mathrm{i}\eta_\ell}\sin\eta_\ell P_\ell(\cos\theta)\right|^2, \tag{2.33b}$$

$$Q(E) = \frac{4\pi}{k^2}\sum_{\ell=0}^{\infty}(2\ell+1)\sin^2\eta_\ell = \frac{\pi}{k^2}\sum_{\ell=0}^{\infty}(2\ell+1)\left|\mathrm{e}^{2\mathrm{i}\eta_\ell}-1\right|^2. \tag{2.33c}$$

This analysis of the scattering processes, based upon the orbital angular momenta ℓ and the phase differences η_ℓ, is called the *method of partial waves*, which was first developed for quantum mechanical scattering processes by the Danish physicists Faxen and Holtsmark (1927). The method has classical analogies, for example in Rayleigh's analysis of sound waves or light scattered by special obstacles. According to (2.33), theoretical calculations of cross sections are reduced to making predictions of phase differences.

If all phases vanish, $\eta_\ell = 0$, or are integer multiple of $\eta_\ell = \pi$, no scattering takes place. An outstanding example for the η_ℓ dependence of scattering processes is the special structure in cross sections associated with the so-called Ramsauer-Townsend effect. This effect manifests itself in the minima of total cross sections and interference structure of differential cross sections. We illustrate the connection between the scattering phases η_ℓ and the total (Q) and partial (Q_ℓ) cross sections in Figs. 2.18 and 2.19 in which we define the relation between these cross sections by $Q = \sum Q_\ell$; the partial cross sections can be represented by $Q_\ell = (4\pi/k^2)(2\ell+1)\sin^2\eta_\ell$, with the specific angular momentum ℓ and the phase η_ℓ.

Calculations of scattering phases η_ℓ, are based upon the following physical arguments. The scattering potential that the electron experiences in a collisional interaction with an atom is composed of several contributions: (1) By the static potential, which represents, both, the direct Coulomb interaction of the projectile electron with the atomic electrons, and also with the nuclear charge,

$$V(\mathbf{r}) = -\frac{Ze^2}{4\pi\varepsilon_0 r} + \frac{Ze^2}{4\pi\varepsilon_0}\int\frac{\rho(\mathbf{r}')\mathrm{d}\tau'}{\left|\mathbf{r}-\mathbf{r}'\right|} \tag{2.34}$$

here $\mathbf{r}$ is the distance between the atomic nucleus and the projectile electron, $\rho(\mathbf{r})$ is the charge density of the electron shell; the calculation for complicated atoms as targets can be carried out with the Hartree-Fock approximation; (2) by Heisenberg's exchange interaction between the projectile electron and an atomic electron; this interaction is identical to the exchange interaction, which is introduced in describing the energy structure of two-electrons atoms; (3) by the so-called polarization interaction which arises because the projectile electron polarizes the atom; this polarization effect is described asymptotically by an α/r^4 potential, in dipole approximation, where α is the electric dipole polarizibility of the atom, i.e., the induced electric dipole moment of the atom becomes $P_i = \alpha F$ with F as electric field strength which the atom experiences from the passing electron.

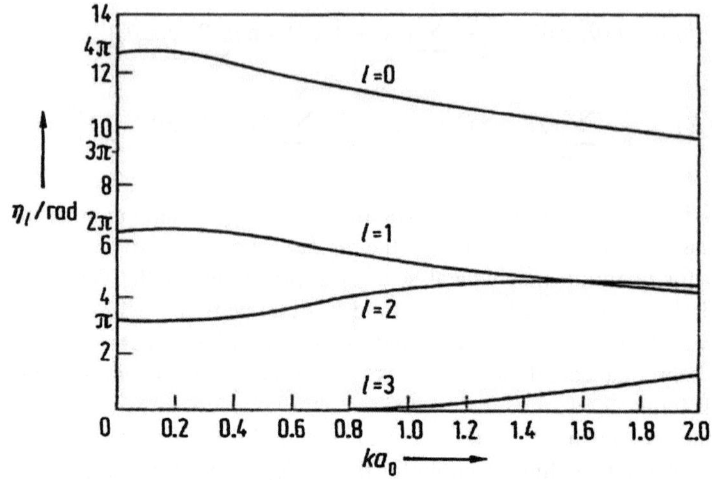

Fig. 2.18 Calibrated scattering phases η_ℓ for $\ell = 0, 1, 2,$ and 3 of elastic scattering on krypton. η_0 tends to 4π at zero energy. The unit of the abscissa $(ka_0)^2 = 1$ corresponds to an energy of 13.6 eV of the incident electrons (Holtsmark 1930)

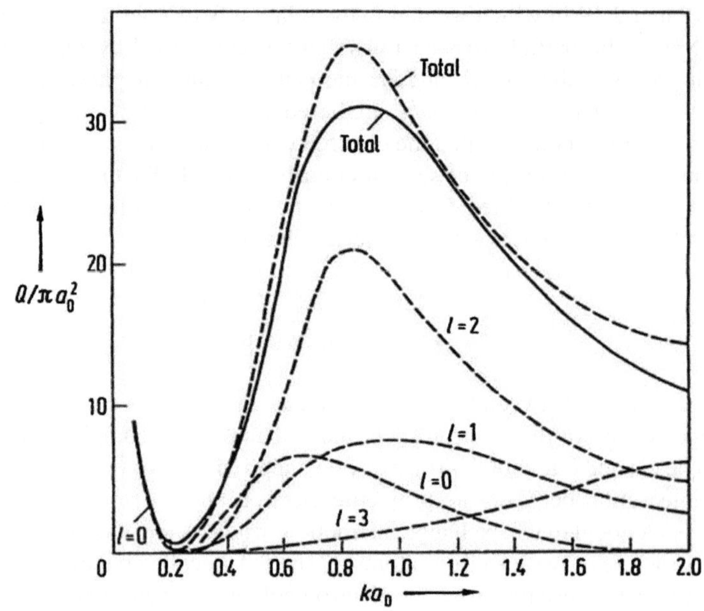

Fig. 2.19 Calculated partial ($\ell = 0$–3) and total cross sections *(dotted line)* for elastic electron scattering on krypton corresponding to the phases of Fig. 2.18 (unit of abscissa as in Fig. 2.18). For comparison the measured total cross section is given by the *full curve* (see Holtsmark 1930)

It has been found that a satisfactory theoretical description of the elastic electron scattering can be achieved only if, in addition to the static Hartree-Fock potential, both the exchange as well as the polarization interaction are taken into account in the calculation.

Figures 2.18 and 2.19 show results of such calculations for elastic scattering of electrons by krypton atoms. Whenever the scattering phase η_ℓ of a partial wave with an angular momentum ℓ approaches an integer multiple of π, the partial cross section vanishes. As can be seen in Fig. 2.18, the $n\pi$ positions of the phases almost coincide at the same electron energies for $\ell = 0, 1$ and 2; this results in a deep Ramsauer-Townsend minimum of the total cross section (Fig. 2.19). The $\frac{n}{2}\pi$ passages of the phases build up the broad maximum of the elastic cross section. This characteristic feature is similar for argon, krypton and xenon atoms, which also show Ramsauer-Townsend minima in the low-energy elastic scattering cross sections (Fig. 2.20b). On the contrary, the light rare gases helium and neon do not have Ramsauer-Townsend minima. This is due to the fact that the scattering potential resulting from the above three contributions leads to phases which lie within a phase interval of π without any $n\pi$ transition. Since the scattering potentials are weaker for helium and neon, so are the scattering phases η_ℓ and cross sections compared to those of the heavier rare gases (Fig. 2.20b).

As mentioned already at the beginning of this section, scattering experiments with positrons have been carried out successfully in recent years; the results of cross section measurements with positrons are compared with those for electrons in Fig. 2.20a. Unlike electrons, positrons cannot be produced by means of tungsten or oxide-cathodes; positrons can be obtained from intense radioactive β^+ emitters such as ^{22}Na (lifetime $\tau \approx 2.6$ years) or from nuclear physics accelerators in which (e^+e^-) pairs are created in bremsstrahlung processes with highly energetic electrons. The energetic positrons from such sources can be decelerated to low thermal energies by a solid state moderator; afterwards they can be accelerated to a given energy to form a beam of positrons.

There are two reaction channels in positron-atom interactions which do not occur in electron scattering; (1) the production of positronium and the annihilation of a positron with an atomic electron in which annihilation radiation occurs. However, since the cross section Q for the latter process is very small ($Q = r_0^2 c/v$ where $r_0 = 2.8 \times 10^{-12}$ cm is the "classical" electron radius and v is the positron velocity); for example, for $Q \geq 10^{-22}\,\text{cm}^2$ and $E = 10\,\text{eV}$ the annihilation process can be neglected. On the other hand positronium production has a cross section which is comparable to that of positron scattering at atoms. The energy threshold E_{thr} for positronium production results from the difference between the ionization energy E_{ion} of the atom that provides the electron and the binding energy of positronium E_{Ps}, i.e.,

$$E_{thr} = E_{ion}(A) - E_{Ps} = E_{ion}(A) - \tfrac{1}{2}E_{ion}(H) = E_{ion}(A) - 6.8\,\text{eV}, \qquad (2.35)$$

in which the binding energy of positronium is half that of atomic hydrogen, which is 6.8 eV. The effect of positronium production is clearly noticeable in the data of

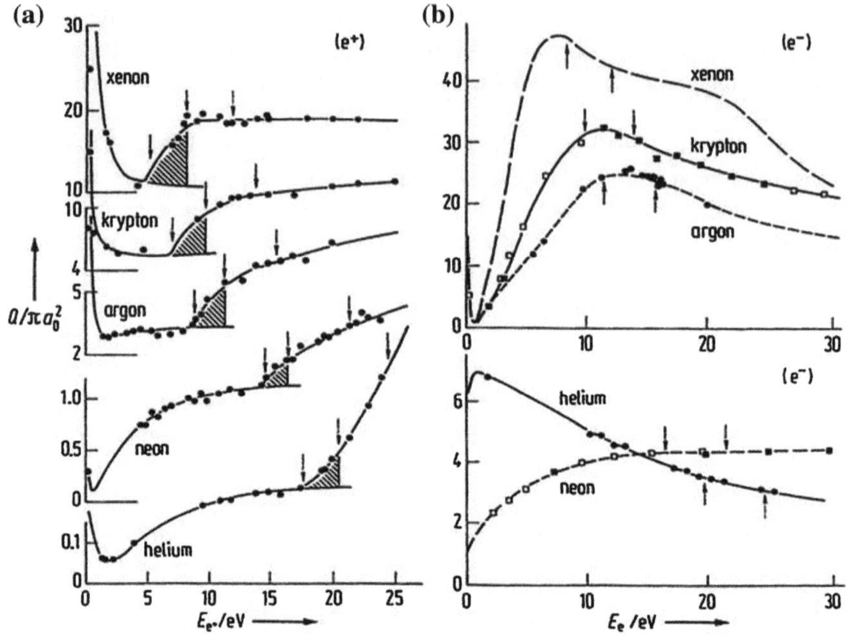

Fig. 2.20 **a** Total cross sections Q of low energy positron scattering on rare gas atoms as a function of the positron energy E_{e^+}; the *three arrows* indicate the onset of the formation of positronium, the first inelastic excitation and the ionization of the atoms. The *hatched areas* represent an estimation for the cross section of the positronium formation, the subtraction of which results in the total elastic cross section (lower part of the hatched area); **b** total cross sections for the scattering of electrons on rare gas atoms as a function of the electron energy E_e; the *two arrows* on each curve indicate the onset of the first inelastic excitation and the ionization of the atoms (after Kauppila and Stein 1982)

Fig. 2.20. Helium and neon atoms display typical Ramsauer-Townsend minima in low energy positron scattering; this is contrary to electron scattering on helium and neon as well as for positron scattering by heavier rare gas atoms. Detailed theoretical description of positron scattering again depends on the scattering phases of the collisional interaction; because of the positive charge of the positron, in comparison to electron scattering, the relevant potentials for positron scattering have a different sign for the static Coulomb interaction. In particular, exchange interaction does not exist in positron scattering since it only takes place quantum mechanically between particles of the same identity. Therefore, only the direct Coulomb interaction remains between the positron and the positive and negative charges of the atom in positron scattering. At very low energies of the incoming positrons, the partial wave with $S = 0$ ("S-scattering" with the symbol S for the quantum number $\ell = 0$ of the partial wave as in spectroscopy) dominates for scattering at helium and neon. Close to zero energy, the attraction between the positron and the atomic electrons dominates, which has the consequence that the scattering phase η_0 becomes positive;

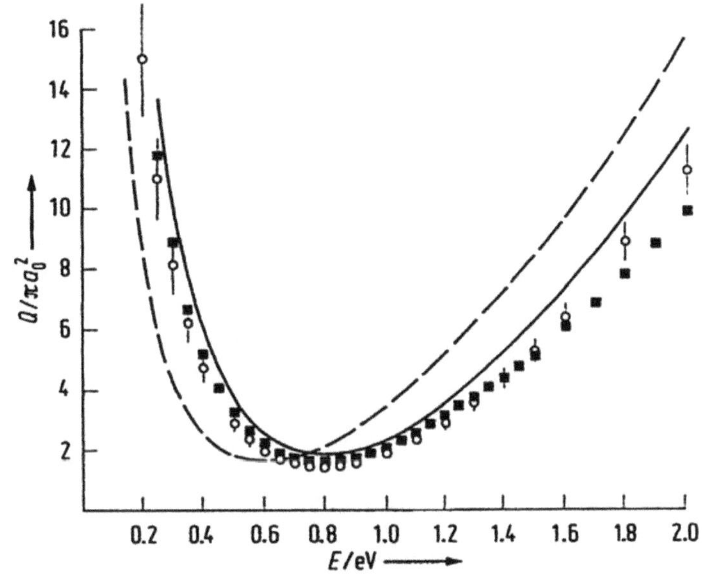

Fig. 2.21 Total elastic cross section Q for electron scattering on xenon atoms; (○) after Jost et al. (1983) by applying a 180° spherical electron monochromator; (■) after Ferch et al. (1987) by applying a time-of-flight spectrometer; (– – –) non-relativistic calculation after McEachran and Stauffer (1984); (——) relativistic theory after McEachran and Stauffer (1987)

however, with increasing positron energy a repulsive potential comes into action and η_0 changes into the negative region and a Ramsauer-Townsend minimum arises for $\eta_0(E)$ passing through zero. Higher partial waves with $\ell \geq 1$ contribute considerably to the positron scattering for heavy rare gas atoms in comparison to the S-scattering. The total cross sections for these atoms show no minima although the S-wave phase η_0 passes through zero. We mention in passing that it was recently possible to measure and to compare the ionization of atomic hydrogen between positron and electron impact (Spicher et al. 1990). The ionization cross section below 450 eV is distinctly larger for positron impact than for electron impact.

New Ramsauer-Townsend measurements have recently been reported; modern electron monochromators were applied, such as special 180° monochromators or time-of-flight spectrometers. The time-of-flight of electrons between a pulsed electrode and a detector electrode is measured in a time-of-flight spectrometer in which the electron energy can be determined with high resolution (for comparison see the scheme of a time-of-flight spectrometer for ions in Fig. 2.4). In a 180° monochromator the electrons move between two concentrically mounted hemispheres in which the electric potentials on the hemispheres and further special properties of the monochromator are arranged so that, after a deflection of 180°, the incoming electrons are focused to the exit slit of the monochromator for a given energy. As an example of the applications of such electron monochromators and spectrometers, we refer to the measurement of the total cross section of elastic electron scattering at xenon atoms (Fig. 2.21). As can be seen in this figure, the agreement between experiment and theoretical data improves when relativistic effects are included in the theoretical calculation.

Angular distributions of electrons, scattered by atoms, can be described by the partial wave method. According to (2.33), diffraction structures of the partial waves arise which are due to the angular dependence of the Legendre polynomials $P_\ell(\cos\theta)$. Such diffraction structures are a quantum mechanical phenomenon that has no classical analogue and does not show up in classical Rutherford scattering due to a Coulomb potential. Diffraction structures in angular distributions of electrons scattered by atoms were first observed by Bullard and Massey (1931) for argon. Figure 2.22 illustrates examples of diffraction structures for various atoms and its relation to the term $[P_\ell(\cos\theta)]^2$ in (2.33) for the dominant phases η_ℓ at 80 eV.

2.3.6 Resonance Structures

Apart from the Ramsauer-Townsend minima and diffraction structures described above, there are resonance structures in electron-atom collisions that were discovered in the 1960s. The expression resonance means that an anomalous structure is observed in total or differential cross section at a sharp electron energy, the so-called resonance energy E_{res}. Resonance structures generally occur in atomic and sub-atomic scattering processes. The description of the quantitative physical process on which the resonance is based requires a complicated theoretical analysis. Qualitatively, the resonance process can be described by the *compound model* which was proposed by N. Bohr and was first applied in nuclear scattering processes. The compound model for electron scattering leads to the production of a short-lived negative atomic ion at the resonance energy E_{res}. We describe this process by means of the reaction equation

$$e(E_{res}) + A \longrightarrow \{A^-\} \longrightarrow A + e(E_{res}). \tag{2.36}$$

The short-lived negative ion $\{A^-\}$ which is produced during a collision by the electron at energy E_{res}, is in a special electron configuration, while the scattered electron is described by the partial waves explained in the previous section. In many cases, the negative ion decays back into the initial state of the atom; however, further decay channels as, for example, into other excited states of the neutral atom are possible. Such inelastic resonance processes are represented by the reaction equation

$$\begin{aligned} e(E_{res}) + A \longrightarrow \{A^-\} \longrightarrow\; & A^* + e(E_{res} - E_0) \\ & \hookrightarrow A + h\nu\,. \end{aligned} \tag{2.37}$$

in which E_0 is the excitation energy.

Formally, resonance processes can be described by a coherent superposition of a potential and a resonance scattering. The potential scattering can be represented by all the phases η_ℓ^{pot} of the partial waves, as described above (see Sect. 2.3.5). The resonance scattering, however, is only associated with a single partial wave with

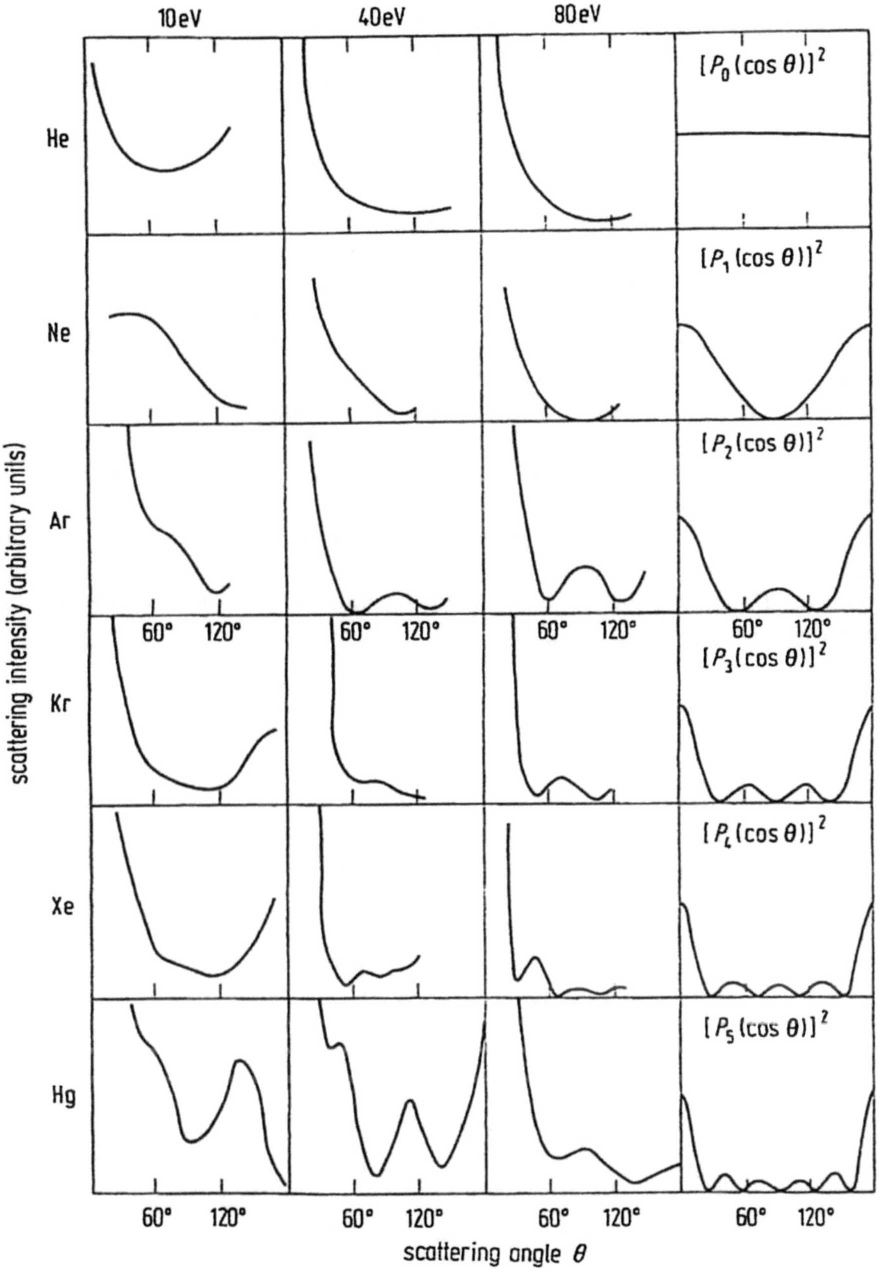

Fig. 2.22 Observed angular distributions for elastic scattering of electrons on rare gas and mercury atoms at 10, 40 and 80 eV; $[P_\ell(\cos\theta)]^2$ is calculated for the dominant phase η_ℓ at 80 eV in each case (after Massey and Burhop 1969)

quantum number ℓ at the resonance energy. It can be shown, by a detailed resonance theory, that the resonance phase η_ℓ^{res} in the finite range of the resonance energy (the resonance width) experiences a phase jump of amount π. This means that the ℓth partial wave possesses a resonance which manifests itself as structure in the cross sections. Since the total phase is composed of the sum $\eta_\ell = \eta_\ell^{pot} + \eta_\ell^{res}$ for which the contribution η_ℓ^{pot} from the potential scattering is almost constant over the range of the resonance, the total phase η_ℓ experiences a phase jump of π from the resonance phase η_ℓ^{res}.

According to (2.33b), the partial cross section for a given value of ℓ is $Q_\ell = (4\pi/k^2)(2\ell+1)\sin^2\eta_\ell$; its resonance structure is illustrated in Fig. 2.23 for various values of η_ℓ^{pot}. If the potential scattering phase η_ℓ^{pot} (also called background scattering phase) is zero or $\pi/2$ the resonance has the shape of a Lorentzian curve in analogy to emission or absorption lines in spectroscopy. The resonance curve for other values of η_ℓ^{pot} has the typical shape of *Breit-Wigner resonances* as first observed in nuclear and particle physics in connection with the compound states mentioned above. The physical process which determines the resonance shape is based on an *interference phenomenon between the potential and resonance scattering*, which we can understand as follows: We consider the term $\mathrm{e}^{2\mathrm{i}\eta_\ell} - 1 = T_\ell(E)$ of (2.33c) as definition of the T-matrix of the scattering theory and transform it with the corresponding scattering phases $\eta_\ell = \eta_\ell^{pot} + \eta_\ell^{res}$ to

$$T_\ell(E) = \mathrm{e}^{2\mathrm{i}\eta_\ell^{pot}}(\mathrm{e}^{2\mathrm{i}\eta_\ell^{res}} - 1) + (\mathrm{e}^{2\mathrm{i}\eta_\ell^{pot}} - 1) = T_\ell^{res} + T_\ell^{pot}. \tag{2.38}$$

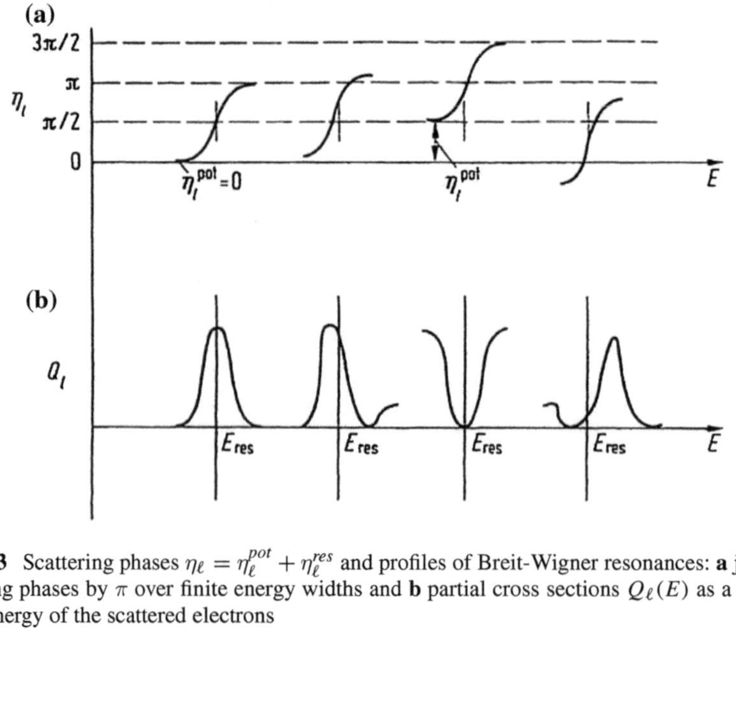

Fig. 2.23 Scattering phases $\eta_\ell = \eta_\ell^{pot} + \eta_\ell^{res}$ and profiles of Breit-Wigner resonances: **a** jumps of scattering phases by π over finite energy widths and **b** partial cross sections $Q_\ell(E)$ as a function of the energy of the scattered electrons

The total and partial cross sections result in

$$Q(E) = \frac{\pi}{k^2} \sum_{\ell=0}^{\infty} (2\ell + 1) \left| T_\ell \right|^2, \tag{2.39}$$

and

$$Q_\ell(E) = \frac{\pi}{k^2} (2\ell + 1) \left| T_\ell^{res} + T_\ell^{pot} \right|^2; \tag{2.40}$$

the latter equation contains the resonance term $\left|T_\ell^{res}\right|^2$, the potential or background term $\left|T_\ell^{pot}\right|^2$ and the interference term $T_\ell^{res*} T_\ell^{pot} + T_\ell^{res} T_\ell^{pot*}$.

By introducing the half-value width Γ as

$$E_1 = E_{res} - \Gamma/2, \quad E_2 = E_{res} + \Gamma/2 \quad \text{and} \quad E_1 - E_2 = \Gamma, \tag{2.41}$$

the exact shape of the resonance cross section in the ℓth partial wave can be calculated according to Breit and Wigner to

$$Q_\ell^{res}(E) = \frac{4\pi}{k^2} (2\ell + 1) \frac{(\Gamma/2)^2}{(E - E_{res})^2 + (\Gamma/2)^2}. \tag{2.42}$$

$Q^{res}(E)$ has its half-value of the maximum for E_{res} at the energies E_1 and E_2.

A further frequently applied representation of the shape of the resonance has been introduced by Fano and Cooper (1965). Equation (2.42) can be rewritten by applying the *Fano parameters* defined by $q = -\cot \eta_\ell^{pot}$ and $\epsilon = -\cot \eta_\ell^{res} = (E - E_{res})/(\Gamma/2)$:

$$Q_\ell(E) = \frac{4\pi}{k^2} (2\ell + 1) \sin^2 \eta_\ell^{pot} \frac{(\epsilon + q)^2}{1 + \epsilon^2}. \tag{2.43}$$

We note that $\sin^2 \eta_\ell^{pot}$ is directly linked to the background scattering, i.e., it becomes $Q_\ell(E) = Q_\ell^{pot} (\epsilon + q)^2/(1 + \epsilon^2)$. Under the assumption that only one partial wave with angular momentum ℓ is present in the scattering process, the total cross section $Q(E)$ will be

$$Q(E) = Q_a + Q_b \frac{(q + \epsilon)^2}{1 + \epsilon^2}, \tag{2.44}$$

with $Q_a = \sum_{\ell \neq \ell'} Q_\ell$ and $Q_b = Q_{\ell'}^{pot}$. With $Q^{pot} = Q_a + Q_b$ we obtain the total cross section

$$Q(E) = Q^{pot} + Q_a \frac{q^2 + 2q\epsilon - 1}{1 + \epsilon^2}. \tag{2.45}$$

Differentiating (2.44) and (2.45) with respect to ϵ results in a maximum at $\epsilon = q^{-1}$ and a minimum at $\epsilon = -q$; both extreme values of the resonance are not located at the resonance energy E_{res} for which $\epsilon = 0$ holds; the maximum and minimum values of the cross sections are $Q^{max}(E) = Q^{pot} + q^2 Q_a$ and $Q^{min}(E) = Q^{pot} - Q_a = Q_b$. The quantities q, Γ, Q_a, and Q_b can usually be taken as constant over the range of the resonance. Figure 2.24 shows the quantity $(q^2 + 2q\epsilon - 1)/(1 + \epsilon^2)$, which determines the shape of the resonance curve as a function of ϵ. The asymmetric Fano profiles of scattering resonances differ from the Lorentzian profile of atomic spectral lines; however, the Fano profiles become identical with the Lorentzian profiles of spectral lines for $q \to 0$ (which represents a *transmission window* for the incoming electrons with a minimum cross section as seen in Fig. 2.24) and for $q \to \infty$ [with functions approaching the form $B/(1 + x^2)$].

The angular distribution of the scattering resonances depends on the Legendre polynomials $P_\ell(\cos\theta)$ (see (2.33)); the characteristics of these polynomials can be used to find out experimentally in which partial wave with angular momentum ℓ the resonance process takes place. Electron scattering resonances were first discovered in helium; the first resonance in helium is called the 19.3 eV *Schulz resonance* after its discoverer (Schulz 1963); shortly after that, elastic and inelastic electron scattering resonances in atomic hydrogen and other atoms were detected. An important characteristic of such scattering is the requirements to achieve high energy resolution. Therefore it is common to apply two electrostatic energy analyzers with energy

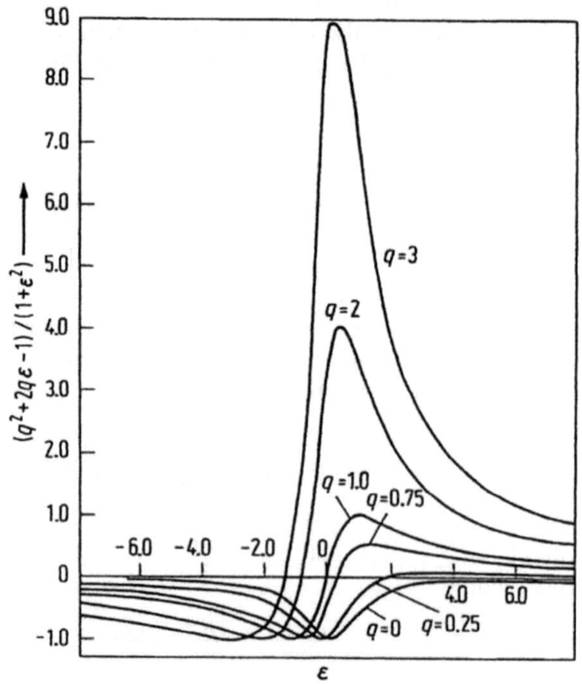

Fig. 2.24 *Fano profiles* of scattering resonances with the q parameter. For negative q values, the signs of the abscissa are reversed (from Fano 1961)

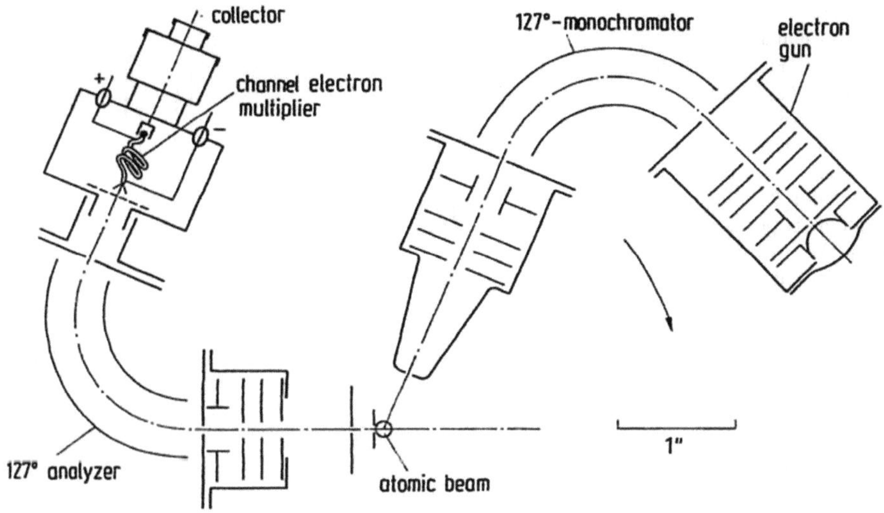

Fig. 2.25 Schematic diagram of a double electrostatic analyzer (after Pavlovic et al. 1972) with two cylindrical 127° electron analyzers (as monochromator for the primary electron beam and as analyzer for the scattered electrons) and electro-optical elements for focussing and extracting the electrons; a channel electron multiplier serves for the detection of the electrons after passing the analyzer. An atomic beam crosses the electron beam of the monochromator perpendicular to the plane of the figure ($1'' = 2.54$ cm)

widths ΔE smaller than about 0.1 eV; one is used for producing mono-energetic electrons and a similar one for analyzing the scattered electrons (Fig. 2.25). The first reported high-resolution measurement with such an experimental arrangement for elastic electron scattering on hydrogen atoms is shown in Fig. 2.26. Observed was a distinct scattering minimum of electrons elastically scattered at atomic hydrogen under a scattering angle of $\theta = 94°$ at an initial electron energy of 9.7 eV. By applying an even higher experimental energy resolution of $\Delta E = 0.07$ eV in the forward direction (i.e. in transmission), further resonance structure below the $n = 2$ energy threshold of atomic hydrogen can be detected (Fig. 2.27).

The physical interpretation of the electron scattering resonances is based upon the compound model described above. The electrons impinging on the atom are bound for the very short times of about 10^{-13} to 10^{-14} s; these times follow from Heisenberg's uncertainty principle, taking into account resonance widths of about $\Delta E_{res} \approx 0.1 \ldots 0.01$ eV in which a doubly excited short-lived negative hydrogen ion is formed. The electron configurations of these negative hydrogen states correspond to singlet and triplet states of two-electron systems. Compound states below the $n = 2$ threshold of atomic hydrogen have the configurations $n_1 = 2$, $\ell_1 = 0, n_2 = 2, \ell_2 = 0$, i.e. $2s^2$ or $n_1 = 2, \ell_1 = 0, n_2 = 2, \ell_2 = 1$, i.e. 2s2p, or $2p^2$, 2s3s, 2sns, 2snp $\pm$ 2pns configurations with the condition that one of the two electrons remains in the $n = 2$ state. Similar to these compound states, there are analogous series of compound states below the hydrogen states $n = 3, 4, 5, \ldots$.

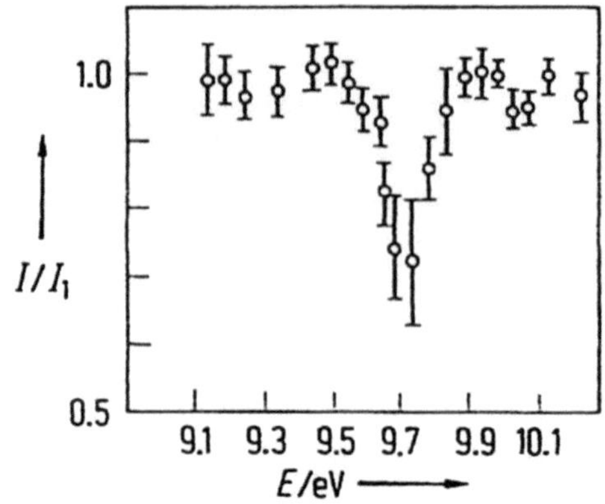

Fig. 2.26 By applying the experimental arrangement of Fig. 2.25, Kleinpoppen and Raible (1965) observed a resonance of elastic electron scattering on atomic hydrogen at about 9.7 eV; angle of observation $\theta = 94°$, energy width of the primary electron beam $\Delta E \approx 0.1$ eV. The scattering intensity I as a function of the electron energy E is given in arbitrary units I_1

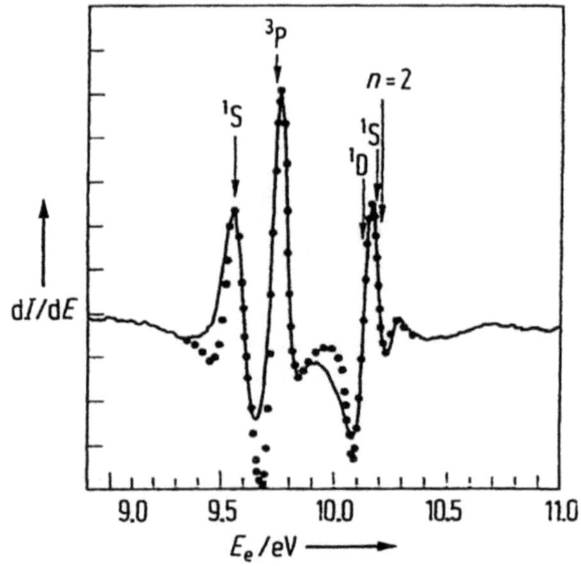

Fig. 2.27 Resonance structure in the transmission of electrons passing through a target of atomic hydrogen; the function represents the energy derivative of the transmission intensity, dI/dE. The *full curve* shows the experimental data. The *points* represent theoretical predictions; $n = 2$ indicates the energy threshold for the $(n = 2)$ state of atomic hydrogen; ^{1}S, ^{3}P, and ^{1}D are configurations of negatively charged ion states H^- (after Sanche and Burrows 1972)

It is common practice to represent a joint level scheme for atomic hydrogen states and negative hydrogen ion states by including the stable ground state of the H^- ion (Fig. 2.28). We can extract the following interesting facts from this figure: The H^- ion has only one stable state, namely the $(1s^2\,{}^1S_0)$ ground state which is 0.75 eV below the ground state of atomic hydrogen. Singly excited H^- states with the configurations $1sn\ell$ do not exist. Doubly excited compound states of the H^- ion below the excited n-states of atomic hydrogen are formed by the binding of a polarization interaction [potential $\sim(-\alpha/r^4)$] between the atom and the electron. There are, however, in addition so-called *shape resonances* or *shape states*, the word "shape" refers to the shape of the special potential curve between the projectile electron and the atom which arises from the term $\ell(\ell + 1)/r^2$ in the Hamilton operator for the

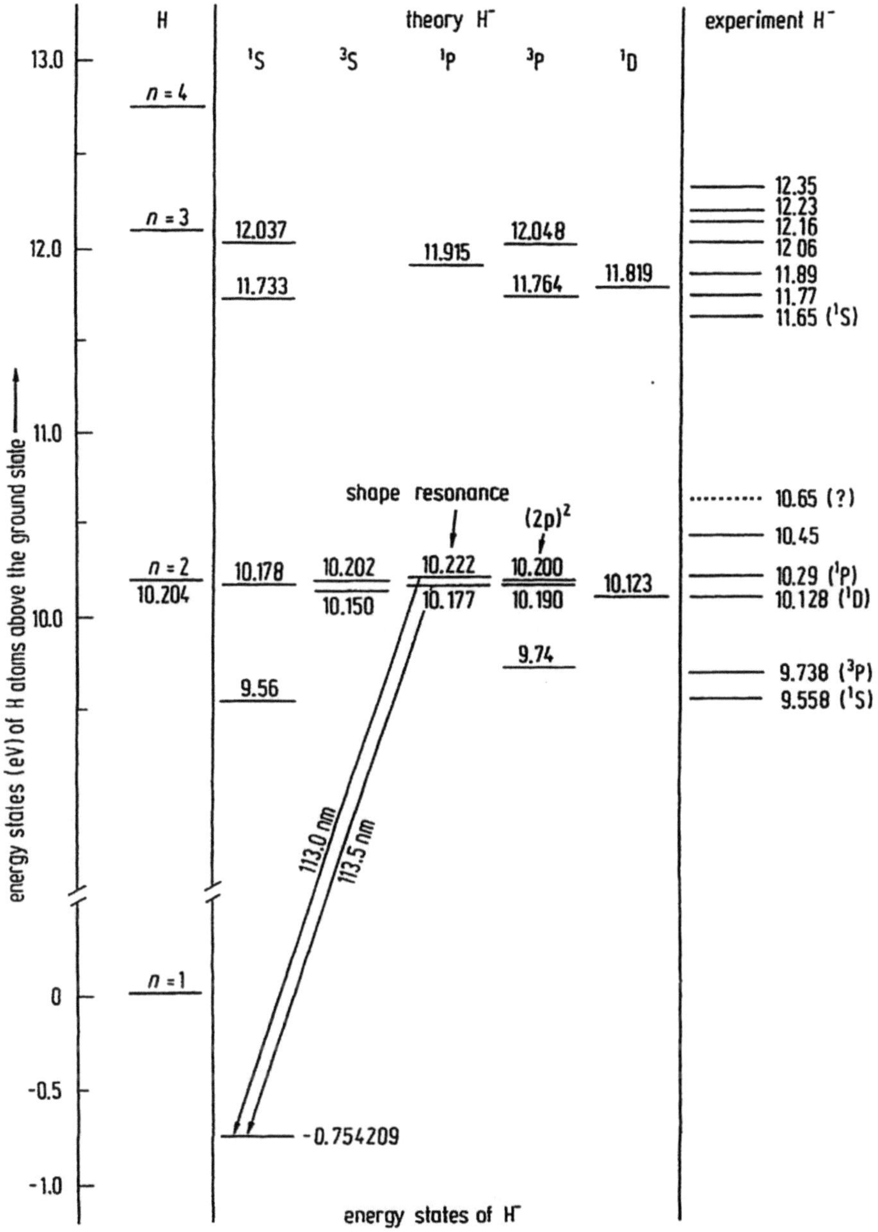

Fig. 2.28 Energy level scheme of the H^- ion compared to the hydrogen states of Bohr's atomic model

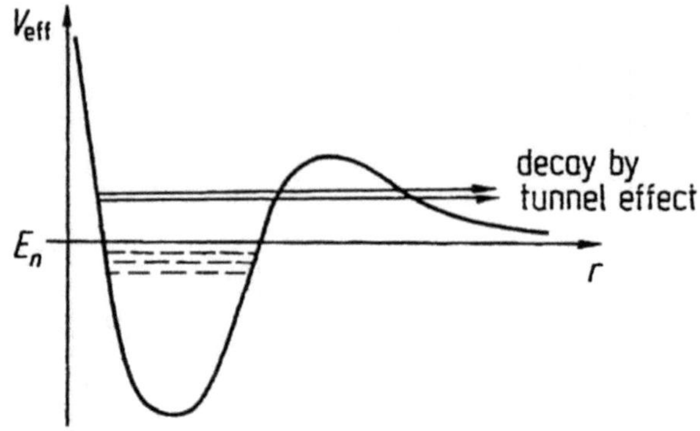

Fig. 2.29 Effective potential V_{eff} of an electron in the field of a hydrogen atom; a repulsive potential dominates for very small values of the distance r between the electron and the hydrogen atom; after that the dipole potential $-\alpha/r^4$ and finally the centrifugal term $\ell(\ell+1)/r^2$ dominate. Below r-axis: hydrogen states with quantum number n and energy E_n are compound states (*dashed lines*); above r-axis: hydrogen states are shape states ("shape resonances", *full lines*) from which the electrons can tunnel through the potential wall

motion of the electron with orbital momentum ℓ. These shape states lie above the excited states of atomic hydrogen. Electrons in shape states can tunnel through the potential barrier of the centrifugal term (Fig. 2.29).

Accordingly we distinguish two types of electron scattering resonances: *Type-I resonances* which are also called *Breit-Wigner* or *Feshbach resonances*, are due to the formation of compound states of the short-lived negative ion. The *Type-II* or *shape resonances* arise from the centrifugal term $\ell(\ell + 1)/r^2$ of the scattering potential. The shape resonance state above the $n = 2$ threshold and the compound states below the $n = 3$ threshold could be detected experimentally in the excitation of the $n = 2$ states (Fig. 2.30).

Feshbach and shape resonances have been observed in electron scattering on many other multi-electron atoms; such resonances often show much more dramatic structure in complex atoms compared to that in atomic hydrogen; this is demonstrated in Fig. 2.31 for the excitation of the first resonance lines of the krypton and mercury atoms. In this connection we draw attention to the steep increase of the cross section of the mercury line $\lambda = 253.7$ nm (transition $6\,^3P_1 \rightarrow 6\,^1S_0$, Fig. 2.31a) which has considerably contributed to the success in the first detection of inelastic energy loss of electrons $\Delta E = hc/\lambda = h\nu$ for the excitation of this spectral line in the historically important *Franck-Hertz experiment*. In this experiment the inelastically scattered electrons in the forward direction are decelerated by an electric field so that they cannot reach the detector electrode. The potential difference ΔV is determined from the difference of the electric potentials between two neighboring minima or maxima of the electron current after traversing the decelerating field. A very slow increase of the cross section of the 253.7 nm Hg line would have been very unfavorable for the detection of the threshold energy $E_{thr} = e\Delta V = h\nu = 4.89$ eV in the Franck-Hertz experiment.

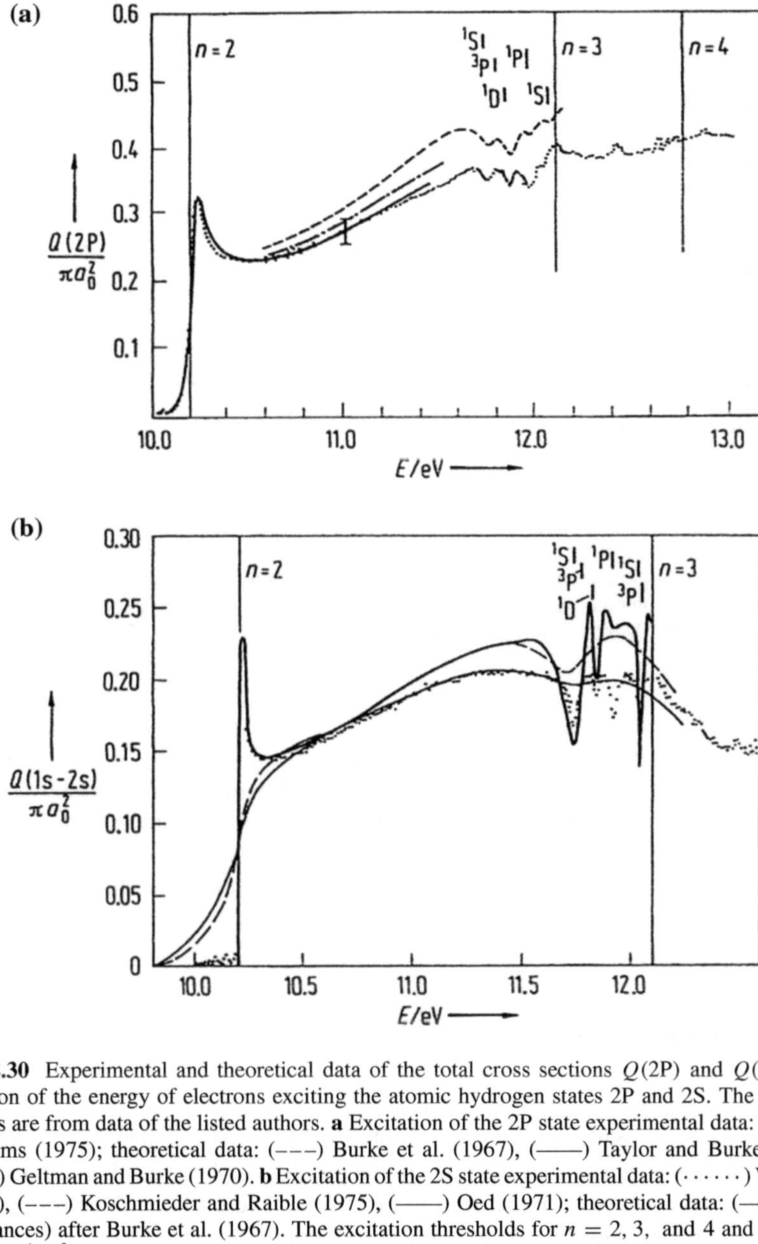

Fig. 2.30 Experimental and theoretical data of the total cross sections $Q(2P)$ and $Q(2S)$ as a function of the energy of electrons exciting the atomic hydrogen states 2P and 2S. The two part figures are from data of the listed authors. **a** Excitation of the 2P state experimental data: (· · · · · ·) Williams (1975); theoretical data: (– – –) Burke et al. (1967), (——) Taylor and Burke (1967), (– · –) Geltman and Burke (1970). **b** Excitation of the 2S state experimental data: (· · · · · ·) Williams (1975), (– – –) Koschmieder and Raible (1975), (——) Oed (1971); theoretical data: (——, with resonances) after Burke et al. (1967). The excitation thresholds for $n = 2, 3,$ and 4 and the level symbols ^{1}S, ^{3}P, ... of the compound states below the ($n = 3$) threshold are marked

A detailed investigation of the Franck-Hertz experiment with mercury has recently been carried out with excitation by polarized electrons and by observing circular polarization of the 253.7 nm mercury spectral line (Wolcke et al. 1983). It was shown

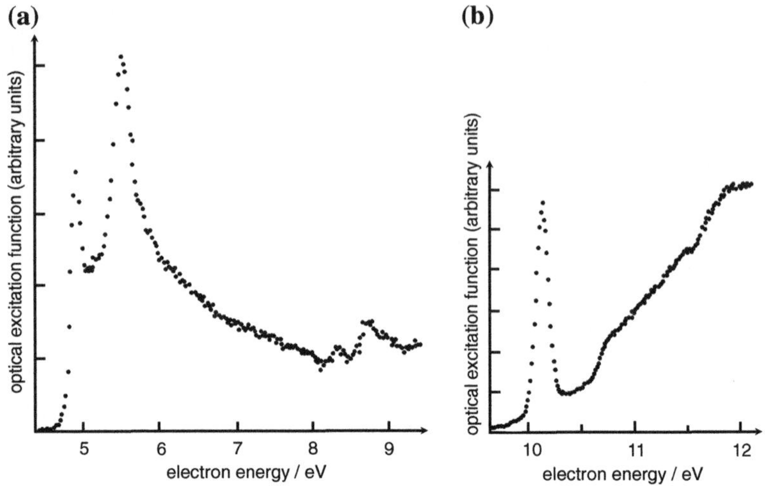

Fig. 2.31 Optical excitation functions, i.e. relative cross sections for the observation of the first resonance lines perpendicular to the direction of the incident electrons: **a** for the 253.7 nm line of the ($6\,^3P_1 \longrightarrow 6\,^1S_0$) intercombination transition of mercury atoms (after Ottley and Kleinpoppen 1975) and **b** for the combined excitation of the 123.5 nm and 116.4 nm lines of krypton $^{3,1}P \longrightarrow {}^1S$ transitions (after Al-Shamma and Kleinpoppen 1978)

that not only the above described resonance effects, but also the amounts and ratios of the cross sections of all three $6\,^3P_{2,1,0}$ states have an influence on the measured potential difference ΔV in the Franck-Hertz experiment. In the interplay of these quantities, the pressure and geometry of a Franck-Hertz experiment have important roles; the measured potential difference may vary between $\Delta V = 4.8$ and 5.15 V in traditional Franck-Hertz tubes as has been shown in a model calculation by Hanne (1988).

2.4 Coincidence Experiments

In the previous sections we have described electron scattering processes in connection with experimental methods in which the observable quantities such as the number of the scattered or transmitted electrons or the number of observed photons are sums of many single events of the scattering processes. These observables result from superpositions or averaging of various types of single processes from which cross sections can be determined as described in the previous sections.

However, because of averaging over the single processes involved, the quality of information from measurements of cross sections is limited. Considerable progress towards more detailed information can be achieved by applying *coincidence techniques* for the study of basic processes of atomic collisions. It has been found that maximum information on the collisional process can be obtained from analyzing special coincidence experiments. In such cases, we talk about *"complete" experiments* which means that maximum information can be extracted from experimental methods, as already discussed in Sect. 2.3.

Coincidence experiments were first introduced by the German W. Bothe in nuclear physics during the end of the 1930s. The terms coincidences, coincidence signals or coincidence counts mean physically that simultaneous or time-correlated signals are registered as pairs or groups of detection signals of the collision process.

2.4.1 Scheme of a Typical Electron–Photon Coincidence Experiment

We illustrate a typical coincidence experiment in connection with the excitation of an excited atomic state by electron impact, which we describe by the reaction equation

$$\begin{aligned} e(E_0) + A \longrightarrow\ & A^* + e(E_0 - E_{thr}) \\ & \hookrightarrow A + h\nu . \end{aligned} \tag{2.46}$$

The excited atoms A^* decay after typical times of about 10^{-9} to 10^{-7} s under the emission of photons of energy $h\nu$ while the scattered electrons lose the threshold energy E_{thr} for the excitation of the atom. The two particles, the photon and the scattered electron, which result from the same excitation process, will leave the same atom after the excitation; the arrival of the two particles at their detectors is correlated in time and can be detected as a coincidence event by coincidence electronics. The electrons produce a detector pulse at the electron detector (electron signal $I_1(t)$), while the photons produce a photon pulse $I_2(t)$ at the photon detector. However, both detector signals arise at different times (t_1, t_2) since photons and electrons have different times of flight and therefore different arrival times at their detectors.

Genuine Versus Statistical Coincidences

It can be expected that *genuine* coincidence signals between the two detector pulses $I_1(t_1)$ and $I_2(t_2)$ may be separated from so-called *chance* or *statistical* coincidence signals by the constant detection time difference $\Delta t = t_1 - t_2$. Statistical or chance coincidences are pairs of detected pulses from events, the origins of which do not arise from the same single elementary process; for example, the photon detector signal may originate from one atom while the electron detector signal from another atom. The trick for the detection of genuine coincidences consists in the following: An electronic time-to-pulse height converter (also called time-to-amplitude converter, abbreviation TAC) transforms the time difference Δt of the arrival of the two particles at the detectors into an electrical pulse; the height ΔH of this pulse coming out of the TAC is proportional to Δt. While the genuine coincidence pulses are expected to have a constant time difference Δt between the two detector signals, the statistical coincidences have variable or statistically distributed time differences Δt. The pulse-height distribution of the genuine coincidences as a function of Δt will have a pronounced peak on a uniform background of the pulse-height distribution of the statistical coincidences (Fig. 2.32). The pulse-height distribution is normally recorded and made visible on the screen of an electronic multichannel analyzer or computer.

A typical characteristic of the coincidence technique is linked to the geometry of incoming and reaction particles (Fig. 2.33). The scattering geometry is normally defined by the directions of the incoming and scattered particles. The mathematical function that describes the angular correlation of the (genuine) count rates of electron-photon coincidences from electron impact excitation depends on the angular coordinates of the observed electrons (θ_e, ϕ_e) and photons $(\theta_\gamma, \phi_\gamma)$. The scattering plane is also a symmetry plane (which is called *planar symmetry*) for the angular correlations of the coincidences and all other possible physical quantities extracted from the coincidence measurements. Electron-atom collision experiments carried out without coincidence analysis (as described above) is normally characterized by a cylindric symmetry which is determined by the direction of the incoming electrons. The ratio of genuine to statistical coincidences can be measured. The real coincidence

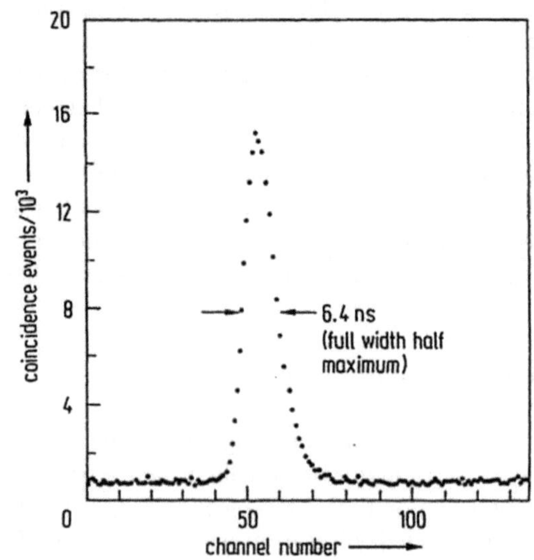

Fig. 2.32 Electron-photon coincidence signal from the excitation and de-excitation $1\,^1S_0 \longrightarrow 2\,^1P_1 \longrightarrow 1\,^1S_0$ of helium at an energy of the impinging electrons of $E_0 = 80\,\text{eV}$ and joint observation of the inelastically scattered electrons at an angle of $\theta_e = 16°$ and the 58.4 nm photon at an angle of $\theta_\gamma = 127°$ in the scattering plane, defined by the directions of the incoming and scattered electrons. Recording time for measuring the coincidence signal was 12 h. Definition of θ_e and θ_γ follows from the next 12 h; after Eminyan et al. (1974)

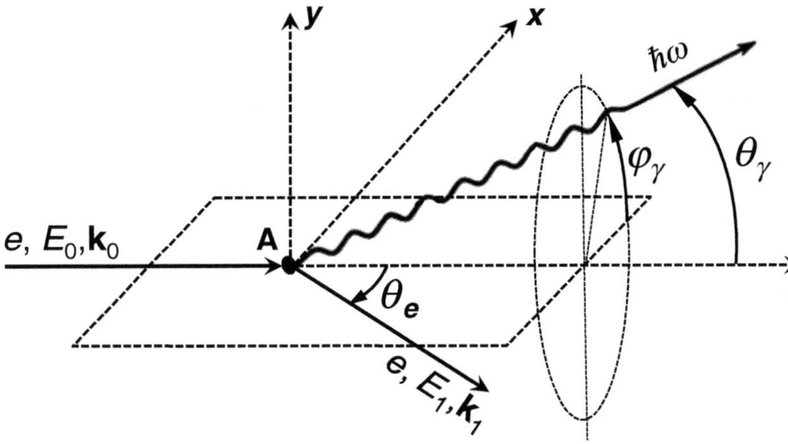

Fig. 2.33 Geometry of the angular correlation between an inelastically scattered electron $(e, E_1, \mathbf{k}_1)$ and the photon $h\omega$. The atom is at the origin of the coordinate system; the incident electrons $(e, E_0, \mathbf{k}_0)$ are propagating parallel to the z-direction

count rate, i.e. the number ΔN_{real} of coincidence signals per unit of time depends on several parameters that we can separate from each other as follows:

$$\Delta N_{real} = f_{\gamma e}(\theta_\gamma, \phi_\gamma, \theta_e, \phi_e) N_0 \eta_e \eta_\gamma \Delta\Omega_\gamma \Delta\Omega_e, \tag{2.47}$$

where N_0 is the total rate of the decaying excited atoms under the emission of the photons; η_e and η_γ are the detector sensitivities for the electrons and photons, respectively; Ω_e and Ω_γ are their solid angles for detection and $f_{\gamma e}(\theta_\gamma, \phi_\gamma, \theta_e, \phi_e)$ the angular correlation function for the coincident detection of the two particles. The statistical coincidence count rate is described by

$$\Delta N_{stat} = f_e(\theta_e, \phi_e) \eta_e N_0 \Delta\Omega_e f_\gamma(\theta_\gamma, \phi_\gamma) \eta_\gamma N_0 \Delta\Omega_\gamma \Delta\tau, \tag{2.48}$$

or with

$$N_e = N_0 f_e \eta_e \Delta\Omega_e \quad \text{and} \quad N_\gamma = N_0 f_\gamma \eta_\gamma \Delta\Omega_\gamma,$$
$$\Delta N_{stat} = N_e N_\gamma \Delta\tau, \tag{2.49}$$

whereby N_e and N_γ are the count rates of the detected electrons and photons, respectively (i.e. the single count rates for the two particles), f_e and f_γ the angular distribution functions of the inelastically scattered electrons and of the photon emission, the product $f_{\gamma e} = f_e f_\gamma$ of which is the combined angular correlation function and $\Delta\tau$ is the time resolution of the coincidence apparatus. The physical meaning of the time resolution is that if, for example, two coincidence events occur within the time interval $\Delta\tau$ they cannot separately be detected with the coincidence apparatus

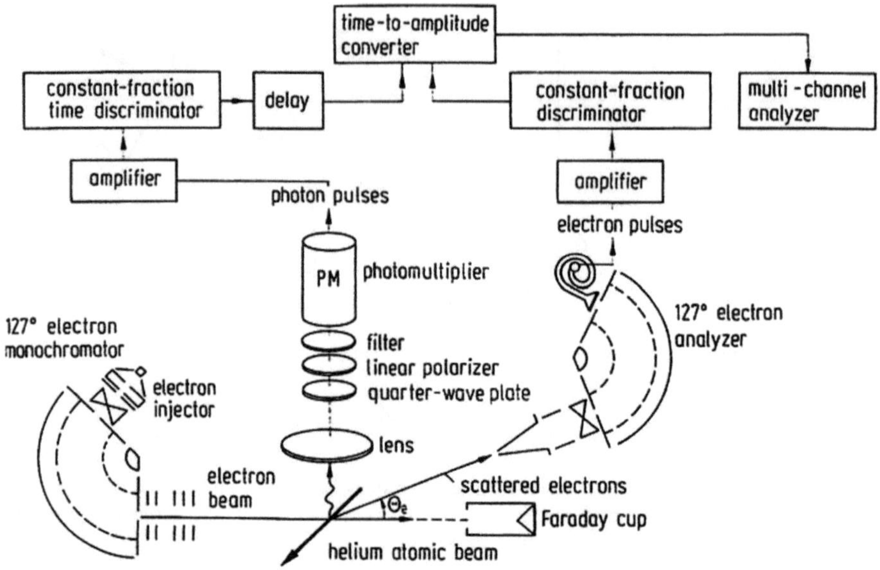

Fig. 2.34 Scheme of a typical electron-photon coincidence apparatus for observing polarized photons perpendicular to the scattering plane, defined by the incoming and scattering electrons; after Standage and Kleinpoppen (1975)

of this limited time resolution. Smaller $\varDelta\tau$ means the time resolution for detecting two consecutive coincident events becomes better; typical values in present coincidence apparatus are in the order of magnitude of $\varDelta\tau \approx 10^{-9}$ s (or shorter). Taking into account of the above relations we obtain for the ratio of genuine to statistical coincidence count rates

$$\frac{\varDelta N_{real}}{\varDelta N_{stat}} = \frac{1}{N_0 \varDelta\tau} = \kappa. \tag{2.50}$$

This ratio of the coincidence count rates increases with decreasing time resolution $\varDelta\tau$ and decreases with increasing decay rate N_0. The $(1/N_0)$ dependence is due to the fact that the statistical coincidences are proportional to N_0^2, while the genuine coincidences are proportional to N_0. Increasing the physical quantity N_0 is only sensible if the ratio κ remains large enough for a realistic coincidence measurement. Relation (2.50) is only correct if no additional or undesired processes influence the coincidence count rates. Note that in the vast majority of physical processes the measurement statistics improves with increasing values for the physical quantity to be measured.

Figure 2.34 shows the scheme of a typical electron-photon coincidence apparatus with electronic devices for amplification and recording of the coincidence signals. The incoming electrons from a 127° electron monochromator are inelastically scattered; the scattered electrons only pass the electron analyzer if it is tuned to a given specific energy. We operate the electron analyzer for the energy $E_0 - E_{thr}$ where

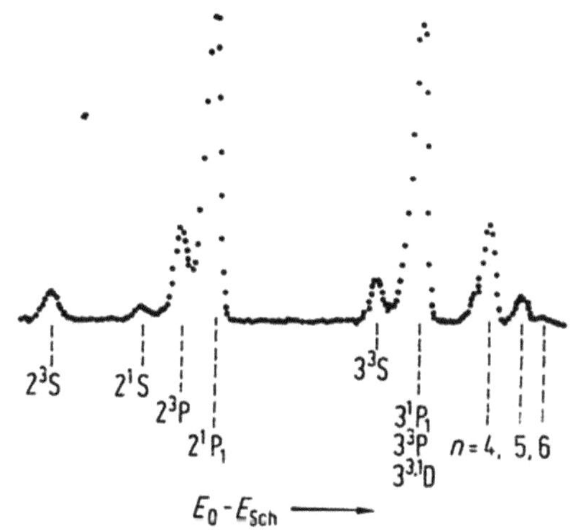

Fig. 2.35 Energy-loss spectrum of electrons having excited states of helium with the principal quantum numbers $n = 2, 3, 4, 5,$ and 6. $E_0 - E_{Sch}$ is the difference between the energy of the incident electrons and the transmission energy of electrons passing the electron analyzer; after Kleinpoppen and McGregor (1980)

E_0 is the energy of the incoming electrons and E_{thr} the threshold energy for the excitation of an atomic state. Figure 2.35 displays a typical energy loss spectrum of electrons; excited energy states of the helium atom can be clearly identified. In an electron-photon coincidence experiment, the electron analyzer is tuned to a sharp energy loss which leads to the required excitation and the subsequent emission of a photon line. As an example we discuss the excitation process of the $2\,{}^1P_1$ state of helium. The electron analyzer is tuned to the energy-loss peak for exciting the $2\,{}^1P_1$ state as in Fig. 2.35 while the photon detector counts the number of photons with wavelength $\lambda = 58.4$ nm from the $2\,{}^1P_1 \longrightarrow 1\,{}^1S_0$ transition.

2.4.2 The λ–χ Representation of the ${}^1S_0 \longrightarrow {}^1P_0 \longrightarrow {}^1S_0$ Excitation/De-excitation Process

The theoretical analysis of the electron-photon angular correlation from this excitation process is based upon the following physical arguments. The excitation of the ${}^1P_1$ state from the ${}^1S_0$ state can be described by introducing the quantum mechanical state vector

$$|\psi({}^1P_1)\rangle = f_0\psi_0 + f_1\psi_1 + f_{-1}\psi_{-1} = \sum_m f_m\psi_m, \tag{2.51}$$

where f_0, f_1 and f_{-1} are complex excitation amplitudes for the excitation of the magnetic sub-states of the ${}^1P_1$ state with the quantum numbers $m = 0$ and $m = \pm 1$ and their eigenfunctions ψ_0 and $\psi_{\pm 1}$, respectively.

By forming Dirac's bracket and taking into account the symmetry and parity of $f_1 = -f_{-1}$ we obtain

$$\langle\psi(^1\mathrm{P})\,|\,\psi(^1\mathrm{P})\rangle = \left|f_0\right|^2 + \left|f_{-1}\right|^2 + \left|f_1\right|^2 = \left|f_0\right|^2 + 2\left|f_1\right|^2. \tag{2.52}$$

where

$$\left|f_0\right|^2 = \sigma_0, \tag{2.53}$$

and

$$\left|f_1\right|^2 = \sigma_1, \tag{2.54}$$

can be defined and normalized as *partial cross sections* for the excitation of the magnetic sub-states $m = 0$ and $m = \pm 1$. The differential cross section then becomes

$$\sigma(E, \theta_e) = \sigma_0 + 2\sigma_1. \tag{2.55}$$

The electron-photon angular correlation can be calculated from the dipole matrix element $P(\theta_\gamma) = \langle\psi(^1\mathrm{P}_1)|e\mathbf{r}|\psi\rangle(^1\mathrm{S}_0)$ by including the above state vector for observing the photons in the scattering plane ($\phi_\gamma = 180°$) and for a fixed electron scattering angle θ_e to

$$\left|\frac{\mathrm{d}P(\theta_\gamma)}{\mathrm{d}\Omega_\gamma}\right|_{\theta_e=const.} = \frac{8\pi}{3} N_{e,\gamma}, \tag{2.56}$$

with

$$N_{e,\gamma} = \lambda \sin^2\theta_\gamma + (1-\lambda)\cos^2\theta_\gamma - 2\sqrt{\lambda(1-\lambda)}\cos\theta_\gamma \sin\theta_\gamma \cos\chi. \tag{2.57}$$

The last term of (2.57) represents an *interference effect* for the excitation of the magnetic substates.

The λ and χ parameters are defined by

$$\lambda = \frac{\sigma_0}{\sigma} = \frac{\left|f_0\right|^2}{\sigma}, \tag{2.58}$$

$$1 - \lambda = 2\frac{\sigma_1}{\sigma} = 2\frac{\left|f_1\right|^2}{\sigma}, \tag{2.59}$$

and χ as the phase difference

$$\chi = \chi_0 - \chi_1 = arg\left(\frac{f_0}{f_1}\right), \tag{2.60}$$

between the excitation amplitudes f_0 and f_1.

2.4.3 *Quantum Mechanical Coherence in the Electron Impact Excitation of Atoms*

The equation for the angular-correlation function $N_{e,\gamma}$ can be transformed as follows.

$$N_{e,\gamma} = \frac{1}{\sigma}\left|f_0 \cos\theta - \sqrt{2} f_1 \sin\theta\right|^2, \tag{2.61}$$

in which $f_0/\sqrt{\sigma}$ is interpreted as a relative amplitude of an electromagnetic Hertzian oscillator parallel to the z-axis (direction of incoming electrons!) and $\sqrt{2}f_1/\sqrt{\sigma}$ as a relative amplitude parallel to the x-axis; this coherent representation of the angular correlation follows directly from (2.57). The two Hertzian oscillators coherently superpose on each other with a phase shift χ, i.e., the physical interpretation of the measurement of the angular correlation is equivalent to an analysis of a *quantum mechanical coherence or interference effect in the electron-impact excitation of the atom.* The relative electric amplitudes and intensities of the electromagnetic radiation of the Hertzian oscillators with reference to the z- and x-direction of Fig. 2.33 are given by

$$E_{z0} \sim \frac{|f_0|}{\sqrt{\sigma}}, \qquad \left|E_{z0}\right|^2 \sim \lambda; \tag{2.62a}$$

and

$$E_{x0} \sim \frac{\sqrt{2}|f_1|}{\sqrt{\sigma}}, \qquad \left|E_{x0}\right|^2 \sim 1-\lambda. \tag{2.62b}$$

Figure 2.36 shows typical electron-photon angular correlations for the electron impact excitation of the $2\,^1\mathrm{P}_1$ state of helium. We note that Born's approximation almost correctly reproduces the observed angular correlation for small angles of the inelastically scattered electrons; however, there is a considerable disagreement between the experimental data and Born's prediction at larger electron scattering angles. The reason for this is that Born's approximation depends on the matrix element $\langle\psi|\mathrm{e}^{(\mathrm{i}/\hbar)\Delta\mathbf{pr}}|\psi_0\rangle$ with ψ_0 and ψ as eigenfunctions of the ground and excited state and $\Delta\mathbf{p}$ as the momentum transfer which the atom experiences by the electron impact $\Delta\mathbf{p} = \mathbf{p}_i - \mathbf{p}_f$, where $\mathbf{p}_i$ and $\mathbf{p}_f$ are the linear momenta of the incoming and scattered electrons, respectively.

Born's approximation is sufficiently valid only for small momentum transfer and with $\mathrm{e}^{(\mathrm{i}/\hbar)\Delta\mathbf{pr}} \approx 1 + (\mathrm{i}/\hbar)\Delta\mathbf{pr}$. The first term does not contribute due to the orthogonality of the $^1\mathrm{S}_0$ and $^1\mathrm{P}_1$ states. This means for our example above that the $^1\mathrm{P}_1$ state will be excited along the momentum transfer $\Delta\mathbf{p}$ (Bethe 1933), i.e. Born's approximation remains finite only for the selection rule $\Delta m = 0$ with respect to the angular momentum transfer vector as axis of quantization. This means for the excitation/de-excitation $^1\mathrm{S}_0 \to {}^1\mathrm{P}_1 \to {}^1\mathrm{S}_0$ that only the $m = 0$ substate of the $^1\mathrm{P}_1$

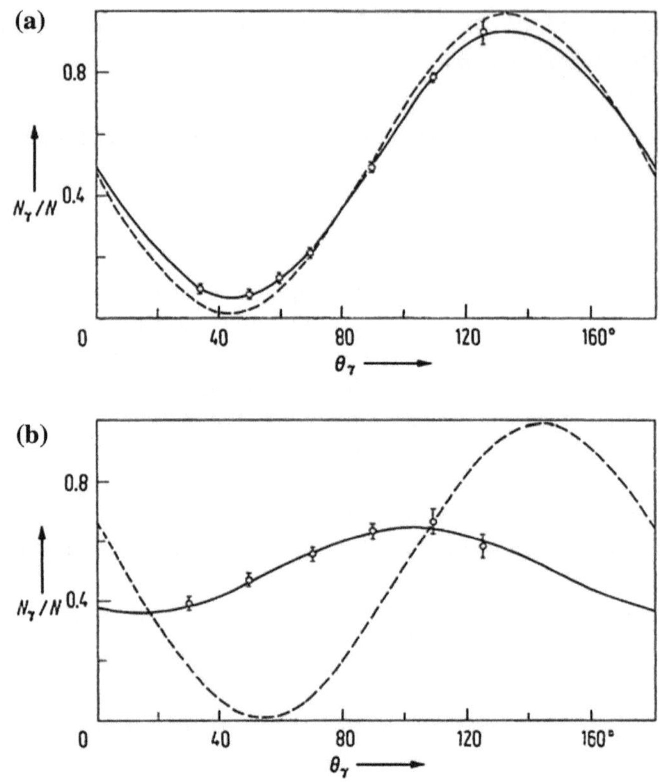

Fig. 2.36 Electron-photon angular correlations from the excitation and de-excitation $1\,^1S_0 \longrightarrow 2\,^1P_1 \longrightarrow 1\,^1S_0$ of helium for electron impact at an energy of 60 eV; the ordinate gives the number N_γ (in arbitrary units N of coincident photons detected in the scattering plane, as a function of the observation angle θ_γ whereby the electron scattering angle θ_e is kept constant: **a** electron scattering angle $\theta_e = 16°$; **b** electron scattering angle $\theta_e = 40°$. The *dotted lines* represent theoretical results of Born's approximation; the *full curves* are least-square fits to the shape of the function for the coincidence rate $N_{e,\gamma}$ of the experimental data, see (2.61); after Eminyan et al. (1973, 1974)

is excited and the emitted photon radiation is that of a dipole oscillating parallel to the momentum transfer $\Delta\mathbf{p}$. Accordingly, the electron-photon angular correlation for a fixed electron scattering angle in the scattering plane is described by a $\sin^2\theta$ relation of the emission of an electromagnetic oscillator (θ is the angle between $\Delta\mathbf{p}$ and the direction of observation: Fig. 2.37). In this approximation, which only takes account of the momentum transfer to the atom, the observed electron-photon angular correlation is described by the coherent superposition of two oscillators parallel to the z- and x-axis discussed above.

In this connection we draw attention to the fact that the above phase difference χ may take on all possible values between 0° and 180° for the case of incoherent excitation; this is equivalent to the average value $\cos\chi = 0$. The interference term of (2.57) would then vanish and the polarization ellipse $\mathbf{E} = \mathbf{E_z} + \mathbf{E_x}$ for the radiation

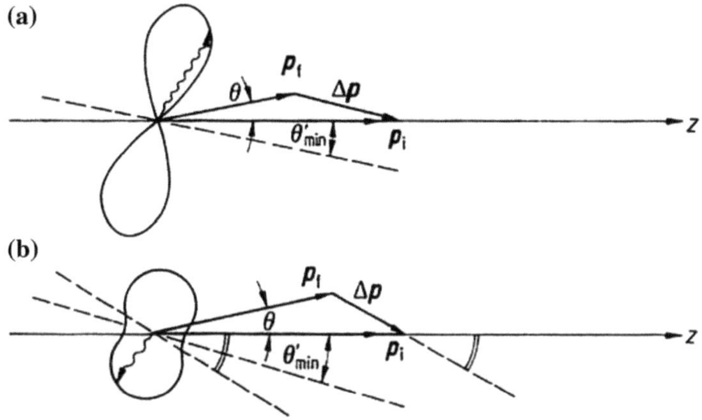

Fig. 2.37 Relation between the electron-photon angular correlation and the momentum transfer $\Delta \mathbf{p}$ for the helium excitation/de-excitation process $1\,^1S_0 \longrightarrow 2\,^1P_1 \longrightarrow 1\,^1S_0$ by electron impact. The electron has initial momentum $\mathbf{p}_1$ and, after the scattering, a momentum $\mathbf{p}_f$. The length of the *wavy arrows* is proportional to the coincident photon intensity in the direction of observation: **a** after Born approximation, $\Delta m_\ell = 0$ refers to $\Delta \mathbf{p}$ as axis where no change of the orbital angular momentum component m_ℓ is possible for the $1\,^1S_0 \longrightarrow 2\,^1P_1$ transition; **b** according to typical experimental observation

of a Hertzian oscillator with amplitudes $\mathbf{E_z}$, and $\mathbf{E_x}$ would lie parallel to the z- and x-axis with regard to its principal axes; the same would be true for the intensity curve $I = \left|E_x + E_z\right|^2 = \left|E_x\right|^2 + \left|E_z\right|^2$ of the coincident photons. However, experiments on electron-photon angular correlation have shown unambiguously that the principal axes of the polarization ellipses are not parallel to the z- and x-axes in general. Therefore, we can assume that the phase χ in the excitation of $^1P_1$ state does not fluctuate statistically. Only a phase difference of $\chi = \pi/2$ would cause the interference term of (2.57) to vanish for finite values of λ.

The very good fit of the experimental coincidence rates (Fig. 2.36) on the above angular correlation function $N_{e,\gamma}$ has encouraged physicists to carry out extended experimental investigations in order to obtain complete information on the excitation process of atoms by electron impact. A possible set of physical quantities which completely describes the excitation of a $^1P_1$ state are the differential cross section σ, the angular correlation parameter $\lambda = \left|f_0\right|^2/\sigma$ and the phase difference χ between the two excitation amplitudes f_0 and f_1. Figure 2.38 shows an example for the excitation of the He($2\,^1P_1$) state at an electron energy of 24 eV, in which a good agreement exists between the theoretical and experimental data for λ and χ. However, we would like to mention that no satisfactory agreement exists between such theoretical and experimental angular correlation parameters at some other, particularly intermediate, electron energies. These discrepancies are unresolved research problems which, on the other hand, imply that the λ, χ parameters (or equivalent parameters) are very sensitive tests for theories of atomic collision processes. We leave these problems,

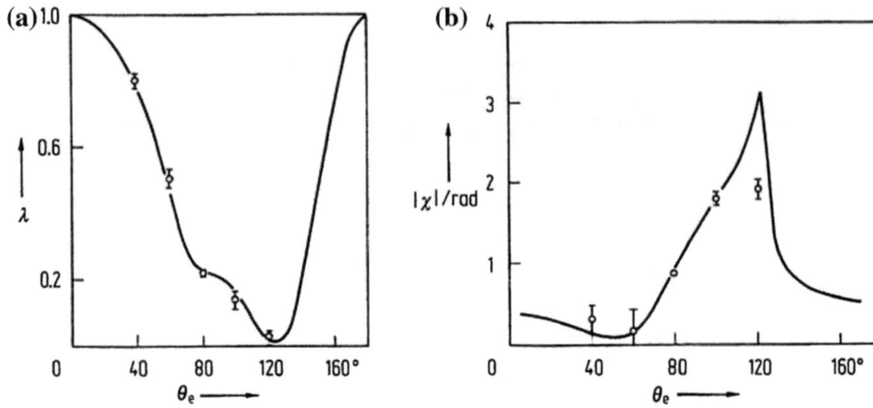

Fig. 2.38 λ- and $|\chi|$-data of the He($2\,^1$P)-excitation by electron impact in dependence of the electron scattering angle θ_e at an electron of 24 eV; experimental data of Crowe et al. (1983); the *full theoretical curve* represents a 5-state **R**-matrix calculation of these authors

however, but will discuss further relevant physical effects that can be extracted from electron-photon coincidence experiments.

The above description of electron-photon angular correlations is based on the *model of coherent excitation* of magnetic substates of the ^{1}P$_1$ state which allows for characterizing the ^{1}P$_1$ state by a wavefunction; see (2.51). The correct interpretation of this coherent excitation can be confirmed by the following arguments and experimental tests. If coherently excited atoms decay by emission of a photon, it can be shown by measuring the polarization of the coincident photon that the photon is completely coherent. Complete coherence of the photons means that two orthogonally oscillating electric vectors of the photon radiation must have a constant phase difference between them. This can indeed be detected for the ^{1}P$_1$ excitation of helium by electron impact. The macroscopically measured phase difference β between two light vectors of the coincident photons oscillating parallel to the z- and x-direction have been proved identical to the phase difference χ between the two excitation amplitudes f_0 and f_1, i.e. $\chi = \beta$. In other words this is an interesting *example in which a macroscopic phase difference* β *is identical to an atomic phase difference* χ.

2.4.4 Alignment and Orientation

Further physical quantities which can be extracted from analyzing electron-photon coincidence experiments are *alignment* and *orientation*. These quantities describe the state of excited states based upon the distribution and orientation of their angular momentum components m_L with reference to a given axis of quantization. Figure 2.39 shows a diagrammatic representation of alignment and orientation of the collisionally excited atomic state. With reference to the above example of ^{1}P$_1$

excitation, the following characteristics of the quantities alignment and orientation can be made: According to Fig. 2.39, orientation means an excess of angular momentum $\Delta \mathbf{L}$ with reference to the z-axis. However, because of the conservation of angular momentum, orbital angular momentum can only be transferred perpendicular to the scattering plane but not in the scattering plane. In other words, only the angular momentum component ΔL_y perpendicular to the scattering plane (Fig. 2.33) can have finite quantized values while the components ΔL_x and ΔL_z vanish because of conservation of angular momentum. The angular momentum transfer $\Delta L_y = \langle L_\perp \rangle$ perpendicular to the scattering plane can be measured directly by the circular polarization of the photons from the transition $^1P_1 \rightarrow {}^1S_0$. For this we set $N(m = 1)$ and $N(m = -1)$ as numbers of atoms in the two magnetic substates $m = \pm 1$ with the normal (parallel to the y-axis in Fig. 2.33) of the scattering plane as axis of quantization. The excess of angular momentum of the excited atoms propagates into an excess of photons with their spins parallel ($N^\uparrow$) and antiparallel ($N^\downarrow$) during the $^1P_1 \rightarrow {}^1S_0$ transition, with reference to the y-direction; the ratio of the difference to the sum of these quantities becomes identical to the negative value of the circular polarization

$$P_3 = P_{circ} = \frac{I(R) - I(L)}{I(R) + I(L)}, \tag{2.63}$$

whereby $I(R)$ and $I(L)$ are the intensity components for right- and left-handed circularly polarized light.

$$\langle L_\perp \rangle = \frac{N(m = 1) - N(m = -1)}{N(m = 1) + N(m = -1)} = \frac{N^\uparrow(h\nu) - N^\downarrow(h\nu)}{N^\uparrow(h\nu) + N^\downarrow(h\nu)} = -P_{circ}. \tag{2.64}$$

The difference of the sign between $\langle L_\perp \rangle$ and P_{circ} is due to the difference in the definition of circularly polarized light in traditional optics and for angular momenta and spins in atomic and particle physics (called helicity of angular momenta). Figure 2.40

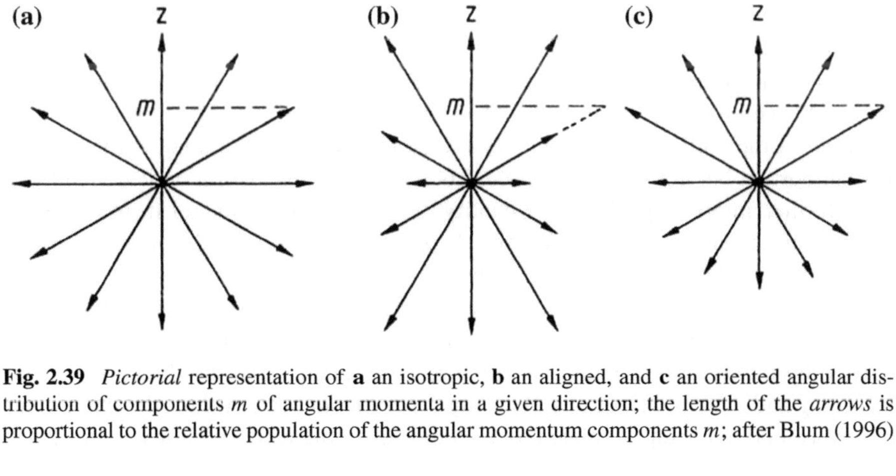

Fig. 2.39 *Pictorial* representation of **a** an isotropic, **b** an aligned, and **c** an oriented angular distribution of components m of angular momenta in a given direction; the length of the *arrows* is proportional to the relative population of the angular momentum components m; after Blum (1996)

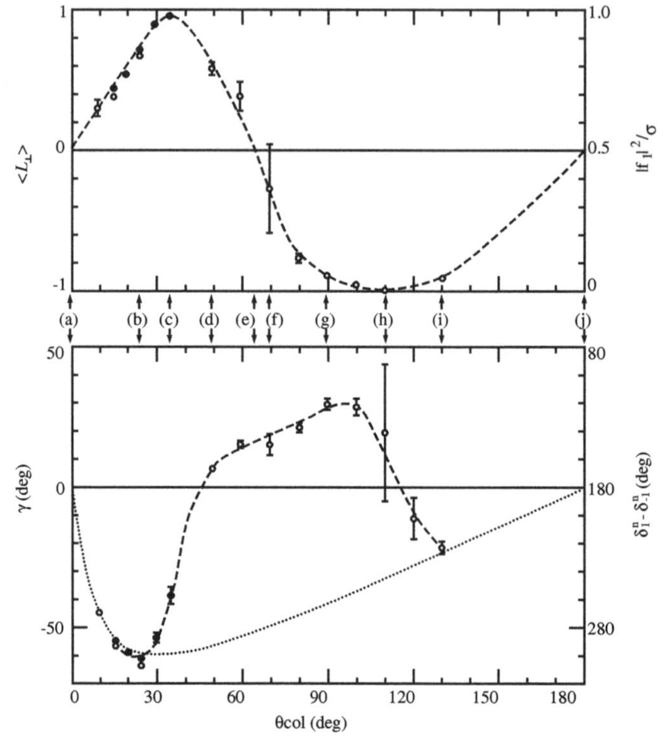

Fig. 2.40 *Upper diagram* angular momentum transfer $\langle L_\perp \rangle$ (in units of $\hbar$) from the He ground state $1\,^1S_0$ to the excited $2\,^1P_1$ state by electron impact at an energy of 80 eV as a function of the scattering angle $\theta_{col} = \theta_e$ of inelastically scattered electrons. Data from Eminyan et al. (1973, 1974) and Hollywood et al. (1979). The *dashed lines* in both figures are drawn to help the eye follow the trend of the experimental data. *Lower diagram* orientation angle γ in degrees of the electron cloud of the $2\,^1P_1$ state (see Fig. 2.41) under the same excitation conditions as in the *upper diagram*

shows experimental data for $\langle L_\perp \rangle$ that have been extracted from electron-photon coincidence measurements of the $2\,^1P_1$ excitation of helium. The orientation of the $2\,^1P_1$ state has optima at the electron scattering angles $\theta_e = 35°$ (maximum) and $\theta_e = 120°$ (minimum) for an initial electron energy of 80 eV. The physical understanding of the collision dynamics in the creation of angular momenta transfer is a very important but not yet fully solved problem of the theory of atomic collisions.

A further interesting physical quantity, which can be extracted from electron-photon coincidence experiments, is the shape and orientation of the electron charge distribution (called the "*electron cloud*" of the excited state). For example, from a knowledge of the state vector $|\psi(^1P_1)\rangle$ (2.51), the charge contribution $e\langle\psi(^1P_1)\,|\,\psi(^1P_1)\rangle d\tau$ of the electron cloud in the volume element $d\tau$ can be calculated by taking into account the initial electron energy and the directions of the incoming and scattered electrons. By using experimental λ and χ data, the shape of an electron cloud has been calculated for a specific example which is shown in Fig. 2.41.

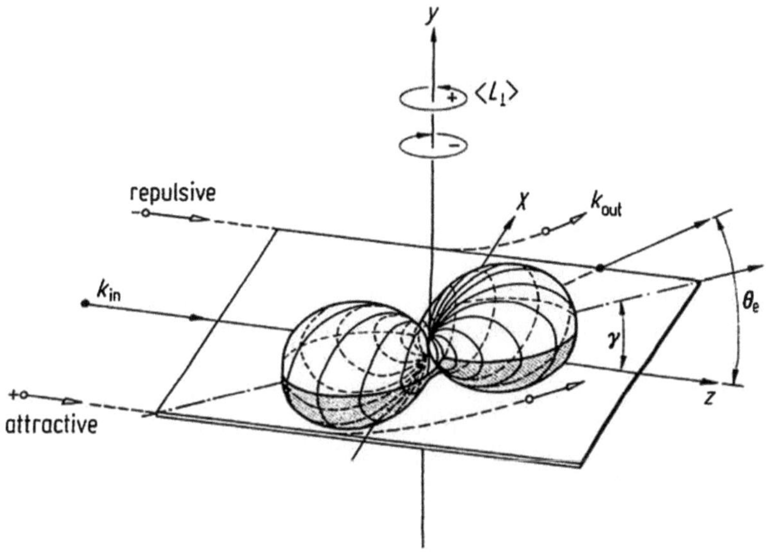

Fig. 2.41 Typical form of the electron cloud for a coherently excited $2\,^1P_1$ state with an orientation angle $\gamma = 45°$ and angular momentum transfer $\langle L_\perp \rangle = 0.75\,\hbar$. The incoming electrons k_{in} are scattered under the angle θ_e (k_{out}) whereby either a *repulsive* and/or an *attractive* potential is effective

It is interesting to note that the special structure and orientation of this electron charge distribution arises only through coherent superposition of the wavefunctions ψ_0 and $\psi_{\pm 1}$ of the magnetic substates with amplitudes f_0 and f_1 and subsequent application of Dirac's bracket involved. We cannot reproduce these types of shape of the electron clouds by incoherent superposition of the charge distributions; they are rotationally symmetric around the axis of quantization. The lower part of Fig. 2.40 shows the orientation angle γ of the charge cloud as a function of the scattering angle for an initial electron energy of 80 eV. The orientation angle γ is related to the linear polarization of photons emitted in the radiative decay of the $^1P_1$ state by

$$\gamma = \frac{1}{2}\arctan\left(\frac{P_2}{P_1}\right), \tag{2.65}$$

where the radiation is propagating perpendicular to the reaction plane (see Fig. 2.41), while

$$P_1 = \frac{I(0°) - I(90°)}{I(0°) + I(90°)}, \tag{2.66a}$$

and

$$P_2 = \frac{I(45°) - I(135°)}{I(45°) + I(135°)}, \tag{2.66b}$$

with $I(\alpha)$ being the intensity component of light polarized under the angle α with respect to the incident electron beam.

The descriptions of the electron impact excitation of the $^1P_1$ state by either the quantities (σ, λ, χ) or $(\sigma, \langle L_\perp \rangle, \gamma)$ are equivalent to each other and each set of such data can be calculated from the other.

It should be emphasized, however, that the primary physical process involved is the *coherence of the excitation of the* $^1P_1$ *state* which manifests itself in an interference effect of coherent oscillations (similar to the Hanle effect) and which is directly detected by interfering amplitudes of the excitation process. We continue the discussion on the parameters of collisionally excited states in Sect. 2.5.3.

2.4.5 Super-Elastically Scattered Electrons

In addition to electron–photon coincidence experiments, we refer to an equivalent experimental method which is based upon *optical pumping* and *simultaneous detection of super-elastically scattered electrons*. As can be proved in atomic spectroscopy, laser radiation intensively induce transitions between atomic states. On average and under condition of saturation between excitation and de-excitation processes a considerable part of atoms can be populated in an excited state by resonance transitions (see Sect. 2.5.5). Electrons which are scattered by atoms in an excited state may induce a superelastic de-excitation process of the atom and gain energy, i.e.,

$$e(E_0) + A^* \longrightarrow A + e(E_0 + \Delta E). \tag{2.67}$$

This superelastic de-excitation process can be read in the opposite direction (called time reversal), which represents an equivalent excitation process. This leads to a comparison of the above electron–photon coincidence experiment with the inverse de-excitation of a laser excited atom (Fig. 2.42). The measurement of angular distributions of super-elastically scattered electrons is then equivalent to the measurement of electron–photon angular correlations since all atoms are expected to be excited by laser photons of identical polarization state. The superelastic collision process has initially been applied to the first resonance transition in sodium (D-line excitation) by Hertel and Stoll (1974a, b, 1977) which provided the above λ and χ parameters or equivalent data. Many reviews and papers with this alternative method have been published (e.g. Campbell et al. 1988): we mention the successful experiments for the super elastic scattering from alkali atoms where laser radiation was available (Karaganov et al. 2001). Since the spectral range of dye lasers is limited, the range of applications of the super-elastic method is somewhat restricted to a few

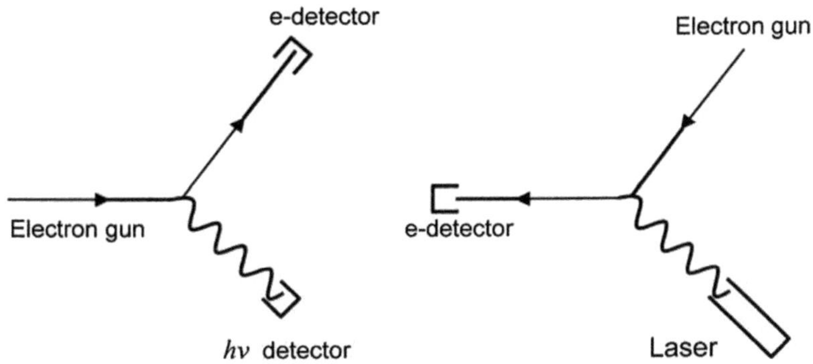

Fig. 2.42 Comparison of electron–photon coincidence experiment (*left*) and the inverse de-excitation of a laser excited atom (*right*); after Hertel (1976)

alkaline and earth alkaline metal atoms so far. On the other hand, the method has the advantage that single count rates of super-elastically scattered electrons are larger than corresponding electron–photon coincidence count rates. Additionally, the use of laser pumping, which is a multiphoton process, allows for accessing a more varied choice of polarization parameters of the laser excited level, including polarization multipoles of higher ranks, not accessible by detecting a single coincidence photon.

2.4.6 (e, 2e) and Multi-ionization Processes

Further coincidence experiments have been carried out in connection with ionization processes of atoms by electron impact. Two electrons or one electron with an ion in a given charge state can be detected in coincidence in such experiments. The relevant events are described by the following reaction processes:

$$e(E_0) + A \longrightarrow A^+ + 2e \tag{2.68a}$$

$$e(E_0) + A \longrightarrow A^{n+} + (n+1)e. \tag{2.68b}$$

The experiments based upon the first reaction are called *(e, 2e) processes*, and those of the second one *multi-ionization processes*. The experimental arrangement for studying such processes is similar to electron-photon coincidence experiments; instead of the photon detector a second electron analyzer is used in $(e, 2e)$-experiments and an ion detector replaces the photon detector in multi-ionization experiments. The ion detector separately records ions of specific charge states A^+, A^{++}, A^{3+}, ..., A^{n+}. The same geometry for angular-correlation measurements can be used, as shown in Fig. 2.33, in which the photon coordinates are replaced by the corresponding electron or ion coordinates.

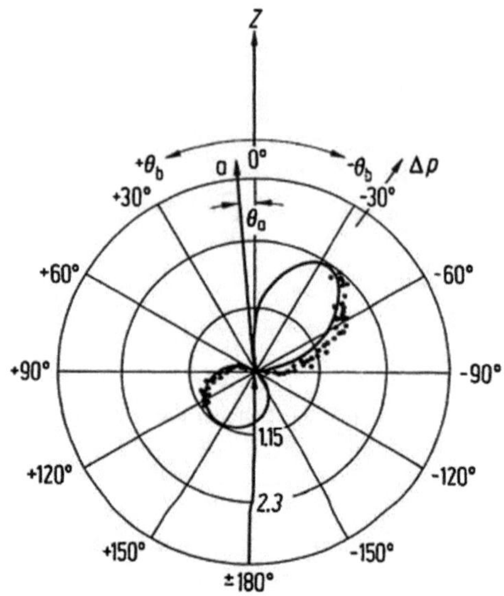

Fig. 2.43 Polar representation of the $(e, 2e)$ angular correlation of helium ionization. The primary electrons enter parallel to the z-axis and collide with helium atoms at the coordinate origin; primary electron energy $E_0 = 256.5$ eV, $E_a = 212$ eV and $E_b = 20$ eV are the energies for the two outgoing electrons. $\theta_a = 4°$ is the scattering angle of electron "a" which is kept constant, while the scattering direction of electron "b" is varied. The joint coincidence count rate for the two electrons a and b is proportional to the length of the radius from the origin to the points which represent experimental data (after Ehrhardt et al. 1980). The *full curve* follows from Born's approximation with planar outgoing waves of the electrons a and b

A typical planar $(e, 2e)$ angular correlation is shown in Fig. 2.43. Planar means that both the outgoing electrons (i.e. the electron resulting from the ionization and the primary electron) and the incoming electron are in a plane defined by these electrons. In this representation the electron detector for the outgoing electron "a" kept fixed, while the second electron "b" is detected in coincidence with the electron a as a function of the angle θ_b. The energies E_a and E_b of the two outgoing electrons can be varied, based upon the energy conservation

$$E_0 = E_{ion} + E_a + E_b, \tag{2.69}$$

with E_0 as the energy of the incoming electron and E_{ion} as the ionization energy of the atom. By assuming $E_a > E_b$ for our example, the momentum transfer $\Delta\mathbf{p} = \hbar\Delta\mathbf{k}$ from the incoming electron with momentum $\hbar\mathbf{k}_0$ to the ion and the electron b becomes

$$\hbar(\mathbf{k}_0 - \mathbf{k}_a) = \hbar(\mathbf{k}_b + \mathbf{k}_{ion}) = \hbar\Delta\mathbf{k}. \tag{2.70}$$

$\Delta\mathbf{p}$ is indicated in Fig. 2.43; we note that one coincidence peak has its maximum in the direction of $\Delta\mathbf{p}$ while the second smaller coincidence peak has its maximum in the opposite direction. The first peak is usually called *binary peak* while the second one *recoil peak*. The binary peak of the ionization interaction has the direction of the momentum transfer in a binary collision; the recoil peak has approximately the direction in which the atom recoils.

Angular correlations of electrons from ionization processes play an important role for our understanding of dynamical two-electron correlations. The ionization of the hydrogen atom, as the most important test atom, is of special interest for testing theoretical models. However, even at larger electron impact energies, the Born approximation is an unreliable method for theoretical predictions, as can be seen in Fig. 2.44. On the other hand, a so-called Coulomb correlation method provides theoretical data which show good agreement with experimental $(e, 2e)$ data. In this connection we refer to the definition of *triply differential cross sections* which can be extracted from the above coincidence measurements of the $(e, 2e)$-process if the coincidence count rate is calibrated absolutely, i.e. $\mathrm{d}^3Q(E_0, E_a, \theta_a, \theta_b, \phi_b)/\mathrm{d}E_b\mathrm{d}\Omega_a\mathrm{d}\Omega_b$ with ϕ_b as azimuthal angle (corresponding to ϕ_γ in Fig. 2.33) and $\mathrm{d}\Omega_a$ and $\mathrm{d}\Omega_b$ as solid angle elements for the detection of the two electrons. Lohmann and Weigold (1981) extended the $(e, 2e)$ method for direct measurements of electron momentum distributions of the hydrogen ground state and of other atoms.

At higher electron impact energies, several electrons can be knocked out of the electron shells of multi-electron atoms. In principle, it would be desirable to

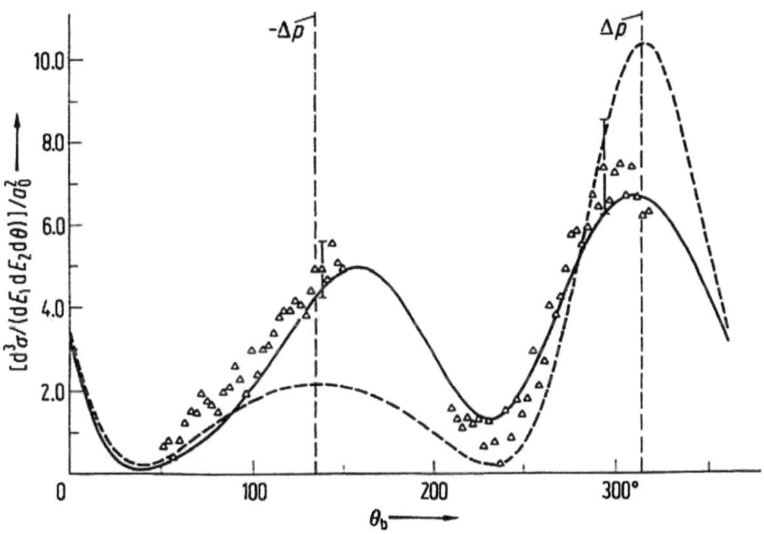

Fig. 2.44 Triple-differential cross section for the $(e, 2e)$ angular correlation of atomic hydrogen in Cartesian coordinates; primary energy $E_0 = 150\,\mathrm{eV}$, $E_b = 5\,\mathrm{eV}$, $\theta_a = 4°$. Experimental data after Klar et al. (1987); (---) Born's approximation; (——) Coulomb-correlation method after Jetzke et al. (1989). The angle on the abscissa for θ_b is measured from the direction of incident electrons from 0° to 360°. The *dotted vertical lines* show the angle for scattering parallel ($\Delta\mathbf{p}$) and antiparallel ($-\Delta\mathbf{p}$) to the vector of the momentum transfer $\Delta\mathbf{p} = \hbar\Delta\mathbf{k}$

detect all knocked-out electrons and the projectile electrons in coincidence (i.e. a $(n+1)$-particle coincidence!). Because of lack of intensities, however, such multiple coincidences ($n > 2$) are very difficult to detect. It has becoming feasible (Senftleben et al. 2010) with the advent and development of *reaction microscopes*. The latter have been reviewed, for example, by Ullrich et al. (2003), Continetti and Hayden (2004) and Vredenborg et al. (2011). On the other hand, coincidences between one electron and multiply charged ions have already been detected in ionization processes of electron impact of atoms. Coincidences between an ion in the charge state A^{n+} and an electron produced by electron impact are described by the doubly differential cross section $\mathrm{d}^2Q^{(n)}/\mathrm{d}E\mathrm{d}\Omega$ where $\mathrm{d}\Omega$ is the solid-angle element and $\mathrm{d}E$ is the energy interval of the electron detected in coincidence with the ion A^{n+}. Figure 2.45 shows schematically the arrangement of a coincidence apparatus for the detection of such coincidences. Figure 2.46 illustrates a typical example of an electron-ion coincidence spectrum from electron impact on xenon atoms. As can be seen in Fig. 2.46 the coincidence count rate associated with the detection of Xe^{2+} ions is considerably larger than that for singly charged Xe^{+} ions. Effects of this kind can be partly related to Auger processes (see Chap. 3) of inner electron shells. We note that Xe ions with up to nine-fold charge state could be detected by electron impact, with a primary electron energy of 6 keV. By means of high energy and synchrotron methods considerably higher charge states can be produced; this, however, is only achieved with much greater financial investments. The electron-impact method is much simpler and cheaper; it can be applied, for example, for the steering and flight corrections of satellites by the recoil of multiply charged Xe ions emitted into space.

2.5 Spin Effects in Atomic Collisions

In general, possible directions of spins of colliding electrons, ions and atoms are statistically distributed. However, spins of these particles can be polarized and an analysis of spin effects after the collision can provide information on collisional interactions, which may not otherwise be obtained. Investigations of such spin effects require complicated apparatus, which we will mention only briefly.

2.5.1 Degree of Polarization for Electrons and Single-Electron Atoms and Ions

We first define the *degree of polarization for electrons* and *single-electron atoms and ions*; their spin components are either $m_s = +1/2$ or $m_s = -1/2$ with reference to an axis of quantization. Unequal distribution of the number of electrons or atoms in the two spin states m_s described by $N_e(\pm 1/2)$ and $N_A(\pm 1/2)$, respectively, leads to the definition of the degree of polarization of electrons and atoms as

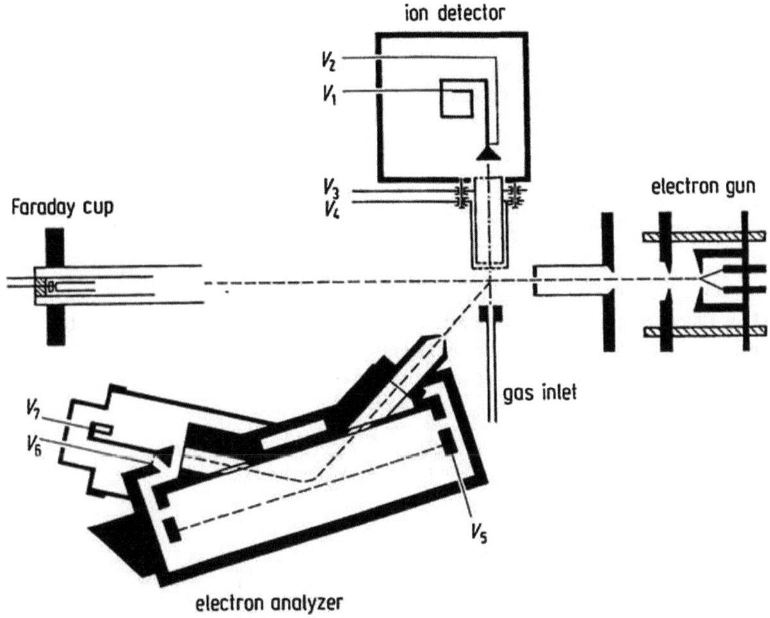

Fig. 2.45 Experimental scheme of an electron-ion coincidence apparatus with a parallel-plate ion analyzer. The ions are accelerated by the potential difference $V_3 - V_4$ and reach the channel electron multiplier (potential V_2) at different times, according to their charge state (after Chaudry et al. 1987). Electrons of selected energy in the electron analyzer are deflected by the negative potential V_5 into a channel electron multiplier at potential V_7

Fig. 2.46 Recorded electron-ion coincidence count rate for scattered electrons and xenon ions with charge states from Xe^+ to Xe^{9+}; observation from the coincidence apparatus described in Fig. 2.45. The coincident 30 eV electrons are scattered through 90° with respect to the 6 keV incident electrons (after Chaudry et al. 1987)

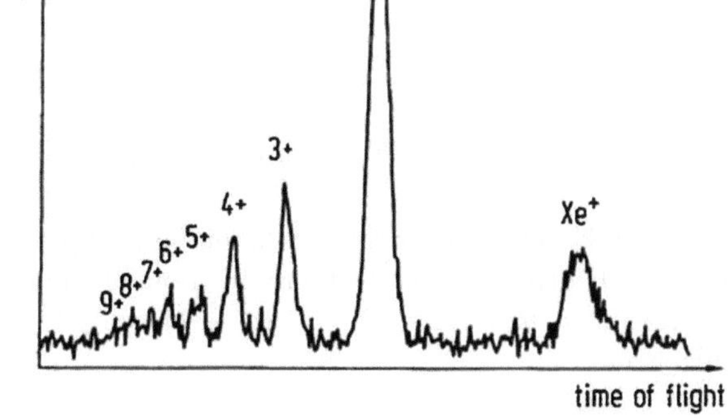

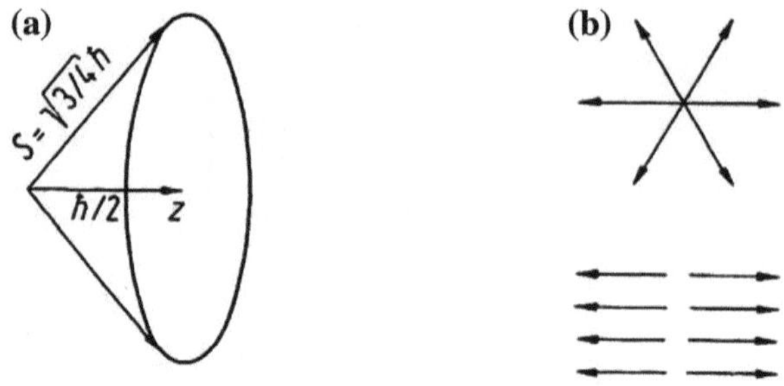

Fig. 2.47 **a** Electron spin which is completely polarized in the z-direction; **b** examples for spin directions of unpolarized electrons; after Kessler (1985)

$$P_e = \frac{N_e(1/2) - N_e(-1/2)}{N_e(1/2) + N_e(-1/2)}, \tag{2.71a}$$

and

$$P_A = \frac{N_A(1/2) - N_A(-1/2)}{N_A(1/2) + N_A(-1/2)}, \tag{2.71b}$$

where $N_e = N_e(1/2) + N_e(-1/2)$, and $N_A = N_A(1/2) + N_A(-1/2)$ are the total number of electrons and atoms. We will use this definition for the description of the following spin experiments.

In principle, the quantum mechanical description of the observable *polarization* is equivalent to macroscopical polarization described above. In quantum mechanics, the spin is represented by an operator which satisfies the characteristic commutation rules for angular momenta. The quantum mechanical expectation value for the spin polarization $\langle \boldsymbol{\sigma} \rangle$, which is identical to the above macroscopical polarization, follows from Schrödinger's equation with Pauli's spin operator $\boldsymbol{\sigma}$ and spin eigenfunction ψ as

$$\langle \boldsymbol{\sigma} \rangle = \langle \psi | \boldsymbol{\sigma} | \psi \rangle = \mathbf{P}. \tag{2.72}$$

The polarization $\mathbf{P}$ is a vectorial quantity, i.e.

$$|\mathbf{P}| = \sqrt{|P_x|^2 + |P_y|^2 + |P_z|^2}, \tag{2.73}$$

in which the above macroscopic definition is applied to each Cartesian coordinate. Figure 2.47 gives a geometric representation for completely polarized electrons in the z-direction $m_s \hbar = 1/2\hbar$, $|S| = \sqrt{s(s+1)}\hbar = \sqrt{3/4}\hbar$ and for unpolarized electrons in general.

2.5.2 Spin and Angular Correlation Experiments Including Photon Polarization Detection

One of the most general scheme for a complete/perfect scattering experiment in atomic collision physics is indicated in Fig. 2.48, which generalizes the scheme in Fig. 2.33 to polarized reaction partners; i.e. a polarized particle, e.g. an electron, collides with another polarized particle, e.g. an atom. In general, coincidence-, spin- and polarized photon analyses will be required to approach a complete experiment. Since around the 1960s of the last century the ideal complete experiment has only been approached in certain steps. However, the way to the basic ideas of complete/perfect experiments for atomic collisions developed through the pioneering research from the Franck-Hertz experiment (Franck and Hertz 1914), the Ramsauer-Townsend effect (Ramsauer 1921; Townsend and Bailey 1922), the Hanle effect (Hanle 1924), electron scattering interference from electron angular distributions by Bullard and Massey (1931), Mott scattering, direct, exchange (Bederson 1969a, b, 1970) and resonance scattering (Schulz 1963). As pointed out by Andersen and Bartschat (2000), the pioneering quantum theories of electron-impact excitation of atoms by Oppenheimer (1927a, b, 1928) and Penney (1932a, b) in the 1920s and 1930s were restricted to the calculation of first-order cross sections; polarization effects of impact line radiation were seen already in experimental investigations (even the magnetic depolarization, called Hanle effect, discovered in 1924) the theoretical understanding of all these effects started only with the theory by Percival and Seaton (1958) and its first experimental verification on alkali resonance lines by Haffner et al. (1965).

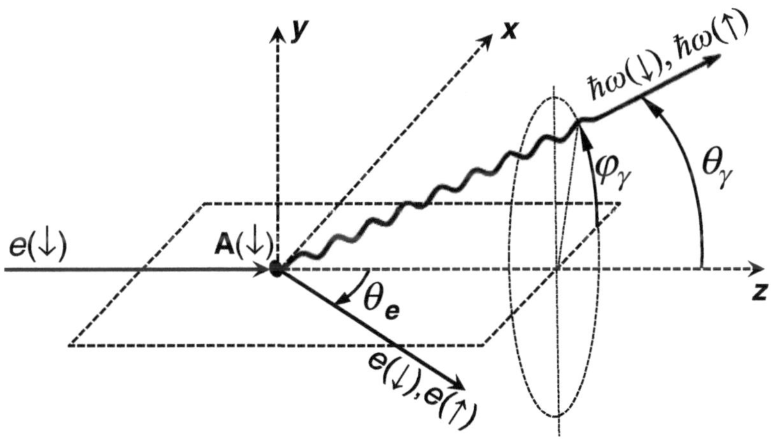

Fig. 2.48 A polarized beam of electrons (*spin up* ↑) is colliding with a polarized beam of one-electron atoms (*spin down* ↓). An electron-photon angular correlation experiment consists of the coincident detection of the scattered electron (including up-down spin analysis) and the polarization of the coincident photon emitted from an excited atom; it represents a *complete/perfect experiment* (Kleinpoppen 1971; Becker and Crowe 2001)

Historically first experimental investigations on the detailed analysis of *scattering amplitudes* started around the end of the 1960s and the beginning of the seventies. It also involved the applications of various fundamental papers by Fano (1957), Fano and Macek (1973), Bederson (1969a, b, 1970), Kleinpoppen (1971), Blum and Kleinpoppen (1974) and Kessler (1985). Comments on direct measurements of phases of quantum mechanical excitation amplitudes and state parameters in atomic collisions were published towards the second half and end of the 1970s (Kleinpoppen 1976, 1980).

On the theoretical side, appropriate formalisms for describing angular correlation and polarization phenomena in collisions were developed in the 1950s, for the needs of the theory of nuclear reactions, and were reviewed for example by Blatt and Biedenharn (1952), Biedenharn and Rose (1953), Devons and Goldfarb (1957), Fano (1957), Fano and Racah (1959) and Fergusson (1965).

It included the development of such concepts as density matrix and statistical tensors of the angular momentum, efficiency tensors, irreducible tensor operators and amplitudes, and many others, which in the seventies started to be widely applied to atomic and molecular processes, as summarized in review papers and monographs such as those by Fano and Macek (1973), Hertel and Stoll (1977), Kessler (1985), Blum (1996), Andersen et al. (1988, 1997), Zare (1988), Balashov et al. (2000), Andersen and Bartschat (2000) and many others.

While applications of the most sophisticated scheme in Fig. 2.48 require still major efforts for complete/perfect scattering experiments we list partial solutions to it. The special case of the $^1P_1$ state excitation can be analyzed in a most straightforward way for helium and light two-electron atoms (Kleinpoppen 1967, 1971 Eminyan et al. 1973). This continues the corresponding discussion of Sects. 2.4.2–2.4.4. In particular, for the helium case the in- and outgoing state vectors are

$$|\psi_{in}\rangle = |\mathrm{He}(1\,^1\mathrm{S}_0)\rangle|p_e\rangle \quad \text{and} \quad |\psi_{out}\rangle = |\mathrm{He}(^1\mathrm{P}_1)\rangle|\mathrm{p}'_e\rangle. \tag{2.74}$$

Using (2.51)–(2.55) and applying the λ–χ representation (see Sect. 2.4.2) with the parameters λ, $1-\lambda$, and χ defined in (2.58)–(2.60), the angular correlation coincidence count rate for observing the photon in the scattering plane at an angle θ_γ and fixed electron scattering angle becomes

$$N_{e,\gamma} = \lambda \sin^2\theta_\gamma + (1-\lambda)\cos\theta_\gamma - 2\sqrt{\lambda(1-\lambda)}\cos\theta_\gamma \sin\theta_\gamma \cos\chi, \tag{2.75}$$

with the last term of (2.75) representing an *interference effect* for the excitation of the magnetic substates.

Figure 2.49, complementary to Fig. 2.36, yields further demonstrations of measured electron–photon angular correlations from the $\mathrm{He}(2\,^1\mathrm{P}_1)$ excitation at various scattering angles (Hollywood et al. 1979). As has been discussed in Sect. 2.4.3, the Born approximation is only close to the experimental data at small scattering angles as already predicted by Bethe (1933) since Born's approximation remains finite only for the selection rule $\Delta M_L = 0$ with respect to the angular momentum transfer

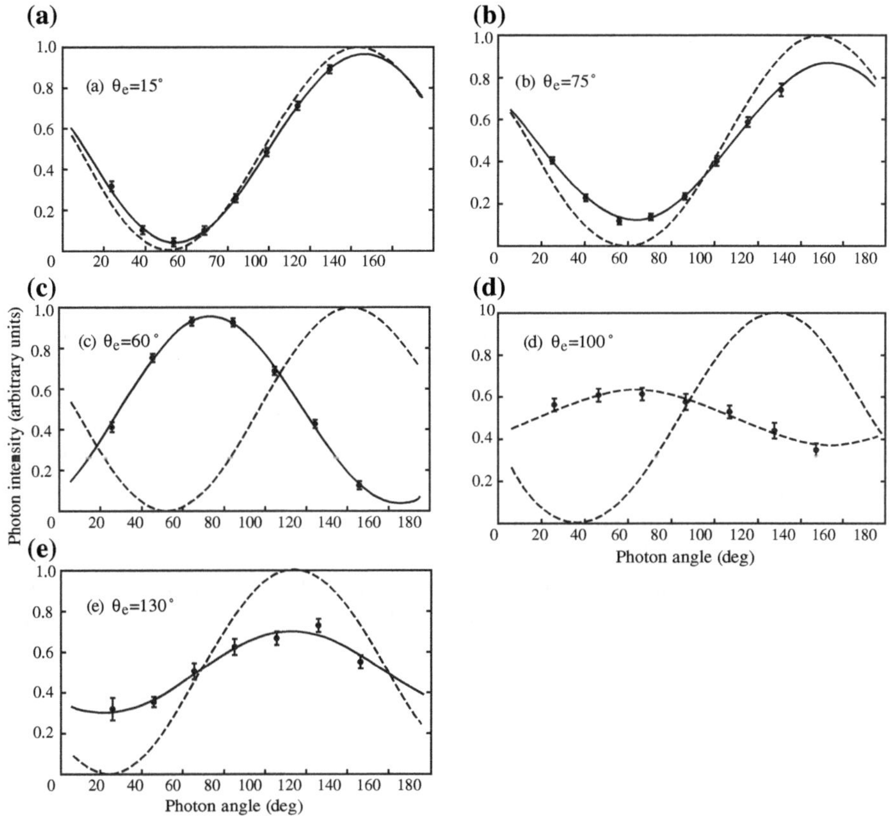

Fig. 2.49 Electron-photon angular correlations from 81.2 eV electron-impact excitation of He ($2\,^1P_1$) at various electron scattering angles. The *full curve* is a least squares of the theoretical representation to the data; the *broken curves* are first Born approximation (data by Hollywood et al. 1979)

vector. Hence, for the excitation/de-excitation $^1S \rightarrow {}^1P \rightarrow {}^1S$, the $M_L = 0$ substate of the ^1P state is excited, only, and the emitted photon radiation is that of a dipole oscillating parallel to the momentum transfer $\Delta\mathbf{p} = \mathbf{p}_i - \mathbf{p}_f$.

An alternative, equivalent interpretation to λ and χ data, outlined in Sect. 2.4.4, is based on the concept of *atomic orientation and alignment* from the electron–photon coincidences, e.g. the He($^1P_1$) excitation (Fano and Macek 1973; Eminyan et al. 1974, 1975). The non-vanishing components of alignment and orientation are determined by the parameters λ and χ as follows:

$$\mathcal{A}_0^{col} = \frac{\langle 3L_z^2 - \mathbf{L}^2 \rangle}{L(L+1)} = \frac{1}{2}(1 - 3\lambda), \tag{2.76}$$

$$\mathcal{A}_{1+}^{col} = \frac{\langle L_x L_z + L_z L_x \rangle}{L(L+1)} = [\lambda(1-\lambda)]^{1/2} \cos\chi, \tag{2.77}$$

$$\mathcal{A}_{2+}^{col} = \frac{\langle L_x^2 - L_y^2 \rangle}{L(L+1)} = \lambda - 1, \tag{2.78}$$

$$\mathcal{O}_{1-}^{col} = \frac{\langle L_y \rangle}{L(L+1)} = -[\lambda(1-\lambda)]^{1/2} \sin\chi. \tag{2.79}$$

Figures 2.50 and 2.51 demonstrate magnitudes of orientation and alignment data of the $2\,^1\mathrm{P}_1$ state as extracted from electron–photon coincidence data. We mention in addition that *alignment* quantities have also been determined in non-coincidence experiments: by the anisotropic angular emission of Auger electrons or the linear polarization of electromagnetic radiation (including x- and γ-rays, Mehlhorn 1983).

After the excitation of an atomic target its state can be represented by various magnetic substates of quantum numbers M_L. The relevant distribution of these substates can be described by state multipoles (Blum and Kleinpoppen 1979), statistical tensors (Balashov et al. 1984), or alignment tensors and an orientation vector (Fano and Macek 1973). Excitation of P states can be particularly simply expressed by *orbital angular momentum transfer* (2.64), which can be calculated from the λ and χ data for the $^1\mathrm{P}_1$ excitation process:

$$\langle L_\perp \rangle = \langle L_y \rangle = -2\sqrt{\lambda(1-\lambda)} \sin\chi. \tag{2.80}$$

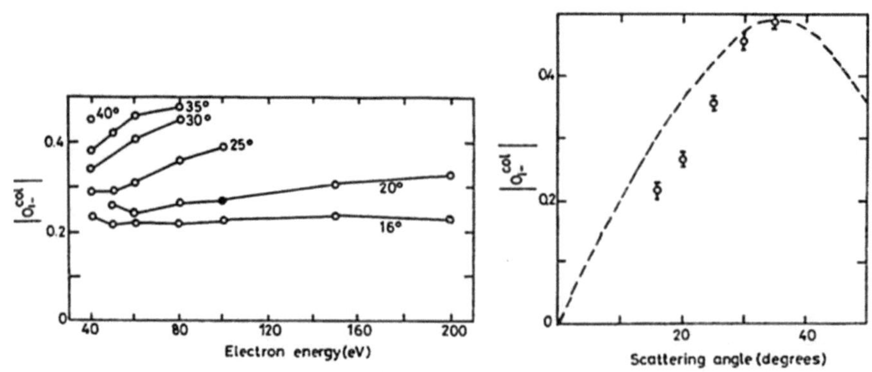

Fig. 2.50 Magnitude of the orientation vector of helium atoms excited to the $2\,^1\mathrm{P}_1$ state as function of the incident electron energy for various electron scattering angles. These data were obtained from the measurement of the angular correlation parameters λ and $|\chi|$ (Eminyan et al. 1973, 1974). The figure on the *right-hand* side shows experimental data at 80 eV electron impact energy; (– – –) theoretical results calculated from the distorted wave approximation of Madison and Shelton (1973) at 78 eV

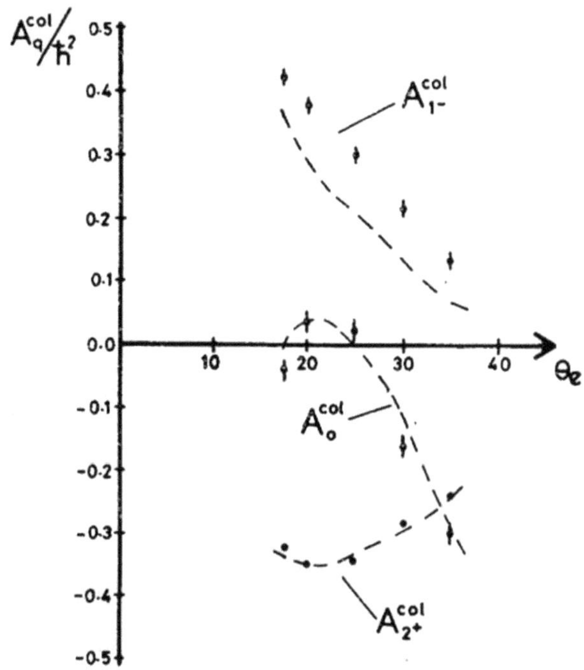

Fig. 2.51 Alignment parameters as calculated from the experimental angular correlation parameters λ and $|\chi|$ of the $2\,^1P_1$ state of helium for various electron scattering angles and an electron impact energy of 80 eV (Eminyan et al. 1975). *Dashed curves* are predictions according to the distorted wave approximation by Madison and Shelton (1973)

According to Fano and Macek (1973) this orbital momentum transfer can be related to the orientation vector $\mathcal{O}_{1-}^{col}$, see (2.79), and the circular polarization P_3 emitted from the P → S photon emission, $\langle L_y \rangle = -P_3$, see (2.63)–(2.64).[1]

The angular and polarization correlation data of electron–photon coincidences can also determine the electron angular distribution of excited states, the so-called *electron cloud* introduced in Sect. 2.4.4, which is the relative charge distribution in the volume element dV, defined as $-e\psi^*\psi dV$. For the $^1S \rightarrow {}^1P$ excitation, the excited state vector can be written as (see Sect. 2.4.2) $\psi = f_0\psi_z - \sqrt{2} f_1 \psi_x$, where $\psi_z = |10\rangle$ and $\psi_x = -1/\sqrt{2}\big(|11\rangle - |1-1\rangle\big)$ denote the corresponding wavefunctions of the p_z and p_x orbitals, respectively. The electron charge distribution depends on amplitudes f_0 and f_1 or λ and χ which can be calculated and graphically represented for a given electron impact energy and scattering angle. A typical picture of the shape of an electron cloud of the excited $^1P_1$ state is shown in Fig. 2.41 (see Sect. 2.4.4); it corresponds to an alignment angle $\gamma = \theta_{min} = 45°$ and $\langle L_\perp \rangle = 0.75\,\hbar$ (in $\hbar$ units). Figure 2.41 also demonstrates the possible repulsive and attractive interactions for atomic orientation by the collision process which has been discussed in detail by Kohmoto and Fano (1981). Figure 2.40 shows experimental data of the orbital angular momentum transfer $\langle L_\perp \rangle$ to the excited 2^1P_1 state and the corresponding alignment angle γ of Fig. 2.41.

---

[1] The sign difference follows from traditional definitions in classical optics and atomic physics.

Beyer et al. (1982) introduced a concept of attractive (f_A) and repulsive (f_R) scattering amplitudes and linked these to the amplitudes $f_{M_L=0}$ and $f_{M_L=\pm 1}$, and λ, χ, and $\sigma = |f_0|^2 + 2|f_1|^2$ for light atoms as follows:

$$f_R = \frac{1}{\sqrt{2}}(f_0 + i\sqrt{2}f_1), \tag{2.81}$$

$$f_A = \frac{1}{\sqrt{2}}(f_0 - i\sqrt{2}f_1), \tag{2.82}$$

with

$$|f_A|^2 = \frac{1}{2}\sigma[1 + 2\sqrt{\lambda(1-\lambda)}]\sin\chi, \tag{2.83}$$

$$|f_R|^2 = \frac{1}{2}\sigma[1 - \sqrt{\lambda(1-\lambda)}]\sin\chi, \tag{2.84}$$

and

$$\tan(\delta_R - \delta_A) = \frac{2\sqrt{\lambda(1-\lambda)}}{2\lambda - 1}\cos\chi, \tag{2.85}$$

where $\delta_R - \delta_A$ is the relative phase of the amplitudes f_R and f_A. Based upon the scattering with repulsive/attractive dynamics for excitation of light atoms a simple classical grazing model by Kohmoto and Fano (1981) and Madison and Winters (1981) leads to the formulation of *orientation propensity rules*. Of particular interest with regard to the classical grazing model and the repulsive and attractive potentials in the electron-atom scattering is a dynamical right-left electron scattering asymmetry (Fig. 2.52) for detecting right or left hand circularly polarized photons of the $3\,^1P_1 \rightarrow 2\,^1S_0$ helium transitions (Silim et al. 1987; Kleinpoppen 1988) in coincidence with the scattered electrons.

As pointed out (Herting et al. 2002; Herting and Hanne 2003) the *orientation propensity rules* should hold for light atoms described in the *LS* coupling scheme. However, for atomic orientation by polarized electron impact excitation of heavy atoms (such as $Hg(6s^2)\,^1S_0 \rightarrow (6s6p)\,^3P_1$) a non-classical interference due to intermediate coupling within the excited state obstructs the interpretation of the orientation propensity rule. The above correlation or deviation from the amplitude relations f_R/f_A and $f_{M_L=0}/f_{M_L=1}$ may help further clarification of the problems with regard to the validity of the orientation propensity rules.

The $He(^1S_0 \rightarrow {}^1P)$ electron impact process at a typically of 50 eV appears to be well understood in connection with detailed electron impact coherence parameters; they serve as most *sensitive tests* of *theoretical models* (see, e.g. Fig. 1.13 in Andersen and Bartschat 2000). The agreement between various experimental data sets, such as differential cross section, orbital momentum transfer, photon polarizations, and charge cloud orientation, on the one hand, and theoretical predictions from *R*-matrix theory with pseudo-states (Bartschat et al. 1996) and the convergent close-coupling (CCC) theory (Fursa and Bray 1995), on the other hand, are most

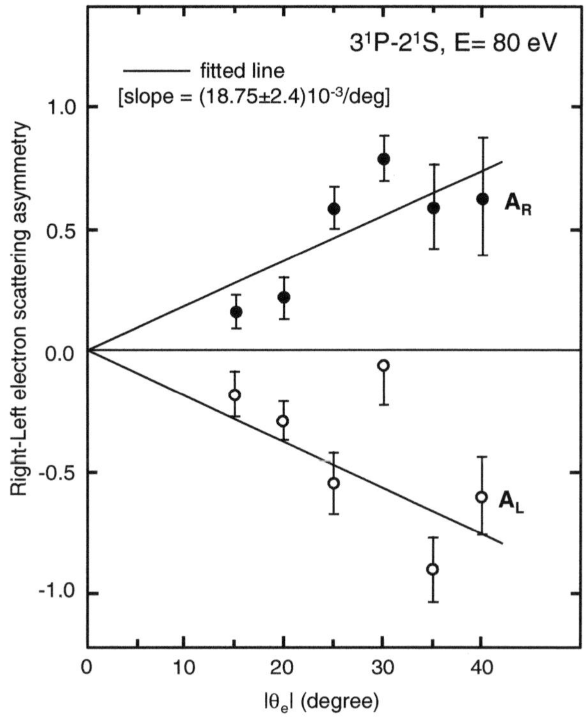

Fig. 2.52 Right-left electron scattering asymmetry as a function of the scattering angle for detecting right (•) or left (∘) hand circularly coincident polarized photons of the helium transition $3\,{}^1P_1 \longrightarrow 2\,{}^1S_0$ at 80 eV. The *full lines* are fitted to the experimental data (Silim 1985)

impressive. We like to draw attention to various reference books (e.g. Andersen and Bartschat 2000; Balashov et al. 2000) and papers (e.g. Andersen et al. 1988) for theoretical and experimental λ, χ, and σ data, they represented so called *complete data* in atomic collision physics for the first time (Eminyan et al. 1973).

The obvious move to investigate the electron impact excitation of atomic hydrogen is complicated by the appearance of two sets of amplitudes for singlet (f^s) and triplet (f^t) scattering for antiparallel and parallel spins of the projectile electron and the atomic hydrogen electron. Neglecting an overall phase, we need to determine 7 independent parameters for a complete experiment with polarized electrons and polarized hydrogen atoms for the 1s $\rightarrow$ 2p $\rightarrow$ 1s excitation/de-excitation. Here we refer to the partial complete experiments by Yalim et al. (1999) for the 2P state excitation of atomic hydrogen using the scattered electron decay photon angular correlation technique at 54.4 eV with the parameters $\lambda = \sigma_0/\sigma$ and $R = Re(f_1 f_0)/\sigma$ as defined by Morgan and McDowell (1975)[2] with the collision frame amplitudes (z in the incident beam direction, see Fig. 2.41). Data of Yalim et al. (1999) are shown in Fig. 2.53 in comparison to other experimental measurements and theories.

[2] As a matter of fact, Morgan and McDowell (1975) defined $R = Re(a_1^c a_0^c)/\sigma$ using a slightly different notation for the amplitudes referring to the collision frame.

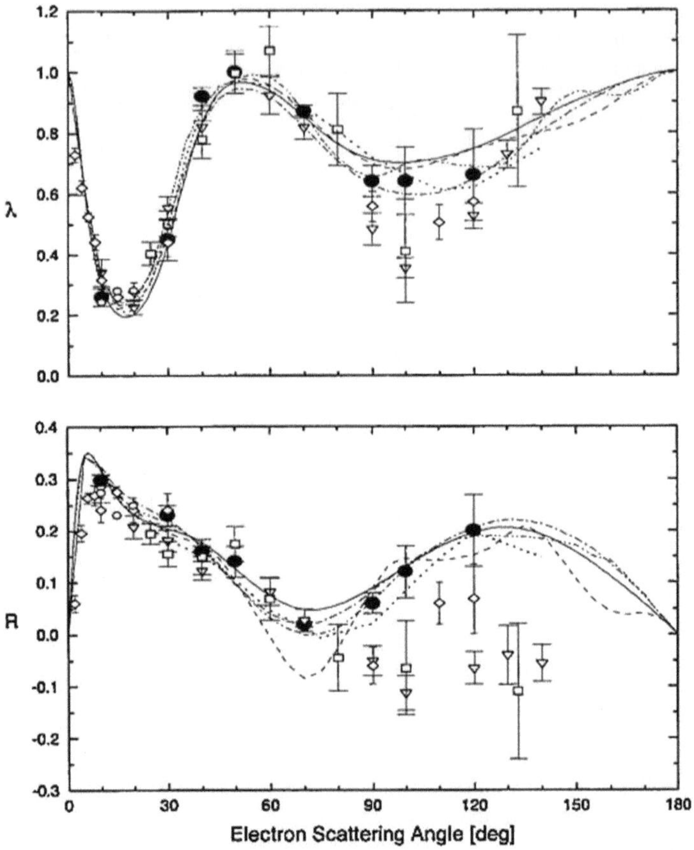

Fig. 2.53 Experimental parameters λ and R: • Yalim et al. (1999); □ Kleinpoppen and Williams (1980); ▽ O'Neil et al. (1998); ◇ Hood et al. (1979); theories: (⋅⋅⋅⋅⋅⋅) van Wyngaarden and Walters (1986); (–⋅–) Madison et al. (1991); (–––) Scholz et al. (1991); (——) Bray and Stelbovics (1992); (– ⋅ ⋅ –) Wang et al. (1994)

2.5.3 Scattering Processes with Polarized Electrons and Polarized Single Electron Atoms

Experiments with polarized electrons and polarized atoms carried out so far can be classified as follows:

1. Scattering of partially polarized electrons by partially polarized single-electron atoms (H, Li, Na, ...);
2. the Mott scattering of unpolarized electrons by heavy atoms and
3. scattering of partially polarized electrons on unpolarized atoms, or alternatively, the scattering of unpolarized electrons by partially polarized atoms associated with polarization measurements of the outgoing electrons or the recoiling atoms.

Spin effects in atomic collisions have been applied to first approaches in *perfect scattering experiments* (Bederson 1969a, b).

In order to understand such experiments, we first develop a set of scattering amplitudes and apply them to the first case (Kleinpoppen 1971). We assume that the electrons and the atoms are completely polarized, i.e. $|\mathbf{P}_e| = 1$ and $|\mathbf{P}_A| = 1$. We then distinguish the following *spin reactions* between electrons completely polarized parallel or antiparallel to a given axis of quantization and characterized by the notations $e(\uparrow)$ and $e(\downarrow)$ on the one hand and polarized atoms with analogous notations $A(\uparrow)$ and $A(\downarrow)$ on the other hand.

The basic spin reactions for elastic electron-atom collisions are:

$$e(\uparrow) + A(\downarrow) \longrightarrow e(\uparrow) + A(\downarrow), \quad \text{direct amplitude } f, \text{ quantity observed } |f|^2, \tag{2.86a}$$

$$e(\uparrow) + A(\downarrow) \longrightarrow e(\downarrow) + A(\uparrow), \quad \text{exchange amplitude } g, \text{ quantity observed } |g|^2, \tag{2.86b}$$

$$e(\uparrow) + A(\uparrow) \longrightarrow e(\uparrow) + A(\uparrow), \quad \text{interference amplitude } f - g, \text{ quantity observed } |f - g|^2. \tag{2.86c}$$

In the first reaction a direct Coulomb interaction takes place, in which the spin directions of both particles are conserved since the projectile electron is not exchanged with the atomic electron. We associate a *direct scattering amplitude* f and its differential cross section $\sigma_{dir} = |f|^2$ with this collisional interaction. In the second reaction an *electron exchange* between the incoming electron and the atomic electron takes place; the amplitude associated with this interaction is the *exchange amplitude* g and its differential exchange cross section is $\sigma_{ex} = |g|^2$. The third reaction is described by a *coherent superposition* or *interference* between the direct Coulomb and exchange interaction both of which take place in the collision but cannot be separated from each other contrary to the first two reactions; the resulting amplitude describing the third process is an *interference amplitude* $f - g$ and its differential cross section $\sigma_{int} = |f - g|^2$. The minus sign between f and g is due to the required antisymmetry of the associated wave function of the scattering process.

Calibrating these reactions with the equivalent opposite spin directions gives the connection between the spin-dependent cross sections and the "normal" differential cross section σ for unpolarized electrons and atoms:

$$\sigma(\theta, E) = \frac{1}{2}(|f|^2 + |g|^2 + |f - g|^2), \tag{2.87}$$

where $|f - g|^2 = |f|^2 + |g|^2 - 2|f||g| \cos\phi$ and ϕ is the phase difference between f and g. We obtain

$$\begin{aligned} \sigma(\theta, E) &= |f|^2 + |g|^2 - |f||g| \cos\phi, \\ &= \sigma_{dir} + \sigma_{ex} - \sigma_{int}. \end{aligned} \tag{2.88}$$

The factor 1/2 in (2.87) takes account of the fact that only 50 % of the possible reactions are of type (2.86a) or (2.86b), and 50 % are of type (2.86c). A simple recalculation results in the familiar relations

$$\begin{aligned}\sigma(\theta, E) &= |f|^2 + |g|^2 - Re(f^*g),\\ &= \frac{3}{4}|f-g|^2 + \frac{1}{4}|f+g|^2,\\ &= \frac{3}{4}|T|^2 + \frac{1}{4}|S|^2 = \frac{3}{4}\sigma_T + \frac{1}{4}\sigma_S,\\ &= \frac{1}{2}\sigma^{\uparrow\uparrow} + \frac{1}{2}\sigma^{\uparrow\downarrow},\end{aligned} \tag{2.89}$$

where $T = f - g$ and $S = f + g$ are the *triplet and singlet scattering amplitudes*, $|T|^2 = \sigma_T$ and $|S|^2 = \sigma_S$ are the triplet and singlet cross sections, and

$$\sigma^{\uparrow\uparrow} = \sigma_T = |f-g|^2, \tag{2.90}$$

and

$$\sigma^{\uparrow\downarrow} = \frac{1}{2}(\sigma_T + \sigma_S) = |f|^2 + |g|^2, \tag{2.91}$$

are further definitions. The interference term

$$\sigma_{int} = Re(f^*g) = |f||g|\cos\phi, \tag{2.92}$$

leads to constructive ($\phi = 180°$) or destructive ($\phi = 0°$) interference contributions to the cross sections. The terms *triplet* and *singlet* are standard notations taken from spectroscopic notations of atomic states (e.g., for the helium atom).

In ideal scattering experiments, which are only possible and successful in a limited way so far, the polarization of the electrons and atoms before (notations $\mathbf{P}_e$ and $\mathbf{P}_A$ for the polarizations of the electron and atom before the scattering) and the polarizations after the scattering (notations $\mathbf{P}'_e$ and $\mathbf{P}'_A$ of the polarizations) are required in order to obtain complete information on the amplitudes f and g (Fig. 2.54). However, the measurement of the polarizations $\mathbf{P}'_e$ and $\mathbf{P}'_A$ after the collision is very difficult for reasons of intensity. On the other hand, combinations of measurements of the following spin reactions are possible (Kleinpoppen 1971), which is shown in Table 2.1. In the first two reactions, only polarized atoms are required; both the polarizations of the electrons and atoms after scattering as well as the differential cross section have to be measured in order to obtain $|f|^2$ and $|g|^2$. In the third reaction the cross section $I(E, \theta)$ for the scattering intensity of polarized electrons or of polarized recoil atoms has to be measured in which, in addition, P_e and P_A and the cross section $\sigma(E, \theta)$ has to be known in order to determine $|f-g|^2$ or the phase term $\cos\phi$.

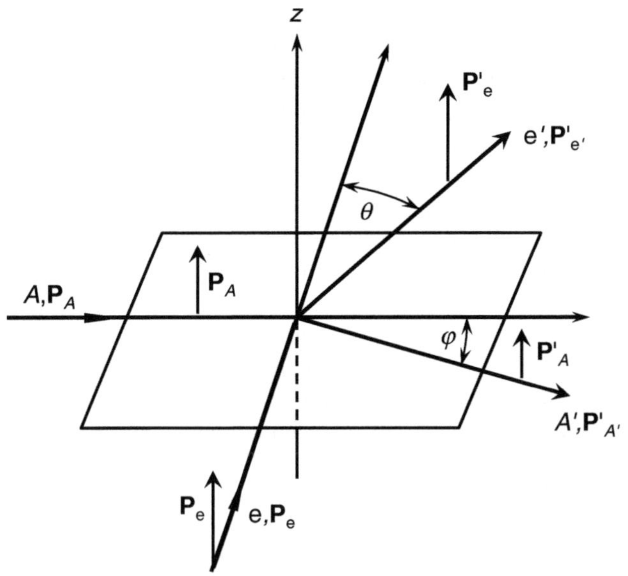

Fig. 2.54 Geometry for scattering polarized electron beams $(e, \mathbf{P}_e)$ on polarized atomic beams $(A, \mathbf{P}_A)$. $\mathbf{P}'_e$, $\mathbf{P}'_A$ are vectorial degrees of polarization after the scattering

Another measurable quantity is the *spin asymmetry*, which is linked to the above quantities and can be defined as follows:

$$A = \frac{1}{P_e P_A} \frac{\sigma^{\uparrow\downarrow} - \sigma^{\uparrow\uparrow}}{\sigma^{\uparrow\downarrow} + \sigma^{\uparrow\uparrow}} = \frac{\sigma_{int}}{\sigma}. \tag{2.93}$$

The same equation is valid for the case in which the differential cross sections σ, $\sigma^{\uparrow\downarrow}$, $\sigma^{\uparrow\uparrow}$, and σ_{int} are replaced by the corresponding total cross sections Q, $Q^{\uparrow\downarrow}$, $Q^{\uparrow\uparrow}$ and Q_{int}.

The pioneering work for the production of polarized beams of electrons is based on the spin-orbit interaction. It utilizes Mott scattering of electrons in the pure Coulomb

Table 2.1 The possible measurements of spin reactions

Polarization before collision	Measurable quantities after collision	Information on the collision process
(1) $P_A \neq 0$, $P_e = 0$	P'_e, $\sigma(E,\theta)$	$\lvert f\rvert^2 = \sigma(E,\theta)(1 - P'_e/P_A)$
(2) $P_A \neq 0$, $P_e = 0$	P'_A, $\sigma(E,\theta)$	$\lvert g\rvert^2 = \sigma(E,\theta)(1 - P'_A/P_A)$
(3) $P_A \neq 0$, $P_e \neq 0$	$\sigma(E,\theta)$	$\lvert f - g\rvert^2 = \sigma + (1 + P_e/P_A)(I - \sigma)$
	$I(E,\theta) = I_e^{\uparrow} + I_e^{\downarrow}$	$I(E,\sigma) = \sigma(E,\theta) - P_e P_A Re(f^* g)$
	$= I_A^{\uparrow} + I_A^{\downarrow}$	equivalent to $\cos\phi$, see (2.88)

field of the nucleus. The Americans Shull et al. (1943) were the first to detect spin polarized electrons from the scattering of energetic electrons (40 keV) in thin gold foils. However, according to theoretical predictions by Massey and Mohr (1941), electron spin polarization also occurs for low energy electron scattering ($\leq$2 keV). This polarization effect is due to an interference between electrons scattered directly in the atomic field without spin flip and those scattered with a spin flip (see below for the connection between the amplitudes h and k in (2.102)). Following a proposal by Kollath (1949), Deichsel (1961) succeeded in detecting electron spin polarization in low energy scattering of electrons by mercury atoms. An electron scattering current of 10^{-9} A and an electron spin polarization of 17 % was measured for elastic scattering of 300 eV electrons at mercury atoms (Steidl et al. 1965). The present most successful method for the production of spin polarized electrons is based on the photoemission in special solids such as gallium-arsenides (GaAs) and gallium-arsenide-phosphides (GaAsP). These materials serve as photocathodes in an ultra-high vacuum system with a minimum pressure of about 10^{-8} Pa. The surfaces of the photoelectron emitting crystals are first treated with cesium and oxygen in order to obtain a negative work function for the electron in the conduction band. When irradiated by circularly polarized light, electron photoemission from such surface films results in longitudinally spin polarized electrons being extracted in the opposite direction to the incoming light. *Longitudinal spin polarization* means that the number of electrons with their spins in the propagation direction is different from the number of electrons with spins in the opposite direction. A cylindrical electrostatic condenser, which deflects the electrons by 90°, leaves the spin orientations of the electrons unchanged, but transforms the longitudinal spin polarization into a transverse one. In this case, the spin polarization refers to the difference in numbers of electrons with their two spin directions perpendicular to the propagation direction (i.e., parallel or antiparallel to a given direction). High-quality photocathodes of the above materials produce spin polarized electrons of a polarization coefficient $P_e \approx 0.35$ to 0.45 and with currents of the order of magnitude of 10 µA or even larger, depending mainly on the laser intensity. Metallo-organic chemical vapor deposition of GaAs–GaAsP photocathodes (*strained-layer photocathodes*) have resulted in a production of over 90 % spin polarized electron beams (Chatwell et al. 1993). The current of polarized electron beams can be as high as that of unpolarized beams and is usually limited by space-charge effects in the electron beam optics.

2.5.4 Production of Polarized Atoms

Polarized atoms can be produced by the following methods.

1. By the spatial separation of atoms into different Zeeman components m by a magnetic hexapole field;
2. by redistribution of Zeeman components m of the atoms by optical pumping and
3. by combinations of both methods (1) and (2).

While Stern-Gerlach and Rabi magnets separate atomic beams in different magnetic substates m from each other, a magnetic hexapole focuses atoms in certain m states and defocusses atoms in other m states. The production of polarized atoms by means of magnets is based on the existence of magnetic moments μ in atoms which experience a magnetic force $\mu\partial B/\partial r$; this force has different signs for the quantum numbers $\pm m$ of the atom. The magnetic field inhomogenity $\partial B/\partial r$ varies proportionally to the distance from the center of a hexapole magnet. The trajectories of atoms of spin component $m_s = +1/2$ are bent toward the center of the hexapole magnet while those which in $m_s = -1/2$ are bent away from it. Single-electron atoms in n^2S states without nuclear spins (which unfortunately do not exist in nature!) would be expected to be polarized to 100 % by the action of a hexapole field. The spins of the focused atoms are oriented parallel to the local B field in the hexapole magnet. When the atoms leave the hexapole magnet they are expected to orient themselves adiabatically (i.e. gradually) into the direction of an external magnetic field. This guiding magnetic field has to be very low in realistic scattering experiments in order to keep the influence of the Lorentz force on the electron to a minimum. This minimal magnetic field results in a reduced effective spin polarization due to the fact that the electron spin is associated with relevant magnetic moments of the hydrogen or alkali atoms. The nuclear spin I and the electron spin S couple to the total spin F so that the electron spin polarization of the atom is reduced from 100 % to $P_A = 1/(2I+1)$ since further magnetic hyperfine components are associated with electron spin components in the opposite direction, i.e. $m_s = -1/2$. In other words the maximum electron spin polarization produced by the hexapole field and the guiding magnetic field becomes 50 % for atomic hydrogen ($I = 1/2$), 33 % for ^{6}Li(I = 1), 25 % for ^{23}Na($I = 3/2$) and only 12.5 % for cesium ($I = 7/2$). An example of a complete photoionization experiment with oxygen atoms polarized by a hexapole magnet is presented in Sect. 4.6.3.

By applying the method of optical pumping with circularly polarized light, substantially larger degrees of polarization can be obtained for the above atoms. Optical pumping with circularly polarized light from the $[F = I + 1/2]$ ground state into the $[F' = F + 1]$-state of the excited $^2P_{3/2}$ state has the consequence that the largest m_F state of the $[I + 1/2]$ ground state is populated preferentially. The nuclear and electron spin in the magnetic substate ($m_F = I + 1/2$) are then oriented in the same direction. However, the electron spin polarization of the atom is not complete since the ($m_F = I - 1/2$) ground state is not altered in its m_F population, and accordingly its spin components m_s remain unpolarized. The resulting electron spin polarizations of the atoms can be calculated as $P_A = (I + 1)/(2I + 1)$ based upon this method, e.g. to $P_H = 75$ %, $P_{6Li} = 66.7$ % and $P_{^{23}Na} = 62.5$ %.

For a further increase in the spin polarization, the two methods of applying hexapoles and optical pumping have been combined. The atoms first pass through a hexapole and subsequently through a laser field for optical pumping. The hexapole eliminates the ($I - 1/2$) states of the ground state so that optical pumping from the ($I + 1/2$) states to the $F' = F + 1$ magnetic substates of the excited $^2P_{3/2}$ state can take place. This should, in principle, result in a complete electron polarization into the ($m_F = +F$) substate of the ground state $F = I + 1/2$. A polarization of about 90 % in sodium atoms was first achieved with this combined hexapole/optical pumping method (Hils et al. 1981).

Simultaneous optical pumping with circularly polarized light from both hyperfine states $I \pm 1/2$ also populates the $(F = I + 1/2, m_F = F)$ substate to a high percentage in which the use of two separate lasers (e.g. two GaAs laser diodes for pumping cesium, see Baum et al. 1989, 1993) is comparable to a method in which by means of electro-acoustic or electro-optical coupling a single laser produces both the required first and the second laser frequency. However, in the latter case, coherent coupling can take place between the states involved, which results in a depolarization of the atomic spin polarization. By using an electro-optically modulated dye laser such depolarization effects could be kept very small in optical pumping with the sodium resonance line; an electron spin polarization of almost 100 % of the ground state of sodium was obtained with this method (e.g. 98 % after Reich (1987) and of 99 % at low atomic density according to Beckord (1989)). Optically pumped cesium with two laser diodes also resulted in high spin polarization of the ground state at low atomic densities (Baum et al. 1989, 1990, 1993). We also mention that, recently, a metastable helium $2\,^3S_1$ atomic beam has been polarized with reference to the $m_j = \pm 1$ Zeeman substates to 90 % by means of a magnetic hexapole (Baum et al. 1988, 1989). Further examples on the production of polarized atoms by optical laser pumping can be found in Sect. 4.6.2. Details of the optical pumping process and producing the atomic polarization are described, for example by Hertel and Stoll (1977), Auzinsh et al. (2010) and Happer et al. (2010).

2.5.5 *Universal Apparatus for Scattering of Polarized Electrons on Polarized Atoms*

As an example of a modern apparatus for the scattering of polarized electrons at polarized atoms, we refer to the schematic arrangement of Fig. 2.55; the instrumental complexity speaks for itself. Apart from investigating elastic and inelastic scattering processes, the setup for such experimental arrangements can also be used for studying ion asymmetries in electron impact ionizations. The *ion asymmetry* A_{ion} follows, in analogy to (2.93), from the difference of the number $N(I)$ of the ions produced in the ionization process in which the arrows refer to initially parallel or antiparallel oriented spins of the electrons and atoms,

$$A_{ion} = \frac{N^{\uparrow\uparrow}(I) - N^{\uparrow\downarrow}(I)}{N^{\uparrow\uparrow}(I) + N^{\uparrow\downarrow}(I)}. \tag{2.94}$$

Out of the considerable number of investigations with polarized electrons and polarized atoms we select, as illustrative examples, measurements of ion asymmetries and the quantities $|f|^2/\sigma$ and $|g|^2/\sigma$ in Table 2.1 for one-electron atoms (Figs. 2.56, 2.57, 2.58).

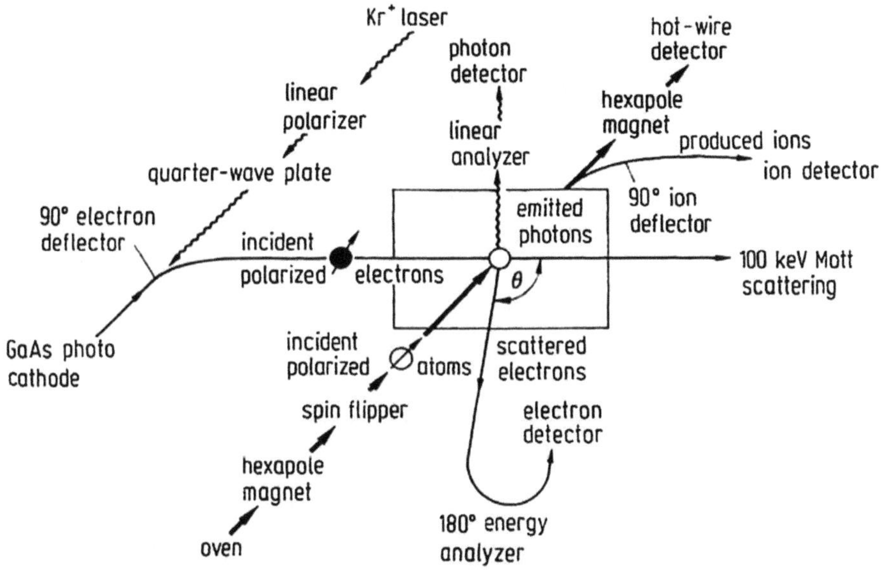

Fig. 2.55 Scheme of a universal apparatus for studying elastic and inelastic scattering and ionization in collisions between polarized electrons and polarized atoms. The photodetector serves for the detection of asymmetries of inelastic excitation processes (after Raith 1988)

Historically, ion asymmetries have been the first spin quantities to be extracted from experiments (Alguard et al. 1977; Hils and Kleinpoppen 1978) with polarized electrons and polarized atoms during the second half of the 1970s. The observation of an astonishingly large ion asymmetry (Fig. 2.56), which shows up as an integral effect in the total ionization cross section, was a surprise in the physics of electron impact ionization.

In a pioneering experiment by Collins et al. (1967) the atomic beam is spin-polarized (polarization P_A) and velocity selected by a Stern-Gerlach magnet and spin-analyzed after electron scattering with unpolarized electrons by a **E–H** gradient balance magnet. Neglecting any other types of interactions the spin polarization of the atom can only vary by an exchange process (see Fig. 2.57). The ratio of the detected beams of atoms with and without spin polarization of the atoms resulted in

$$\frac{\sigma_{ex}(\theta)}{\sigma(\theta)} = \frac{|g(\theta)|^2}{\frac{1}{2}(|f|^2 + |g|^2 + |f - g|^2)} = \frac{|g(\theta)|^2}{\sigma(\theta)} = 1 - \frac{P'_A}{P_A}. \tag{2.95}$$

As an example of spin effects in elastic electron-atom scattering, a few years later a paper was published by Hils et al. (1972) on the measurement of the *direct* differential elastic scattering of unpolarized electrons on polarized potassium atoms. The scattering process polarizes the electrons and their degree of polarization (P'_e) was measured by Mott scattering. The relation between the scattering amplitude $f(\theta)$

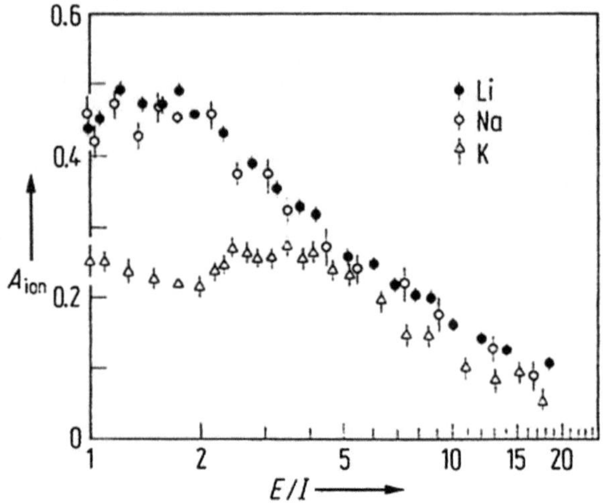

Fig. 2.56 Ion asymmetries $A_{ion} = [N^{\uparrow\uparrow}(I) - N^{\uparrow\downarrow}(I)]/[N^{\uparrow\uparrow}(I) + N^{\uparrow\downarrow}(I)]$ from scattering of polarized electrons on polarized light alkali metal atoms. E/I is the ratio of the energy of the incident electrons to the ionization energy of the relevant atom (after Baum et al. 1985)

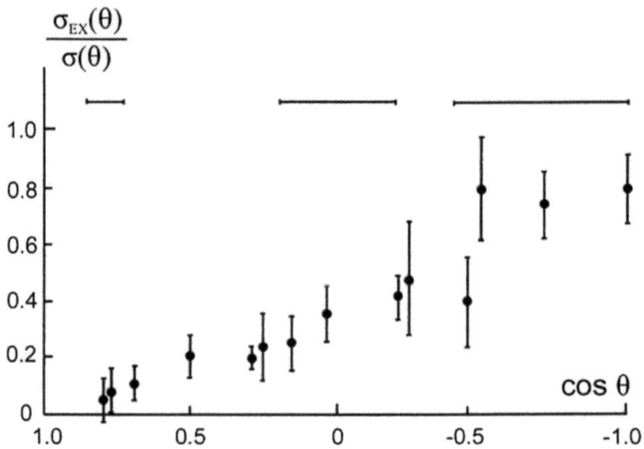

Fig. 2.57 Experimental data for $\sigma_{ex}(E,\theta)/\sigma(E,\theta)$ of Collins et al. (1967) for electron scattering on polarized potassium atoms at an energy of 0.5 eV. *Horizontal error bars* indicate uncertainties in $\cos\theta$ at three different angles

and the measured electron polarization is given by

$$\frac{\sigma_{dir}(\theta)}{\sigma(\theta)} = \frac{|f(\theta)|^2}{\sigma(\theta)} = 1 - \frac{P_e'}{P_A}. \tag{2.96}$$

Figure 2.58 shows a distinctive interference structure in the intensity of the electron scattering on polarized potassium atoms, which manifests itself in the *direct* Coulomb

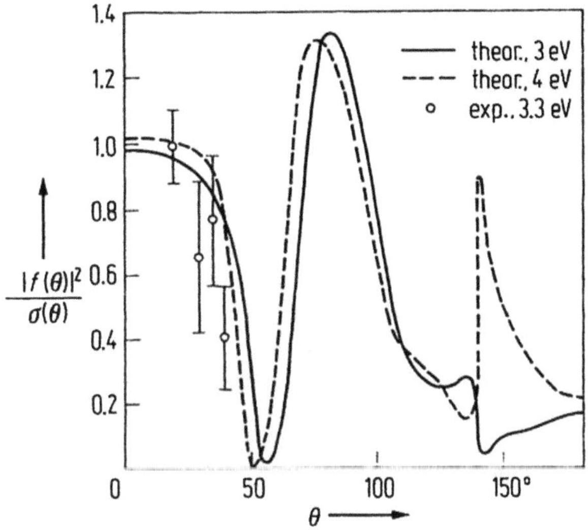

Fig. 2.58 Experimental (∘) and theoretical data for the square of the direct elastic scattering amplitude f divided by the differential cross section $\sigma(\theta)$ as a function of the scattering angle θ with polarized potassium as target atoms. The experimental data follow from the measurement of the polarization of initially unpolarized electrons scattered elastically on partially polarized potassium atoms ($P_A = 20\,\%$). Process (1) of Table 2.1; after Hils et al. (1972). Theoretical values from the work of Karule and Peterkop (1973) (two-state close-coupling approximation) at 3 eV (——) and 4 eV (– – –)

interaction (amplitude f). The basic experiment for the underlying scattering process consists of the measurement of the polarization of the scattered electrons, which were initially unpolarized; see reaction (1) of Table 2.1. The data clearly demonstrated the direct scattering process, although the accuracy of the measurement was limited.

Measurements of exchange amplitudes have also been carried out by the usual electron–photon coincidence experiment with electron impact excitation of the $3\,^3$P state of helium (Silim 1985; Silim et al. 1987) with emission of light at wavelength at 388.9 nm and the following process:

$$\begin{aligned} e + \mathrm{He}(1\,^1\mathrm{S}_0) &\longrightarrow \mathrm{He}^*(3\,^3\mathrm{P}_{0,1,2}) + e \\ &\qquad \hookrightarrow \mathrm{He}^*(2\,^3\mathrm{S}_1) + h\nu. \end{aligned} \tag{2.97}$$

Only exchange amplitudes g_0, g_1 contribute to the excitation of the ^{3}P states so that a parameter

$$\lambda = \frac{|g_0|^2}{\sigma} = \frac{|g_0|^2}{|g_0|^2 + 2|g_1|^2}, \tag{2.98}$$

and a phase χ can be defined by $g_1 = |g_1| \exp(i\chi)$, $g_0 = |g_0|$ which is similar to the ^{1}P excitation of helium. Experimental data for excitation energy of 60 eV showed total polarization and coherence for the $3\,^3$P excitation by electron impact.

Passing on to electron scattering by heavy atoms such as rubidium and cesium, spin-orbit interaction between the projectile electron and the target atom takes place in addition to the direct Coulomb and exchange interaction; therefore many more amplitudes are necessary for describing the scattering. This situation is similar to the description of the normal fine structure of excited atoms or the photoionization of heavy alkali atoms where spin orbit interactions increase with the larger masses of atoms involved. Six amplitudes are necessary for the description of elastic electron scattering on heavy alkali atoms, which means that 11 independent quantities, i.e. 6 moduli and 5 phase differences have to be determined for a complete analysis of the scattering process (Khalid and Kleinpoppen 1983).

The complication due to a large number of amplitudes is reduced by using target atoms without a resulting total spin (*spinless atoms*), as for example with rare gas atoms or two-electron atoms with opposite spins. Consequently, two spin reactions can be defined for the scattering of polarized electrons on spinless atoms A:

$$\begin{aligned} &e(\uparrow) + A \longrightarrow A + e(\uparrow), \quad \text{with amplitude } h, \quad \text{modulus } |h|^2, \\ &e(\uparrow) + A \longrightarrow A + e(\downarrow), \quad \text{with amplitude } k, \quad \text{modulus } |k|^2; \end{aligned} \tag{2.99}$$

we denote, as before, the first process as a *direct* one with amplitude h and the second one as a *spin-flip* process with amplitude k.

We note that the direct process can be superposed coherently with an electron exchange process. Both, the direct Coulomb and the exchange process, cannot be separated from each other due to their indistinguishability in the experiment. In order to measure the amplitudes h and k, partially polarized electrons are scattered by atoms; the change of the spin polarization of the electrons after scattering determines the moduli $|h|$ and $|k|$ and their phase difference $\Delta\phi = \gamma_1 - \gamma_2$ based on the following relations:

$$S = -\frac{2|h||k| \sin \Delta\phi}{\sigma}, \tag{2.100a}$$

$$T = \frac{|h|^2 - |k|^2}{\sigma}, \tag{2.100b}$$

and

$$U = \frac{2|h||k| \cos \Delta\phi}{\sigma}, \tag{2.100c}$$

where $\sigma = |h|^2 + |k|^2$ is the differential cross section; the quantities S, T, and U are connected to the components of the spin polarization of the scattered electrons (see Figs. 2.59 and 2.60) as follows

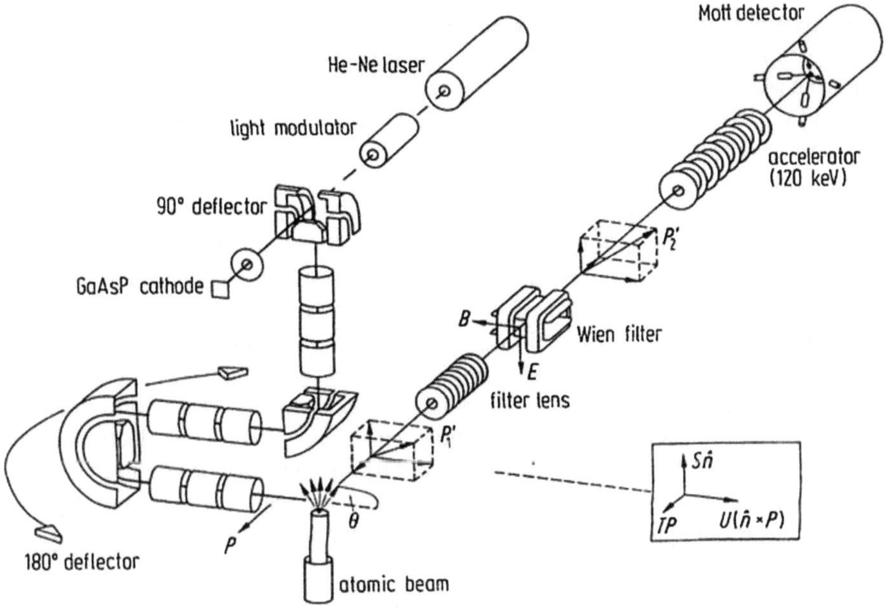

Fig. 2.59 Scheme of an apparatus for scattering partially polarized electrons on spinless atoms (after Berger and Kessler 1986)

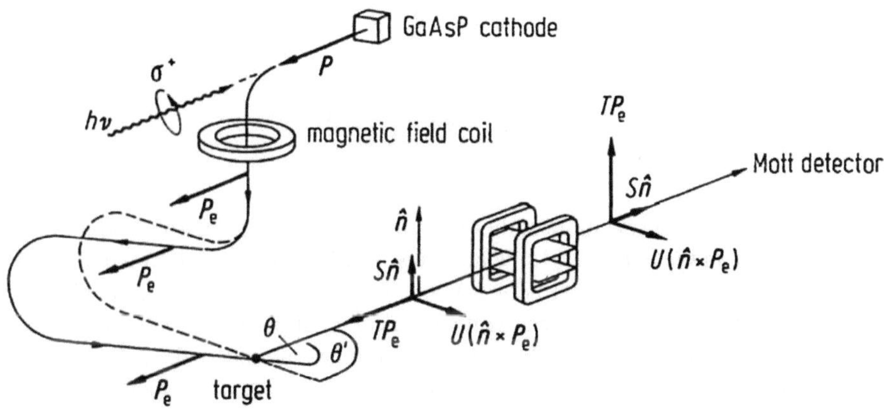

Fig. 2.60 Orientation of the polarization $\mathbf{P}_e$ of electrons from the GaAsP cathode as a source of polarized electrons into a Mott detector in the apparatus of Fig. 2.59

$$\mathbf{P}_e' = S\mathbf{n} + T\mathbf{P}_e + U(\mathbf{n} \times \mathbf{P}_e). \tag{2.101}$$

$\mathbf{P}_e'$ is the vector polarization of the scattered electrons under the condition that the initial spin polarization $\mathbf{P}_e$ of the incoming electrons is in the scattering plane. The quantity $S\mathbf{n}$ is the component of the spin of the scattered electrons perpendicular to the scattering plane (i.e. in the direction of the scattering normal), $T\mathbf{P}_e$ is parallel or antiparallel to $\mathbf{P}_e$ and $U(\mathbf{n} \times \mathbf{P}_e)$ is rotated by 90° with reference to $\mathbf{P}_e$ in the scattering plane. Figure 2.59 displays the schematic layout of such experiments; the method for producing polarized electrons, for the deflection of the spin direction and for the Mott detector are already known to us. The *Wien filter* has two orthogonally superposed magnetic and electric fields and acts on the electron beam as follows: Electrons of a given, fixed energy pass through the Wien filter only if the Lorentz force due to the magnetic field compensates the Coulomb force of the electric field; in this way the Wien filter acts as an energy or velocity filter. In addition, the spin components $\mathbf{P}_e$ and $S\mathbf{n}$ oriented perpendicular to the magnetic field, can carry out Lamor precessions, i.e., these spin components may be rotated by 90° according to the choice of the magnetic field strength. In this way it is possible to one which is required for the measurement of the polarization by means of the Mott detector. Figure 2.60 shows schematically the various spin directions of the electron on its long journey from the GaAsP cathode to the Mott detector according to the layout of the apparatus shown in Fig. 2.59. From the measurements of the quantities S, T, and U, the relevant amplitudes and their phase differences can be calculated as

$$|h| = \left(\sigma \frac{1+T}{2}\right)^{\frac{1}{2}}, \tag{2.102a}$$

$$|k| = \left(\sigma \frac{1-T}{2}\right)^{\frac{1}{2}}, \tag{2.102b}$$

and

$$\gamma_1 - \gamma_2 = \tan^{-1}\left(\frac{-S}{U}\right), \tag{2.102c}$$

according to (2.100).

Figure 2.61 shows an example of these quantities for elastic electron-krypton scattering. As one can see in Fig. 2.61, the modulus $|h|$ of the direct scattering amplitude shows a distinctive diffraction structure, which is due to the superposition of several partial waves of scattered electrons with various angular momenta. This structure is determined by the dipole and exchange interaction, as previously described in connection with the Ramsauer-Townsend effect (see Sect. 2.3.5). The modulus of the spin-flip amplitude $|k|$, which originates from the spin-orbit interaction is considerably smaller than that of the direct amplitude $|h|$; the spin-flip amplitude is primarily determined by the ($\ell = 1$) partial wave of the scattered electrons, which has the result that the diffraction structure is hardly discernible.

We note that the measurement of the complex amplitudes h and k signifies a *complete experiment* in the meaning defined in Sect. 2.2. "Complete" means, for our example, a measurement based on a physical method which allows one to determine the complex amplitudes h and k; however, because of coherent superposition of the Coulomb-direct, the Coulomb-exchange, and the spin-orbit interaction of this collision process, it is not possible to determine the amplitudes of these interactions.

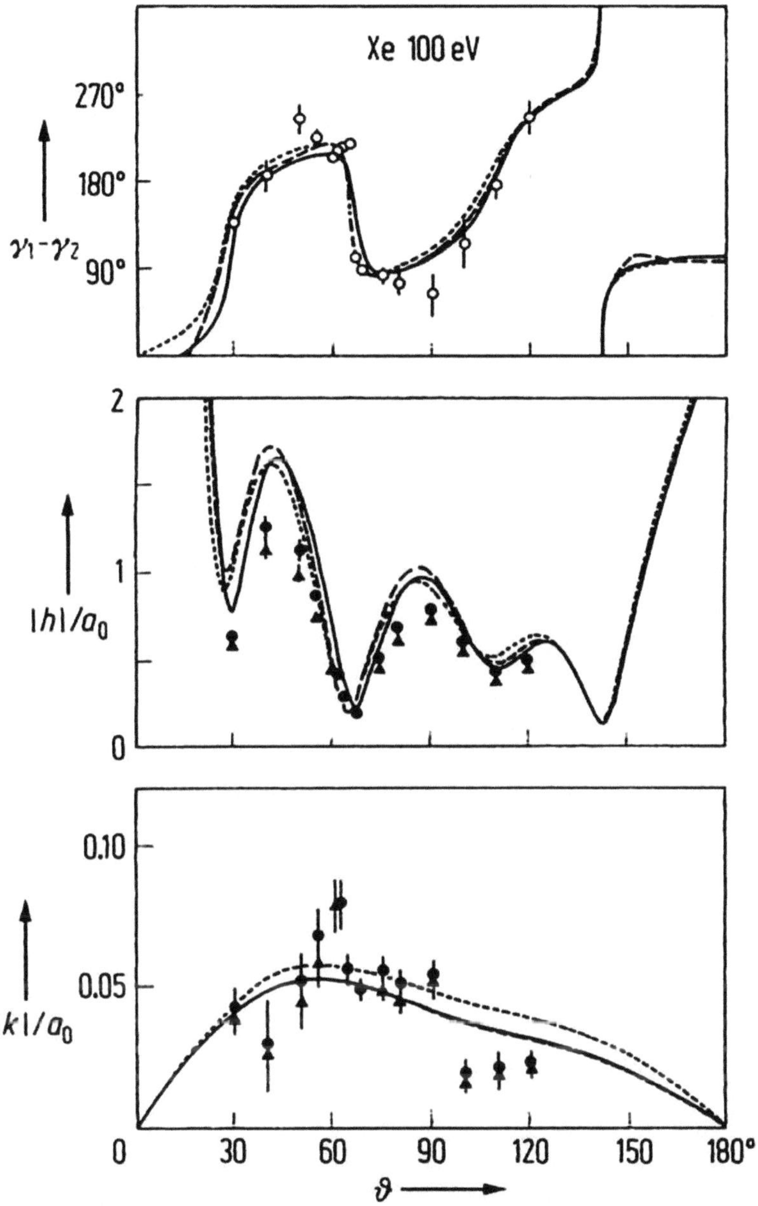

Fig. 2.61 Moduli of amplitudes $|h|$ and $|k|$ and phase differences $\gamma_1 - \gamma_2$ between the two amplitudes for elastic scattering of polarized electrons on xenon atoms, as a function of the scattering angle ϑ at an energy of 100 eV. Experimental data points with error bars after Berger and Kessler (1986). The *dotted* and *full curves* represent various theoretical predictions: (– – –) after Haberland et al. (1986), (· · · · · ·) after McEachran and Stauffer (1986), (——) after Awe et al. (1983). The data for $|h|$ and $|k|$ are given in units of the Bohr radius a_0 and are normalized to the measured differential cross section $\sigma = |h|^2 + |k|^2$

In the research literature in electron-atom collisions, the above amplitudes h and k for the direct and spin-flip process are denoted by the letters f and g. This leads to a confusion with regard to the description of the direct Coulomb interaction (normally associated with the amplitude f) and the Coulomb-exchange interaction (normally associated with the amplitude g) in electron scattering by atomic hydrogen and light alkali atoms; see (2.87).

The aim in extending these experiments was applying spin polarized electrons in collisions with spin polarized atoms. First approaches at the end of the 1970s aimed at spin-asymmetry effects of ionization cross sections with spin polarized electrons and polarized target atoms (Alguard et al. 1977; Hils and Kleinpoppen 1978; Hils et al. 1981, 1982).

More recent experiments with spin polarized electrons ($|\mathbf{P}_e| = 0.65$), and polarized cesium atoms ($|\mathbf{P}_A| = 0.90$) showed dramatic asymmetry effects in the differential cross section σ_0 with unpolarized collision partners; based upon the theory of Burke and Mitchell (1974) the differential cross section for scattering of the spin polarized beams is given by

$$\sigma = \sigma_0\big[1 + A_1(\mathbf{P}_A\hat{\mathbf{n}}) + A_2(\mathbf{P}_e\hat{\mathbf{n}}) - A_{nn}(\mathbf{P}_A\hat{\mathbf{n}})(\mathbf{P}_e\hat{\mathbf{n}})\big], \tag{2.103}$$

(Baum et al. 1999, 2002).

Figure 2.62 shows impressive experimental and theoretical data for the differential elastic cross section and the various spin asymmetry parameters A_1, A_2, and A_{nn} of low-energy electron scattering from cesium atoms. The physical meaning, with respect to the reaction plane, of these spin-asymmetries are as follows: A_1 and A_2 correspond to spin-up, spin-down asymmetries in the differential cross section for scattering of unpolarized electrons on unpolarized atoms (A_1) or polarized electrons on unpolarized atoms (A_2); A_{nn} represents an *antiparallel–parallel* asymmetry.

Data on differential cross sections and spin asymmetries have also been reported for spin polarized electron impact excitation of spin polarized cesium atoms (Baum et al. 2004). The optically allowed excitation

$$\mathrm{Cs(6s)}\,{}^2\mathrm{S}_{1/2} \longrightarrow \mathrm{Cs(6p)}\,{}^2\mathrm{P}_{1/2,3/2}, \tag{2.104}$$

and the optically forbidden transition

$$\mathrm{Cs(6s)}\,{}^2\mathrm{S}_{1/2} \longrightarrow \mathrm{Cs(5d)}\,{}^2\mathrm{D}_{3/2,5/2}, \tag{2.105}$$

were measured and data are compared to theories of the non-relativistic convergent close-coupling (CCC) and semi-relativistic R-matrix with pseudo-states (RMPS). Satisfactory agreements between experiments and theories were notable.

By applying both, spin polarized projectile electrons and spin polarized hydrogen atoms, Fletcher et al. (1985) were able to fully determine the various scattering processes for direct, exchange and interference, and also triplet- and singlet interactions; see (2.89)–(2.92). The spin asymmetry A, see (2.93), for the elastic 90° scattering of polarized electrons on polarized hydrogen atoms has been reported by

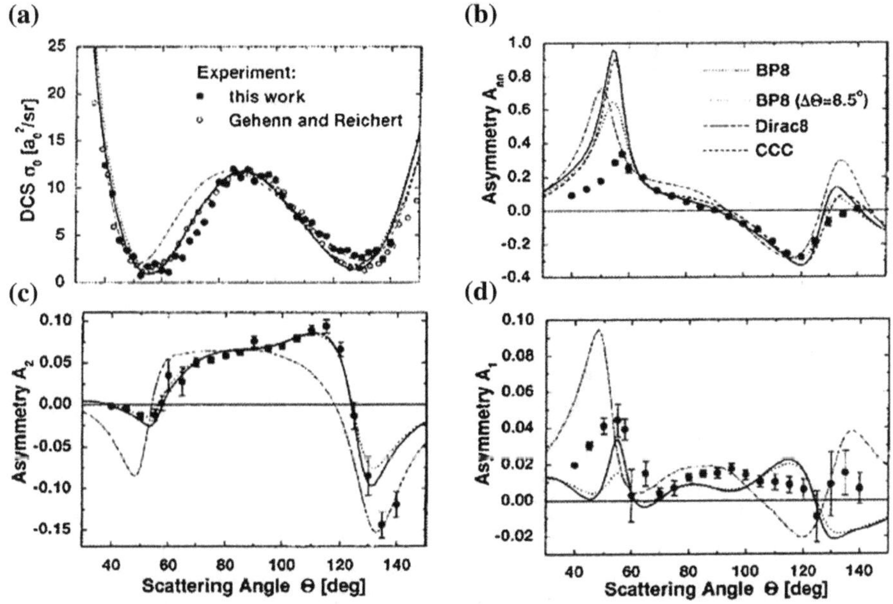

Fig. 2.62 **a** Differential cross section σ_0 for scattering of electrons by cesium atoms (normalized to theory at $\theta = 90°$, impact energy of 3 eV); (•) Baum et al. (1986); (○) Gehenn and Reichert (1977); **b** the asymmetries A_{nn}, **c** A_2, and **d** A_1 are compared to theoretical predictions: Breit-Pauli R-matrix approaches BP8 (including after convolution with the experimental angular resolution of $\Delta\theta = 8.5°$), a relativistic Dirac 8-state R-matrix model (Dirac 8) and a non-relativistic convergent close-coupling calculation (CCC). Many more such data for projectile energies from 5 to 25 eV by Baum et al. (2002) have been published

Fletcher et al. (1985); see Fig. 2.63. Similar measurements, including theoretical predictions, were reported on different atoms by the Bielefeld group on lithium (Baum et al. 1986) and the NBS group on sodium atoms (McClelland et al. 1987).

2.5.6 Advanced Implementations and Developments

Since the first relevant experiment by Ehrhardt et al. (1969) the physics of low-energy electron impact ionization with completely determined kinematics has had an outstanding development both, experimentally and theoretically, in atomic, molecular, cluster and surface physics (see, e.g. Weigold and McCarthy 1999; Ehrhardt and Morgan 1994; Neudatchin et al. 1999). Proceedings of Conferences on $(e, 2e)$ and double photoionization link the process of electron impact and photodouble ionization together, respectively (Madison and Schulz 2001; Hanne et al. 2003; Lahmam-Bennani and Lohmann 2005; Martin and de Harak 2010):

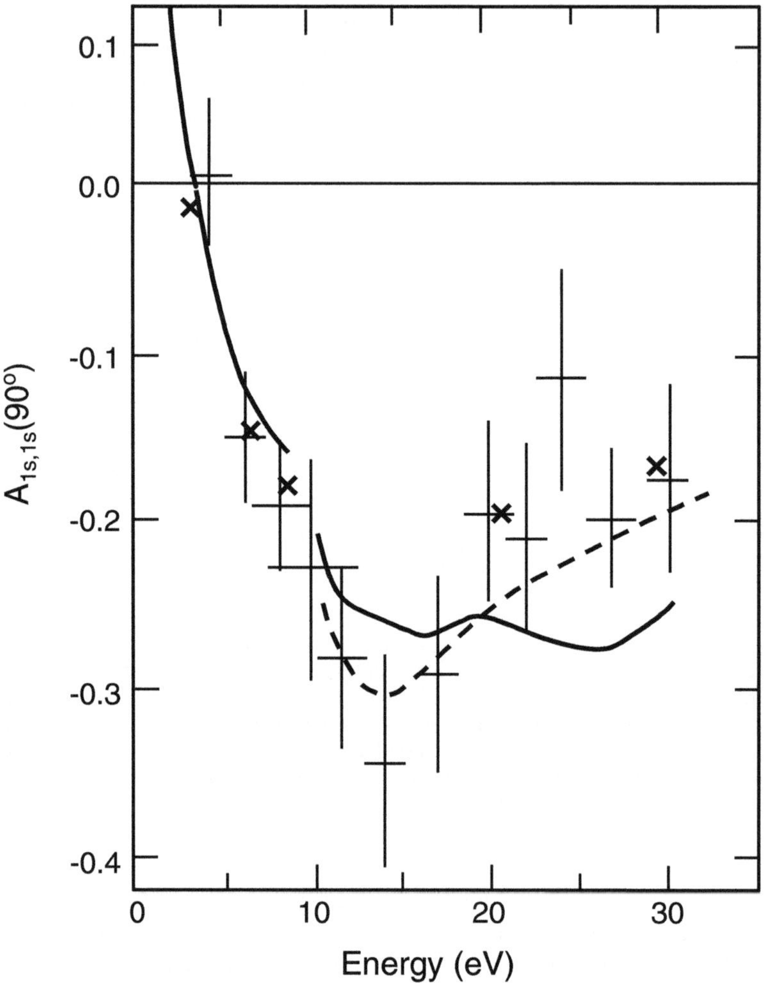

Fig. 2.63 Polarization spin asymmetry A for the elastic 90° scattering on atomic hydrogen. (*Error bars*) experimental results of the Yale group (Fletcher et al. 1985); (*crosses, solid and dashed curves*) theoretical results from van Wyngaarden and Walters (1986)

$$\begin{aligned} e + A &\longrightarrow A^{+} + 2e, \\ h\nu + A &\longrightarrow A^{++} + 2e. \end{aligned} \tag{2.106}$$

In a paper by Lower et al. (2004), schematically, a quantum mechanically complete experiment, in addition to the kinematically complete experiment on the $(e, 2e)$ process would require to determine a scheme of measurements as illustrated in Fig. 2.64. As pointed out by these authors the ultimate objective to determine experimentally these physically variables lies beyond the present technologies although

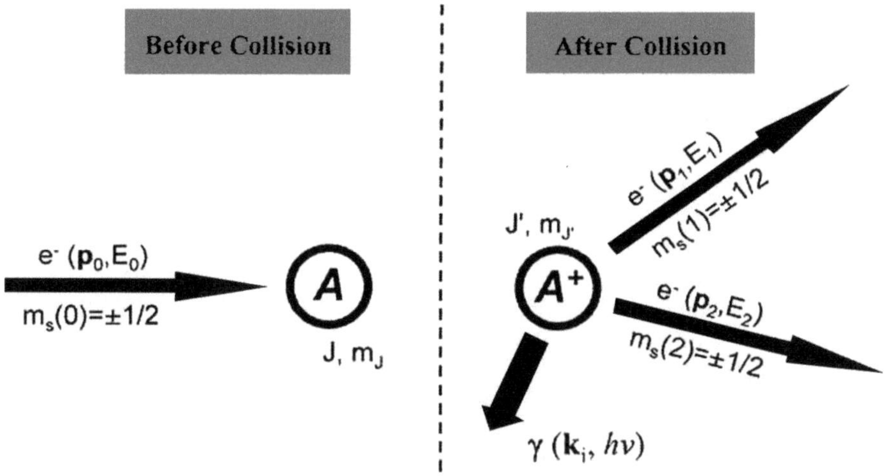

Fig. 2.64 Kinematical and quantum mechanical variables of the $(e, 2e)$ process: momenta of the continuum electrons $\mathbf{p}_0$, $\mathbf{p}_1$, $\mathbf{p}_2$, their spin projections m_s and angular momentum states of the atomic target and the residual ion (after Lower et al. 2004)

significant progress has been made recently on the quantum states of the projectile electron and the target atom.

Another type of quantum mechanically complete experiment is the analysis of photoionization of polarized atoms according to the reaction

$$|\psi_{in}\rangle = |A\rangle|h\nu\rangle \longrightarrow \mathcal{H}_{\text{lin. operator}} \longrightarrow |\psi_{out}\rangle = \sum_i c_i|A^+\rangle|e_i^-\rangle; \qquad (2.107)$$

the photoprocess transfers the initial state into a pure, final state described by amplitudes and phases (Klar and Kleinpoppen 1982). We particularly refer to the pioneering experiment by Siegel et al. (1983) with polarized metastable neon atoms. We also mention the photoionization of polarized atomic oxygen by Plotzke et al. (1996) and by Prümper (1998). Godehusen et al. (1998) reported a complete photoionization experiment with laser polarized (aligned or oriented) europium atoms; reviews on such experimental data have been published by Sonntag and Zimmermann (1992, 1995). For more detailed discussions on atomic photoionization experiments we refer to Chap. 4.

2.6 Ion–Atom and Atom–Atom Collision Processes

The atom(ion)-atom collision problem, which, in its "simplicity", approaches the previously described electron–hydrogen atom collision process, is the scattering of protons on atomic hydrogen. However, one has to realize that even the scattering of atomic hydrogen by atomic hydrogen is a much more complicated process since it takes place between two "*composite particles*". In addition the multitude of possible atom(ion)-atom collisions is almost without limit, particularly if one realizes that higher charge states of ions with sufficient beam intensity or target densities are available nowadays; taking into account all the isotopes of each element, the number of long-lived nucleides are of the order of a thousand for heavy particle-atom collision processes. Last but not least, a further complication arises through the fact that the colliding ions or atoms may undergo a short-lived molecular bonding building-up thereby a quasi-molecular state with the target-atoms. Of course, stable molecules based upon a chemical reaction may also be formed during the collision. However, such processes belong strictly to the subject of chemistry, which is beyond the scope of this section.

For the reasons mentioned above, the content of this section is more fragmented and selective than that of previous sections.

2.6.1 Impact Parameter Representation in the Classical Approximation

The de Broglie wavelength $\lambda = h/p$ of atomic particles in the eV to MeV energy range is small compared to the dimensions of atoms and of their scattering potentials. In accordance with this, the classically defined impact parameter is often used for the description of atom(ion)-atom collisions. This contrasts with electron scattering by atoms in which, for example, the de Broglie wavelength of the electron is $\lambda = 0.1$ nm at a scattering energy of about 150 eV; the corresponding de Broglie wavelength for protons is reduced by the ($m^{-1/2}$) dependence (m = mass) to $\lambda = 2.3$ pm at the same energy. In this case, the wave packet of the proton and other similar heavy particles moves on an approximately classical trajectory during the collision process. The relationship between the dimension a of the atom or its scattering potential,the impact parameter b and the orbital angular momentum L of a projectile is shown in Fig. 2.65.

According to this figure, scattering only takes place for $b < a$ but not for $b > a$. By taking into account the orbital angular momentum $L = pb$, particles with $L > pa$ are not scattered according to the classical arguments. We write $L = \ell\hbar$ and $p = \hbar k (k = 2\pi/\lambda)$ in semi-classical and quantum mechanical approximation so that the above inequality becomes $\ell > ka$. In this approximation scattering processes with $\ell > ka$ only take place with low probability. By knowing the force $F = -\partial V/\partial r$, or the potential $V(r)$ responsible for the scattering, an associated scattering angle θ and subsequently a differential cross section can be calculated for each impact parameter b. Alternatively, if the scattering potential and the differential cross section are known, the impact parameter can be calculated. However, only a few potentials are sufficiently well known for an analytical solution of the impact parameter. In most applications only numerical approximations are possible for calculating the required impact parameters.

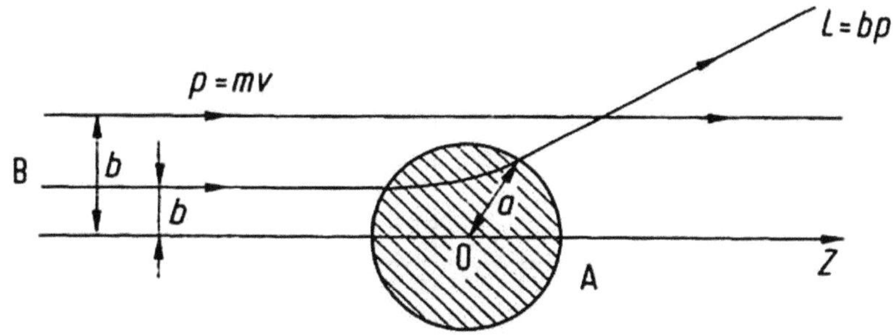

Fig. 2.65 Impact parameter *b* for the scattering of a particle *B* on particle *A*; the scattering potential of *A* has the limited range *a*

2.6.2 Quasi-Molecular Bonds

As already mentioned above, collisions between atoms or ions and atoms can result in a mutual penetration of their electron clouds and a short-lived molecular bonding may take place, in which *quasi-molecules* between the collision partners may be formed. Figure 2.66 demonstrates how an incident proton in colliding with a hydrogen atom gradually distorts its electron cloud and attracts part of the electron cloud of the hydrogen atom. If the incident proton carries the whole electron cloud (i.e. the electron of the target atom) with it at the end of the collision, an electron capture process has taken place, which can be represented by the following reaction equation: $\mathrm{p} + \mathrm{H} \rightarrow \mathrm{H} + \mathrm{p}$. The description of this process is based upon the model of formation of quasi molecules when the proton approaches the hydrogen atom (see Fig. 2.67). This figure represents a *correlation* or *Fano-Lichten diagram*; the hydrogen states are indicated for the case that the distance R between the projectile proton and the hydrogen atom is infinitely large. When the proton and the hydrogen atom approach each other in the collision process, molecular bound states are formed, which are described by the electron configurations $1\mathrm{s}\sigma$, $2\mathrm{s}\sigma$, $2\mathrm{p}\sigma$, $2\mathrm{p}\pi, \ldots$ and so forth of the H_2^+ quasi molecule; the Greek letters σ and π denote the magnetic quantum numbers $m_\ell = 0$ and $m_\ell = \pm 1$ of the orbital momentum $\ell = 1$ with reference to the axis connecting the two protons. For the limiting case of zero distance between the proton and the hydrogen atom, an He^+ ion is formed. This ion is often called a *united atom* (a name which is clearly misleading!). The energy states of the ion for $R = 0$ are helium-ion like, with the usual notations 1s, 2s, 2p, ...; they are lower than the corresponding ones for atomic hydrogen because $Z = 2$.

United atoms are the subject of research projects, particularly in high-energy heavy-ion physics carried out at high-energy accelerators (e.g., at the Darmstadt GSI [Gesellschaft für Schwerionenforschung]). By bombarding a uranium target ($Z = 92$) with highly charged uranium ions, a united highly charged atom with $Z' = 184$ can be produced for a short time; this enables physicists to study atomic physics for at least inner shells of a super-heavy atom. Such studies are particularly important for the further development of relativistic atomic physics, based upon the Dirac

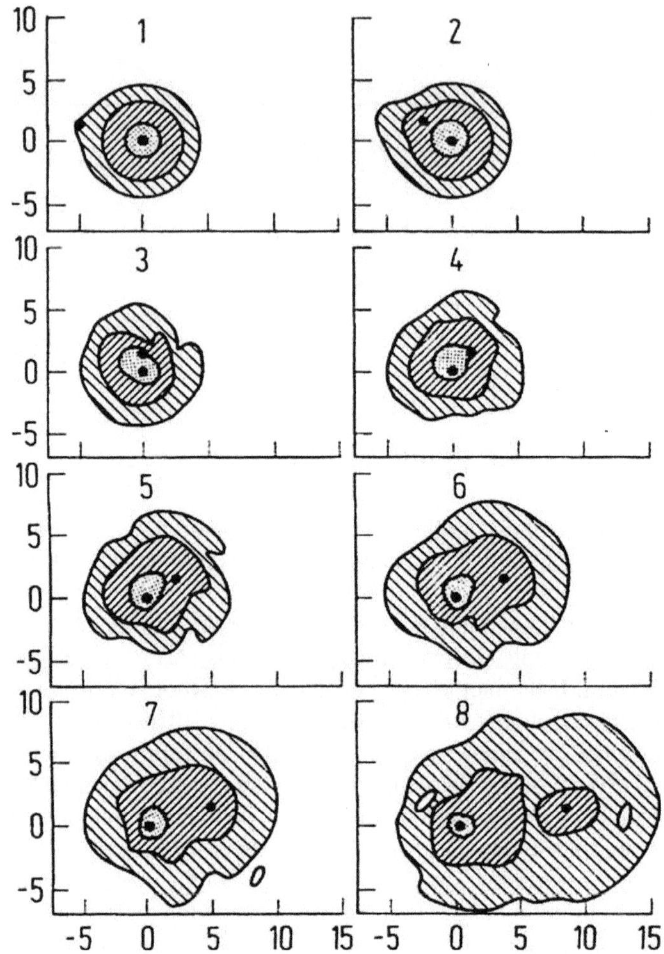

Fig. 2.66 Charge distribution of an electron in atomic hydrogen at the approach of a 40 keV proton; the two protons are represented by black dots. Events 1 to 8 follow in time sequence; the hatched areas illustrate the charge density distribution of the electron in atomic hydrogen. Distances are measured in atomic units (theoretical calculation of Shakeshaft 1978)

theory, beyond the critical nuclear charge ($Z \sim 173$), and quantum electrodynamics in strong fields (Pieper and Greiner 1969; Zeldovich and Popov 1971; Greiner and Reinhardt 1995; Labzowsky et al. 2006).

The two states 1sσ and 2pσ interfere with each other in the intermediate range between $R = 0$ and $R = \infty$ of the correlation diagram of Fig. 2.67 via radial coupling. The averaged, energy separation $E(1\mathrm{s}\sigma) - E(2\mathrm{p}\sigma) = h\nu$ is related to the frequency ν by which the electron oscillates between the two protons. When the proton, as a projectile, leaves the target area it has either captured the electron or not. The oscillatory characteristic of the charge capture can be detected by a measurement of

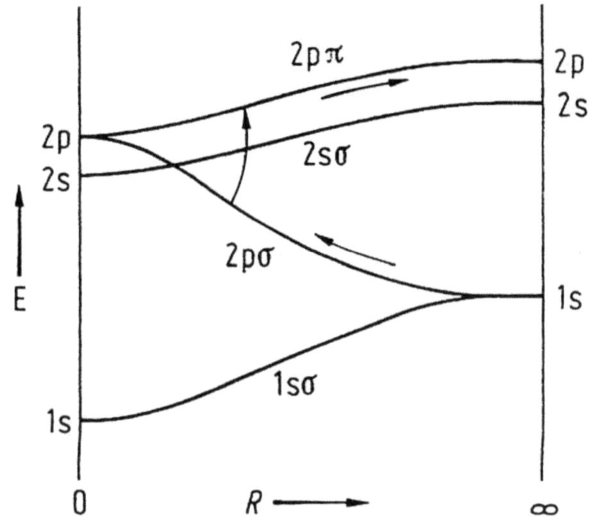

Fig. 2.67 Correlation diagram for the collision system p + H. R is the distance between the proton and the hydrogen atom; E is the energy of the collision system, i.e. for the atomic hydrogen states at $R \rightarrow \infty$ and of He^+ at $R = 0$. Radial $1s\sigma - 2p\sigma$ coupling and rotational $2p\sigma - 2p\pi$ coupling (see Fig. 2.71 and text later on) is indicated by *arrows* along the electron path. For clarity, the ℓ-degeneracy has been removed

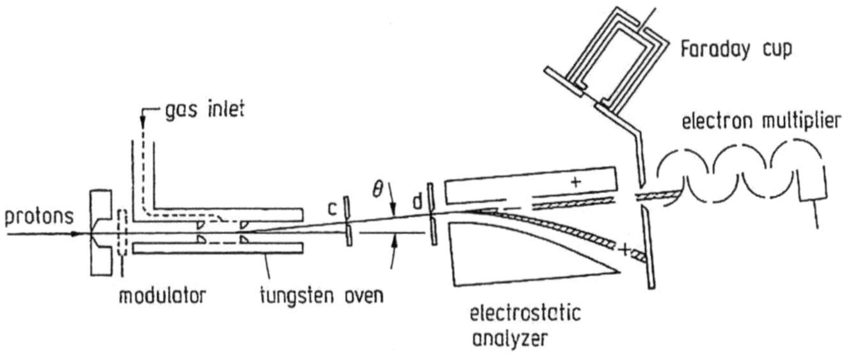

Fig. 2.68 Scheme of an apparatus for the detection of the charge capture ($p + H \longrightarrow H + p$) as a function of the scattering angle θ. The tungsten oven is at a temperature of $T = 2400$ K at which molecular hydrogen is dissociated into atomic hydrogen to a high degree of dissociation (≈95 %); after Lockwood and Everhart (1962)

the differential cross section. The experimental apparatus required is shown schematically in Fig. 2.68. Protons pass through a target of atomic hydrogen; the scattered particles, which are either protons or hydrogen atoms from the charge capture, pass through an electrostatic energy analyzer. The undeflected neutral hydrogen atoms are detected by an open electron multiplier in which electrons, are released from its surface. The deflected protons are detected by a rotatable Faraday cup. Figure 2.69 demonstrates the oscillatory structure of the energy dependence of the charge capture of the proton in atomic hydrogen.

Similar distinctive interference effects between direct and charge exchange processes can also be detected in the total cross sections. As a typical example, we refer to the collision process

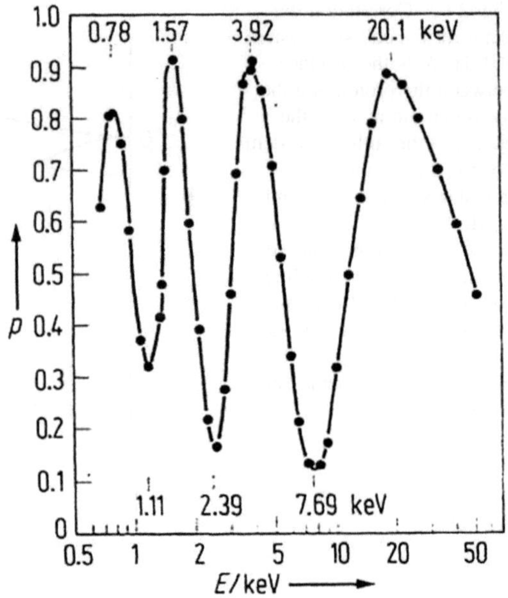

Fig. 2.69 Electron capture probability p for the collision process $p + H \longrightarrow H + p$ as a function of the energy of the incoming protons; the hydrogen atoms produced by this process are recorded at an angle of $\theta = 3°$ (see Fig. 2.68); after Lockwood and Everhart (1962)

$$Na^+ + Ne \begin{array}{l} \nearrow Na^+ + Ne^*(3p) \\ \searrow Na^*(3p) + Ne^+ ; \end{array} \tag{2.108}$$

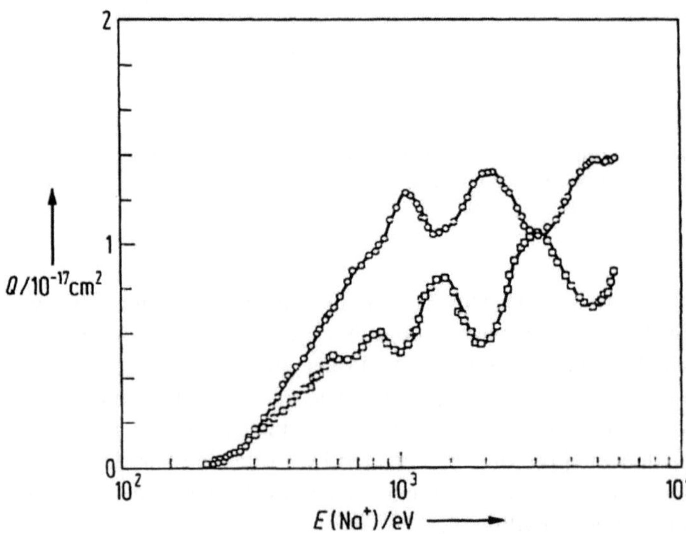

Fig. 2.70 Absolute total cross section Q for the excitation of the Ne(3p) state (□) and Na(3p) states (○) for collisions between Na^+ ions of energy $E(Na^+)$ and neon atoms (after Tolk et al. 1967)

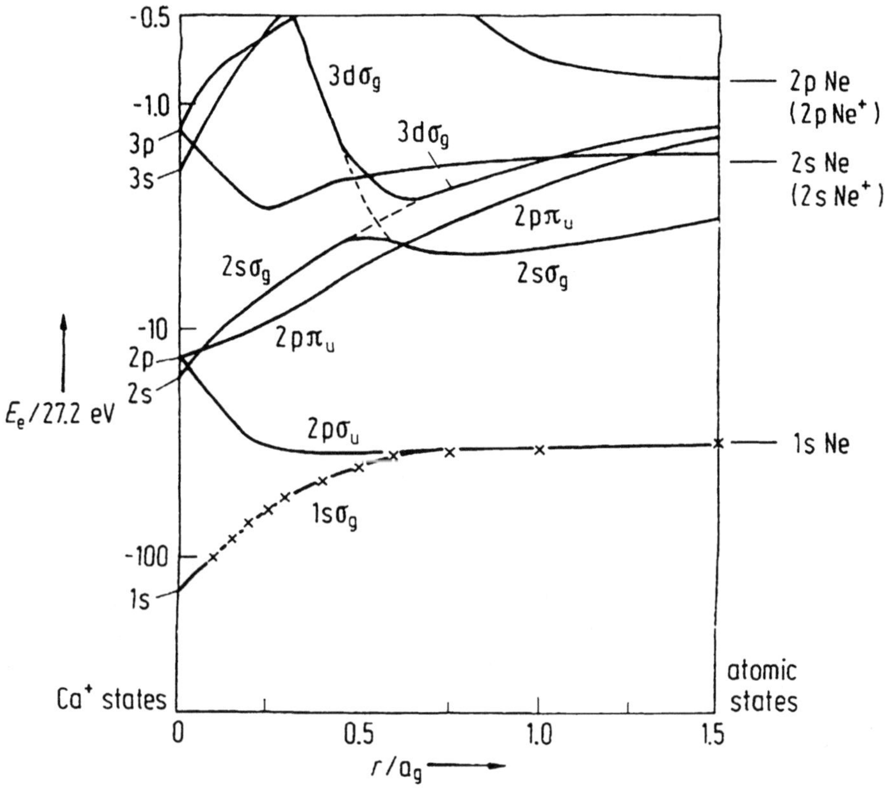

Fig. 2.71 Correlation diagram of the symmetric ($Ne^+ + Ne$) collision process; r = internuclear distance between the colliding particles, a_0 first Bohr radius (after Larkins 1972)

the first reaction is a *direct excitation process*, the second one is a *charge exchange process* including simultaneous excitation. Both processes interfere with each other, this follows from an observation of the oscillations of the line intensities from the excited 3p states (Na D lines and NeI 622.6 nm line) of the two neutral atoms (Fig. 2.70). A quasi-molecular interference model after Rosenthal and Foley (1969) and Bobashev (1970) correctly predicts the observed anti-coincidences between the maxima and minima of such processes in Fig. 2.70.

A further important example for the application of short-lived quasi-molecular bonds in atomic collisions is associated with the production of electron holes in inner shells. We illustrate this process in connection with the $Ne^+ - Ne$ collision process which leads to the emission of a K X-ray line or the corresponding KLL Auger electron. Consider the relevant correlation diagram of the $Ne^+ - Ne$ system in Fig. 2.71. At vanishing small distances between the Ne^+ ion and the Ne atom, a calcium nucleus with the electrons of the colliding ion and atom is formed (a *united atom* or, better, a *united ion*). Two of the original four 1s electrons of the two neon particles occupy the 1s state of the united atom. The Pauli principle prevents further occupation of electrons in the 1s state; instead, the two remaining electrons are promoted into the molecular 2pσ state of the united atom (ion). Of course, the state of the united atom cannot be reached in practice because of the Coulomb repulsion of the nuclei; if, for example, the projectile ion has an energy greater than 50 keV, the minimum distance between the ion and the target atom for head-on collisions becomes smaller than the K-shell (5×10^{-10} cm) of the neon atom. In this case the energy separation between the 2pσ and 2pπ molecular states becomes very small.

If, for example, an electron hole were directly produced in the 2p subshell during a ($Ne^+ + Ne$) collision, the hole can be "shared" between the states 2pπ and 3dσ according to the correlation diagram in Fig. 2.71. Since a transition 2pπ $\rightarrow$ 2pσ can take place during the process of closer approach of the collisional partners, a hole in the 1s state of the separated atoms will be produced by the transition 2pσ $\rightarrow$ 1sσ at large distances between the atoms. In other words, an initial electron hole in the 2p subshell is transferred into a hole of the 1s shell by the dynamics of the collision process and the formation of quasi-molecular states. It can be shown theoretically that the Schrödinger equation for the description of the motion of the atomic nuclei can be transferred into a set of coupled equations which comprise dynamical coupling terms and determine transition probabilities between the quasi-molecular states. These coupling terms are reduced to the expressions $v_R\langle i|P_R|j\rangle$ and $\Omega\langle i|L_z|j\rangle$ in semi-classical approximation in which $v_R = \partial R/\partial t$ are the relative velocity, P_R the relative momentum and Ω the rotational angular momentum of the nuclear motion; $|i\rangle$ and $|j\rangle$ are the eigenvectors of the quasi-molecular states and L_z the angular momentum perpendicular to the scattering plane. The first of the above expressions

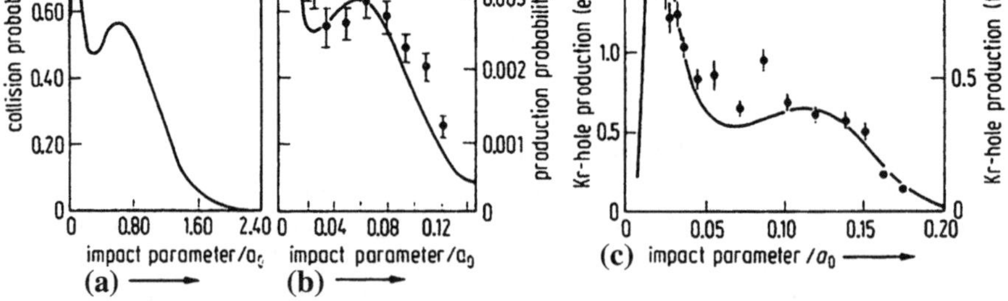

Fig. 2.72 Impact parameter dependence of **a** the Lyman-α excitation of the ($H^+ + H$) collision system according to theoretical predictions by Bates and Williams (1965), **b** the production of a K-electron hole in a 363 keV ($Ne^+ + Ne$) collision (experimental data and theoretical curve after Sackmann et al. 1974), and **c** a L-electron hole production from a 1.75 MeV ($Kr^+ + Kr$) collision (after Shanker et al. 1984)

is called the *radial coupling term*; this term relates to transitions between states of the same rotational symmetry, with reference to the axis connecting the two nuclei. For example, even symmetry exists if the sign of the eigenfunction of the quasi-molecular state remains unchanged on rotation of 180° around the connecting nuclear axis (notation of the states with the symbol g from the German word "gerade"). Odd (German "ungerade" with symbol u) symmetry is valid if the sign of the wavefunction changes under the above rotation operation. For example, transitions between the quasi-molecular states $2s\sigma_g$ and $3d\sigma_g$ can be induced under the influence of the radial coupling term, if the two particles approach each other at a distance of $R = 0.5$ atomic units. The matrix element $\Omega\langle i|L_z|j\rangle$ of the rotational coupling connects states of different angular symmetry as, for example, the quasi-molecular states $2p\sigma_u$ and $2p\pi_u$; the finite transition probability between these states which follows from this rotational coupling is the dominant process for the production of the 1s electron hole in the ($Ne^+ + Ne$) collision. As an example Fig. 2.72b shows experimental results for the probability of the production of a K-shell electron hole in the ($Ne^+ + Ne$) collision process.In the experiment of Sackmann et al. (1974) K_α-X-ray photons are detected in coincidence with the scattered Ne^+ ions. The probability $P(b)$ as a function of the impact parameter b follows from the ratio

$$\frac{N}{I_0} = nQL, \tag{2.109}$$

with N as number of the produced K_α photons, I_0 the number of the ions hitting the target of density n along the length L and the total cross section Q in which the fluorescence yield and the detection sensitivities of the X-ray and ion-detector have to be taken into account. The kinetic maximum of the probability $P(b)$ at very small impact parameter is attributed to the above ($2p\sigma - 2p\pi$) rotational coupling, i.e. to the fast rotation of the inter-nuclear axis of the two collisional partners. The lower second maximum at larger impact parameters can be explained by the adiabatic *Massey criterion* which follows from Heisenberg's uncertainty relation

$$\Delta E \Delta t > \hbar. \tag{2.110}$$

The interaction time $\Delta t = d_w/v$ of the two collision partners Ne and Ne^+ is determined by their relative velocity v and the spatial interaction range of an extension d_w. With these quantities the Massey criterion can be written as

$$\frac{\Delta E d_w}{\hbar v} > 1. \tag{2.111}$$

A significant transition probability between two quasi-molecular states only arises for such separations between the colliding partners for which their energy differences is within the interval ΔE determined by the Massey criterion. Transitions between states for which their energy difference is large, compared to the energy interval according to Massey's criterion, are very unlikely. In such cases, the number of coupling quasi-molecular states which may contribute to inner-shell excitation is substantially reduced.

This double structure in the probability for producing electron holes in inner shells (Fig. 2.72), i.e., the principal maximum based on the $(2p\sigma - 2p\pi)$ rotational coupling and the side maximum due to the adiabatic Massey criterion has been observed experimentally in many examples of inner-shell excitation. Rotational couplings of the type $2p\sigma - 2p\pi$ were first postulated by the British physicists Bates and Williams (1965) in connection with the H(2P) excitation in the proton–hydrogen collision process (Fig. 2.72a). The qualitatively good agreement between theory and experiment in the double structure of the excitation probabilities implies the existence of the rotational coupling and an adiabatic process, which induce transitions between quasi-molecular states.

2.6.3 Potential Scattering and Quantum Mechanical Structure Effects

A series of structure effects show up in measurements of a cross section of scattering processes between neutral atoms. In approaching each other, the electron clouds of the colliding atoms are deformed whereby two electric dipoles may be induced for a short time. The potential energy between the two dipoles is proportional to $(-1/r^6)$ which results in an attractive force, the *van der Waals force* between two atoms (r is the distance between the atoms). However, as expected at very small separations, the atoms repel each other so that the total potential has the form

$$V(r) = \frac{A}{r^n} - \frac{C}{r^6}, \tag{2.112}$$

of which the *Lennard-Jones potential* represents a typical case with $n = 12$. The special characteristics of such potentials leads to a series of *rainbow* and *glory structures* which are interference effects in cross sections of low-energy collisions between atoms. A typical apparatus for such studies is shown schematically in Fig. 2.73. The incoming atoms pass through a set of rotating discs in the form of toothed wheels acting as a velocity or energy selector; the slits and rotational velocity of the discs are correlated so that only atoms of a specific energy reach the target area.

As examples of the above structure effects, we refer to Figs. 2.74, 2.75 and 2.76. Glory undulation effects are shown in Figs. 2.74 and 2.75 for collisions between potassium and krypton atoms and also between sodium and mercury atoms. In both experiments the projectile atoms sodium or potassium first pass through the velocity selector before colliding with the target atoms. Figure 2.76 shows measured angular distributions for five different energies of Na–Hg scattering; in analogy to the rainbow scattering of sunlight, the maxima in the scattering intensities are interpreted as rainbow maxima. The broad maximum at the largest scattering angle is the primary rainbow and the remaining ones are secondary rainbows. Based on such rainbow data, Buck and Pauly (1971) determined the following parameters and the shape of the potential curve (Fig. 2.77) of the collision dynamics of colliding sodium and mercury atoms:

$$r_m = (4.72 \pm 0.02)10^{-8}\,\text{cm}, \quad \varepsilon = -(5.49 \pm 0.17)10^{-2}\,\text{eV}. \tag{2.113}$$

An interesting interference effect arises in the scattering of atoms of the same kind. An atom A may be elastically scattered through a given angle θ by an atomic target of identical atoms. Because of conservation of momentum and energy, the target atom will experience a recoil through 90° with respect to the direction of the scattered atom, i.e. the scattering angle of the recoiling atom is $-(\pi/2) + \theta$; $f(\theta) = f(-\theta)$ and

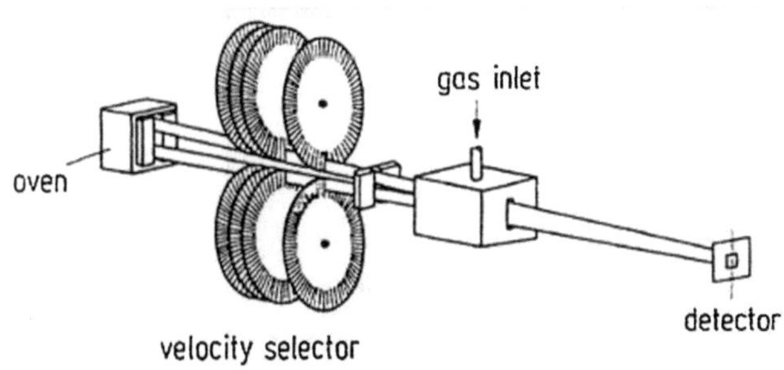

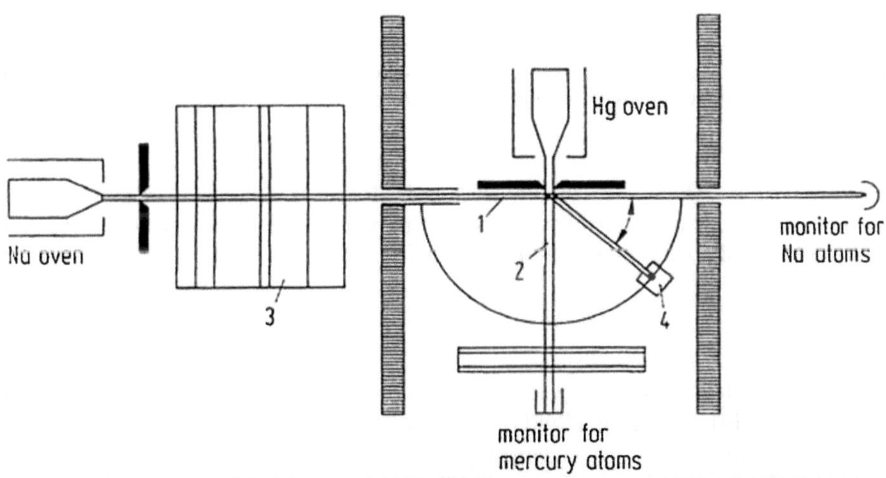

Fig. 2.73 **a** Scheme of an apparatus for measuring total cross sections in collisions between atoms; a velocity selector (after von Busch et al. 1967) varies the velocity of the incoming atoms; **b** crossed atomic beams (1, 2) with a velocity vector (3) for the energy selection of the primary sodium beam and an atomic detector (4) for the measurement of differential cross sections (after Buck and Pauly 1968)

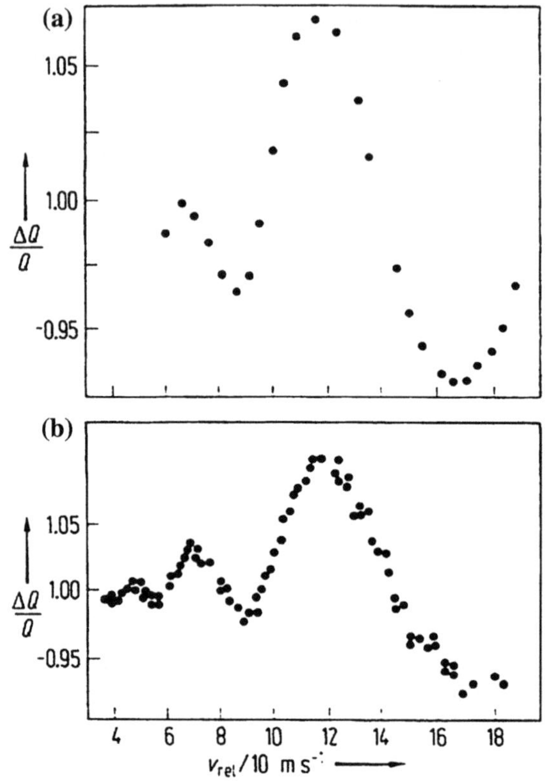

Fig. 2.74 Relative variation $\Delta Q/Q$ of the averaged total cross section Q for the (K+Kr) scattering as a function of the relative velocity v_{rel} between the colliding atoms. **a** Experiment after von Busch et al. (1967); **b** experiment after Beck and Loesch (1966)

$f[(\pi/2)-\theta] = f[\theta-(\pi/2)]$ are the amplitudes for the scattered and recoiling atom. In view of the indistinguishability between the two colliding atoms, the differential cross section follows as an interference effect

$$\sigma(E,\theta) = \left|f(\theta) + f\left(\frac{\pi}{2} - \theta\right)\right|^2, \tag{2.114}$$

since the amplitudes coherently superpose on each other and are expected to have a certain phase difference. This kind of symmetry interference effect has indeed been observed in the (^{4}He + ^{4}He) scattering (Fig. 2.78) but not with different isotopes such as in the scattering process ^{4}He + ^{3}He.

The existence of this quantum mechanical interference effect due to the indistinguishability between the scattered and recoil atom was first recognized and predicted by Mott (1929, 1932). Chadwick (1929) detected this effect a year later in the scattering of α particles (i.e. He^{++} ions) on He nuclei. 40 years later Pauly (1988)

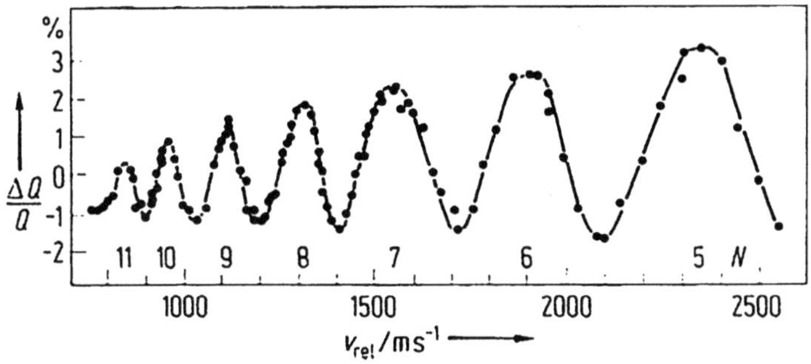

Fig. 2.75 Identical quantity to that of Fig. 2.74 for scattering of sodium on mercury atoms (after Buck et al. 1971)

Fig. 2.76 Angular distribution $J(\theta)$ of sodium atoms scattered by mercury atoms for various collision energies. $J(\theta)$ as a function of the scattering angle is given in arbitrary units J_1 (after Buck and Pauly 1971)

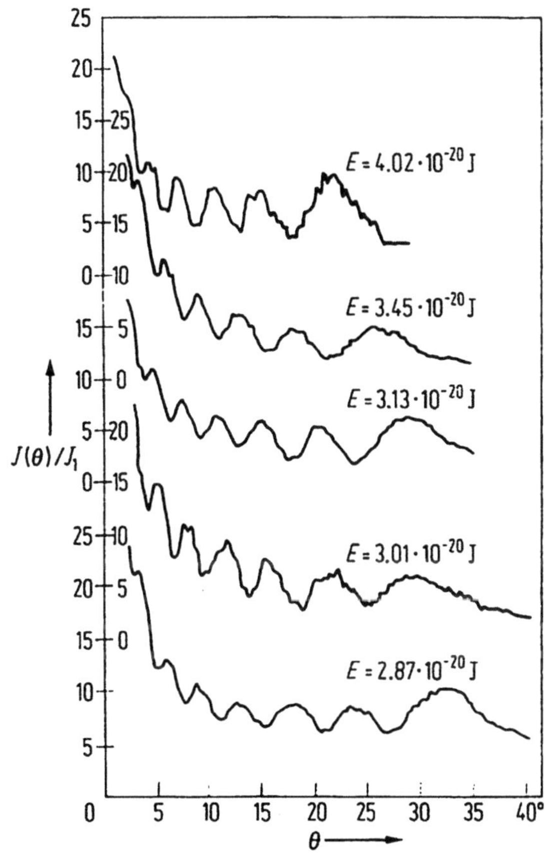

succeeded in verifying this interference effect in low-energy scattering of atomic particles.

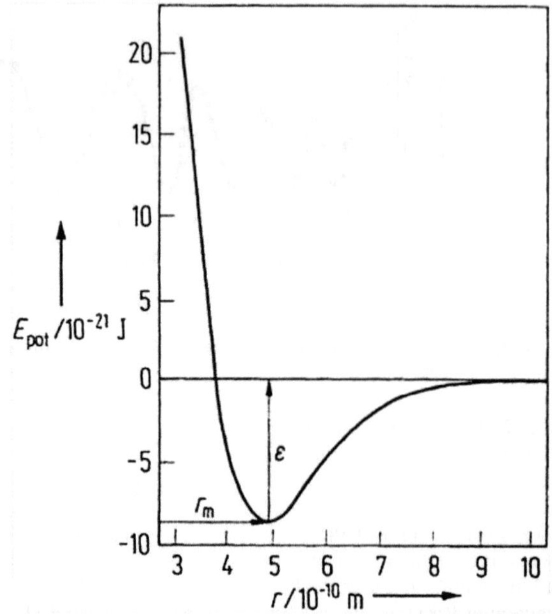

Fig. 2.77 Typical form of the interaction potential energy $E_{pot}(r)$ as a function of the distance r between colliding Na and Hg atoms (after Buck and Pauly 1971)

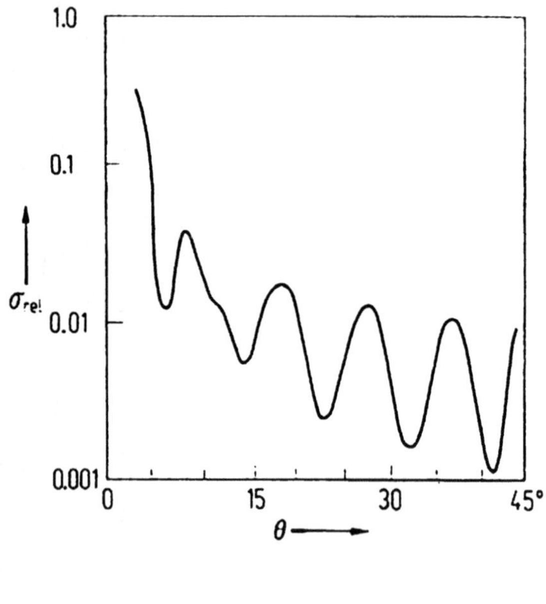

Fig. 2.78 Relative differential cross section σ_{rel} for the scattering of ^{4}He atoms on ^{4}He atoms as a function of the scattering angle θ at a collision energy of $E = 63$ meV (after Siska et al. 1979)

2.6.4 Coincidence and Spin Experiments

We have already explained in Sects. 2.4 and 2.5 how coincidence and electron-spin methods can be applied to studies based upon "complete experiments" in electron–atom collisions. Such coincidence and spin methods have also been applied for ion and atom scattering on atoms.

Though quantum mechanically complete collision experiments with ions and atoms and projectiles have been approached, its large amount of possibilities goes far beyond electron impact and photoabsorption processes. Accordingly, the large selectivity of possibilities is by far more with heavy particle impact. Experiments have been carried out specifically for coincidence measurements between photons, electrons, ions and atoms and also for spin effects in collisions between spin-polarized ions and atoms. While coincidence experiments have already provided an impressive set of data, spin experiments for heavy-particle-atom collisions are only at the beginning of their development.

As an example we again discuss the excitation of Lyman-α radiation by a *charge transfer process*:

$$\begin{aligned} p + He \longrightarrow\ & H(2p) + He^+ \\ & \llcorner\!\!\longrightarrow L_\alpha + H(1s). \end{aligned} \tag{2.115}$$

In relevant experiments by Hippler et al. (1987), Lyman-α photons and hydrogen atoms in the ground state are detected in coincidence; the excited H(2p) atoms decay into the ground state on their way from target to detector. The directions of the incident protons and the detected hydrogen atoms define the planar reaction plane of the above process. The analysis of the count rates of the coincident detection of the L_α photons and the hydrogen atoms can be based upon the excitation amplitudes f_0 and f_1 for the magnetic substates $m_\ell = 0$ and $m_\ell = \pm 1$ of the P state of

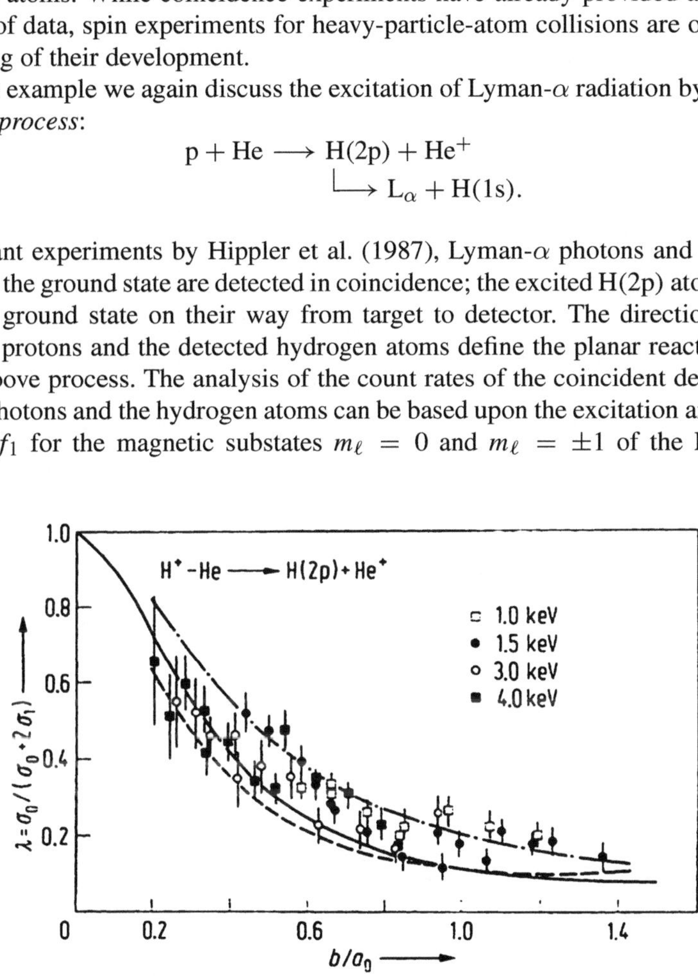

Fig. 2.79 λ parameter for the relative population of the H(2P, $m = 0$) state as a function of the impact parameter b (in atomic units a_0) of the charge exchange process $H^+ + He \longrightarrow H(2p) + He^+$ at various energies; experimental data with error bars after Hippler et al. (1987), theoretical predictions (——) by Macek and Wang (1986) and Fritsch (1986) for the other two curves: (– – –) for 1 keV and (– · – · –) for 4 keV of the incident protons

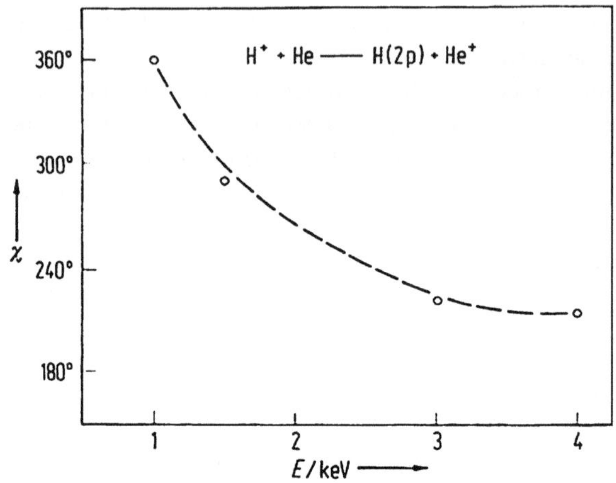

Fig. 2.80 Relative phase χ between the amplitudes f_0 and f_1 as a function of the energy for the same process as in Fig. 2.79 and an impact parameter $b \simeq 0.7a_0$ (data from Hippler et al. 1987)

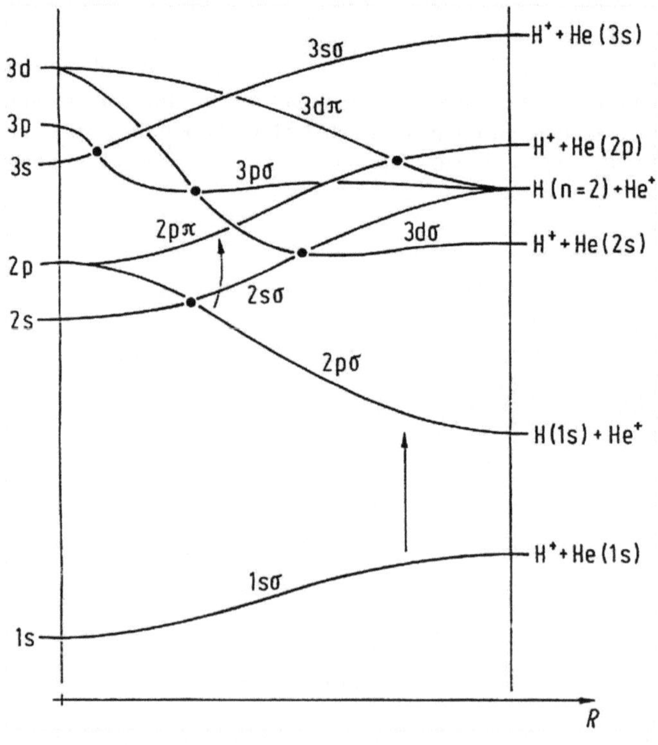

Fig. 2.81 Correlation diagram (schematic) for the collision process $H^+ + He \longrightarrow H + He^+$ (R = internuclear distance). Note that the nucleus at $R = 0$ corresponds to a lithium nucleus, which is surrounded by two electrons (following Hippler et al. 1987)

atomic hydrogen which is identical to electron–photon coincidences from the electron impact excitation of $^1P_1$ states of helium described in Sects. 2.4 and 2.5. The fine-structure splitting $2\,^2P_{1/2,3/2}$ results in a certain depolarization of the L_α-emission which can be accounted for theoretically. The description of the coherent excitation of the H(2p) state by charge transfer requires only three quantities. These are the differential cross section σ, the angular correlation parameter $\lambda = |f_0|^2/\sigma$ and the phase difference χ between the amplitudes f_0 and f_1, as for the corresponding He(2p) electron impact excitation process $^1S_0 \rightarrow {}^1P_1$ (Sect. 2.5). As an alternative, the angular momentum transfer $\langle L_\perp \rangle$ into the 2p state and the alignment angle γ of the charge state distribution of the excited state can be applied for the interpretation of the coincidences (again in accordance with the electron impact excitation in Sects. 2.4 and 2.5). Figures 2.79 and 2.80 show examples for the measurement of the angular correlation parameters λ and χ. The dependence of these quantities on the impact parameter b can be understood by means of the correlation diagram of Fig. 2.81 and appropriate theoretical approximation methods. The quasi-molecular states 1sσ and 2pσ are correlated with the H(1s) and He(^{1}S) states when the two colliding partners approach each other. Excitation into the H(2p) state by charge transfer takes place in two stages: The first stage occurs at the relatively large internuclear separation of a few atomic units of the length a_0 (see arrow in Fig. 2.81); the quasi-molecular state 1sσ couples radially to the next higher state 2pσ. During a further process of closer approach of the two particles, the 2pσ state radially interacts with the 2sσ or 3pσ state and promotes the H(2p) state by the (2pσ − 2pπ) rotational coupling. According to the data of Fig. 2.79, the excitation of the magnetic sublevel $m_\ell = 0$ of the H(2p) state takes place preferentially below the value $b = 0.3$ of the impact parameter, which appears to be in good agreement with theoretical predictions.

We are pleased also to refer to the method of *Momentum Imaging in Atomic Collisions* which has dramatically been applied for many kinematically complete studies of three- and more particle atomic collision systems. The technique emerged form the work of Schmidt-Böcking and his students and colleagues at Frankfurt a. M. and is also named as COLTRIMS, i.e. *Cold Target Ion Momentum Spectroscopy* (Cocke 2004). The technique first applied to ion–atom collisions is now frequently used for a wide range of experimental areas including, both, charged particles and photonic collisions. Important reviews about COLTRIMS have been published (Ullrich et al. 1998; Dörner et al. 2001; Schmidt-Böcking et al. 2002; Ullrich et al. 2003).

Atom–photon coincidence measurements of atom–atom excitation have also resulted in detailed information on the λ and χ parameters, alignment and orientation and the shape of the electron charge distribution of excited states. In such experiments, monoenergetic projectile atoms are produced either by charge exchange of monoenergetic ions passing through a gaseous target (see Sect. 2.6.2) or by neutralization on a metallic surface. The scattered atoms are detected by surface ionization. In modern experiments of this type, position-sensitive surface detectors are used as indicated in Fig. 2.82. The scattered fast atoms are detected by a chevron arrangement of two micro-channel plates. The fast neutral atoms release secondary electrons from the front side of the first micro-plate. The position of each resulting electron

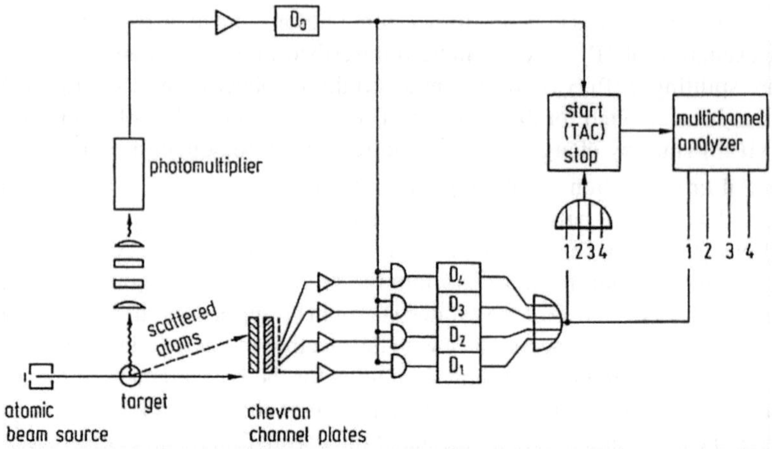

Fig. 2.82 Scheme of a multiple-coincidence apparatus for coincidence measurements with atoms and photons (after Hermann et al. 1980)

avalanche through the plates is accurately detected by 16 separate anodes fixed behind the second micro-plate. The resulting pulses are amplified and stopped by a time-to-amplitude converter, which has been started by a photon produced by the decay of the excited atom. The time-of-flight spectra of the coincidence events are registered for each scattering angle by an appropriate computer program. Hermann et al. (1980) and Menner et al. (1981) have studied with this technique the impact excitation of the first resonance lines of alkali atoms in collisions with rare gas atoms. Figure 2.83 shows an example for the measurement of relevant λ and χ parameters as defined in Sect. 2.4.2.

We have already drawn attention to the fact that spin experiments for scattering of ions (or atoms) by atoms are only at the beginning of developments of relevant research programmes. The main reason for this situation is the fact that spin polarized ions cannot be separated and focused by a magnetic hexapole, because the Lorentz force acts on the ion charge with the result that the spatial separation and focusing of ions in selective spin states is completely destroyed. Even using very small apertures for the ion beam does not provide spin selection because of diffraction effects encountered with such small apertures in accordance with Heisenberg's uncertainty principle. However, such difficulties can be overcome by applying the technique of optical pumping of one-electron ions with circularly polarized light. This has already been successful in polarizing ion beams of considerable intensities (Reihl 1993; Lutz 1992).

Successful collision experiments with unpolarized ions and polarized atoms have already been reported with impressive results. An experimental scheme for such investigations is shown schematically in Fig. 2.84; ions of rare gas atoms or protons cross a sodium atomic beam which has been partially polarized by a hexapole magnet. As a result of the collisions, the sodium D lines were excited. Their radiation

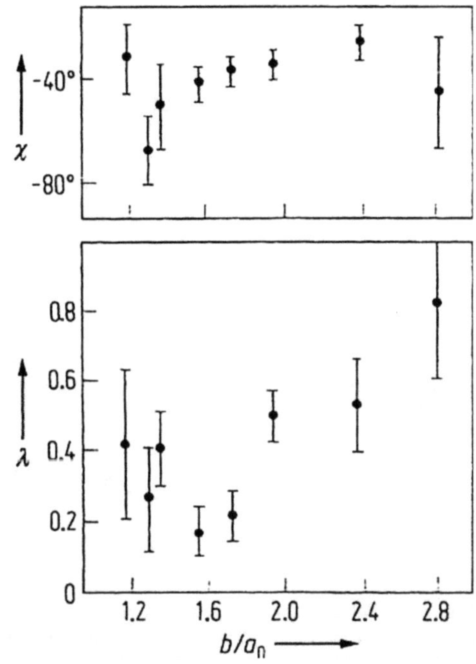

Fig. 2.83 Experimental λ and χ data for the excitation of the first resonance lines of lithium atoms in collisions with xenon atoms (after Menner et al. 1981)

was observed parallel to the polarization direction of the spin of the sodium and the linear and circular polarization of the D lines was measured. Of special interest in this experiment was the detection of a circular polarization of the D lines, which would be expected if spin conservation were maintained during the excitation process. The focused atomic beam with its spin component $m_s = +1/2$ in the $3\,^2S_{1/2}$ ground state is transferred into the $(m_j + m_s)$ components of the $3\,^2P_{1/2,3/2}$ states in which the m_j-components with positive sign are more strongly populated than the ones with negative sign. This then leads to more intense σ^+ components in the $3\,^2P \rightarrow 3\,^2S$ transition than for the σ^- components, which explains the circular polarization of the D-line radiation. In the quasi-molecular system $(Ne^+ + Na)$, however, electron exchange between the polarized sodium atoms and the excited $Ne^+(2p^5 3s)$ or $Ne^+(2p^5 3p)$ can take place, which leads to a depolarization of the circular polarization of the D-line radiation. The observed circular polarization of the D lines in Fig. 2.85 for high Ne^+ energies ($\sim$5 keV) agrees well with the value $P_{circ} = 17.2\,\%$ which follows from a knowledge of the polarization of the sodium beam of $P_{Na} = 20.8\,\%$ and of spin conservation during the collision process. The maximum depolarization caused by the above electron exchange interaction results in a lower limiting value of $P_{circ} = 11.7\,\%$ for the circular polarization. The experimental data in Fig. 2.85 oscillate between the two limiting values of the circular polarization. The electron exchange comes into play if the primary energy of the Ne^+ ions is very large since the interaction time involved in this process is too short

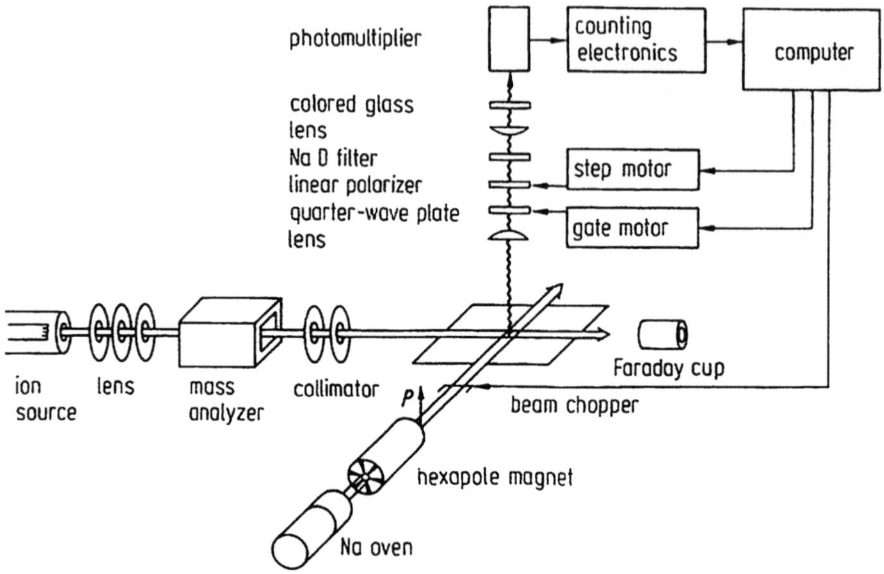

Fig. 2.84 Scheme for the excitation of the Na D resonance radiation by H^+, He^+, and Ne^+ ions, colliding with partially polarized Na atoms (after Osimitsch et al. 1989)

at small energies. However, if sufficient time for the electron exchange process is available, depolarization of the D-line radiation occurs.

2.6.5 Antiproton–Atom Collisions

Anti-protons as the anti-particles of protons can be produced in high-energy interactions of protons with matter. For example, at the European Organization for Nuclear Physics Research (i.e., CERN = Centre European de Rechèrches Nucléaires), anti-protons with energies between 5.9 MeV and 1.3 GeV can be produced with intensities of 10^5–10^6 particles per second. Anti-protons of lower energies are obtained by deceleration of the primary beam after passing through thin foils.

The first scattering experiments of protons (p) and anti-protons ($\overline{p}$) as projectiles resulted in interesting data for single (Q^+) and multiple (Q^{n+}) ionization of atoms. As an example of such experiments, we discuss the ionization of helium. It has been proved experimentally that the cross sections of the reactions

$$\mathrm{p} + \mathrm{He} \longrightarrow \mathrm{He}^+ + \mathrm{p} + \mathrm{e}, \tag{2.116}$$

and

$$\overline{\mathrm{p}} + \mathrm{He} \longrightarrow \mathrm{He}^+ + \overline{\mathrm{p}} + \mathrm{e}, \tag{2.117}$$

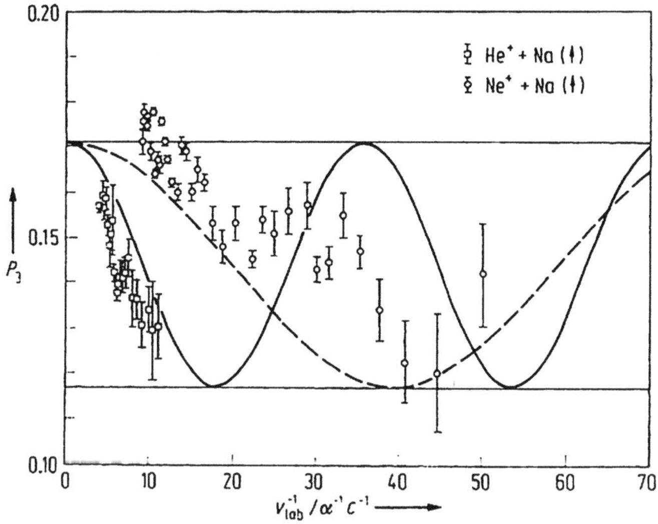

Fig. 2.85 Measured and calculated circular polarization P_3 of the Na D line radiation as a function of the inverse velocity $1/v_{lab}$ of He^+ and Ne^+ ions colliding with sodium atoms (Osimitsch 1989; Osimitsch et al. 1989). The *upper horizontal line* corresponds to spin conversion, the *lower one* to maximum depolarization by electron exchange during the collisional excitation; (α = Sommerfeld's fine structure constant, c = light velocity). The theoretical curves have been calculated according to a simple model for a phase shift between singlet and triplet states under the assumption of a straight line motion of the nucleus for small impact parameters; (---) for Ne^+, (——) for He^+

agree with each other within an error of a few percent. On the other hand, the cross sections for double ionization (Q^{++}) of the reactions

$$p(\overline{p}) + He \longrightarrow He^{++} + p(\overline{p}) + 2e, \tag{2.118}$$

are different for protons (p) and anti-protons ($\overline{p}$) as shown in Fig. 2.86, for the ratios Q^{++}/Q^+ of double to single ionization. This surprising Q^{++}/Q^+ effect has been interpreted as an interference effect due to the following two processes. The first process is called "*shake-off*" process; it arises from the fact that the electron shielding of the atomic nucleus is changed when the first electron is knocked out. During the second process the two knocked out electrons interact with the projectile one after another with different probability (a two-step process). Accordingly both processes and their interference terms for the production of the He^{++} ions are different for protons and anti-protons. However, the details of a clear-cut theory for this Q^{++}/Q^+ effect are not yet completely understood.

For high energies it has been found experimentally (Fig. 2.87), that, as projectiles, the ratio $Q^{++}/Q^+ = R_{p^-}$ for anti-protons is the same as that for electrons (R_{e^-}) and for protons (R_{p^+}) is the same as for positrons (R_{e^+}).In the high-energy range the

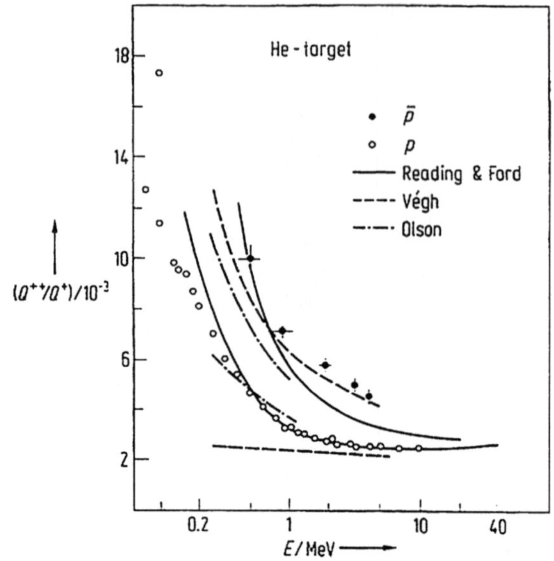

Fig. 2.86 Experimental (∘, •) and theoretical (——, – – –, · – · –) data for Q^{++}/Q^{+} ratios between double and single ionization of helium atoms by proton and antiproton collisions (after Elsener 1989)

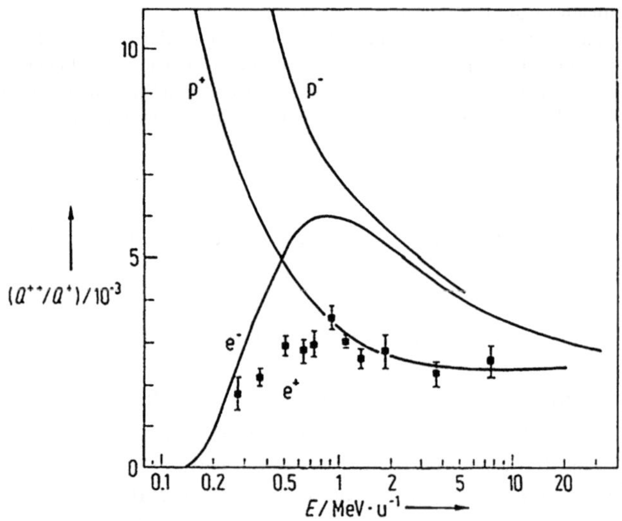

Fig. 2.87 Q^{++}/Q^{+} data for helium atoms as target with protons and anti-protons as projectiles. The *full curves* represent experimental data for electrons (e^{-}), protons (p^{+}), and anti-protons (p^{-}) while the data (■) are the positron ratios $R_{e^+} = Q^{++}/Q^{+}$ according to measurements of Charlton et al. (1988) $\left(1\,\text{u} = \frac{1}{12}m(^{12}\text{C})\right)$

Q^{++}/Q^{+} ratios depend on the projectile charge, but not on the projectile mass. This effect could only be detected by actual scattering experiments with anti-particles in high-energy accelerators.

Chapter 3
Auger Emission and Inner Shell Hole Experiments

3.1 Basic KLL Auger Transitions

Let us consider an ionization process of the atomic K-shell. Unanimously, the K-shell of atoms consists of s-electrons, only. Having no angular momentum available in their initial state but only spin, KLL Auger transitions form the fundamental transitions in the field of Auger emission.

The Auger emission is usually described within a two-step model, the so-called *two-step ansatz*, (Wentzel 1927; Åberg and Howat 1982; Mehlhorn 1990) i.e., the primary ionization/excitation process is dealt with indenpendently of the subsequent Auger decay. Particular processes of ionization or excitation by photons or light particles like electrons will be discussed later on; we refer to discussions in Sects. 4.3.2 and 4.6.2, also.

Considering an Auger transition from a singly ionized state with total angular momentum J to a final state with total angular momentum J_f

$$A^{+}(J) \longrightarrow A^{++}(J_f) + e_{Auger}, \tag{3.1}$$

the Auger rates and the relative intensities, respectively are proportional to the squared transition matrix elements.

$$I_0 = \frac{1}{|\mathbf{k}|^2} \frac{1}{2J+1} \sum_{\ell j} \left| \langle J_f \| V \| (Jj) J_f \rangle \right|^2, \tag{3.2}$$

where $\mathbf{k}$ is the momentum of the emitted Auger electrons, V denotes the Coulomb interaction, and j refers to the total angular momentum. KLL Auger transitions are prominent in almost all elements of the periodic table though can be particularly best observed in comparatively light atoms like neon which are best described in the LS coupling scheme. Applying LS coupling, the Auger transition matrix elements become independent of the total angular momenta, which solely enter the geometrical

H. Kleinpoppen et al., *Perfect/Complete Scattering Experiments*,
Springer Series on Atomic, Optical, and Plasma Physics 75,
DOI: 10.1007/978-3-642-40514-3_3, © Springer-Verlag Berlin Heidelberg 2013

factors, and can be expressed as a function of the orbital angular momenta of the transition matrix elements, only,

$$I_0 = \frac{1}{|\mathbf{k}|^2} \frac{1}{2J+1} \sum_{\ell j} (2j+1) \begin{Bmatrix} J & j & J_f \\ L & \ell & L_f \\ S & 1/2 & S_f \end{Bmatrix}^2 \left| \langle L_f \| V \| (L\,\ell) L_f \rangle \right|^2. \tag{3.3}$$

Here, the quantum numbers ℓ, L, L_f refer to the Auger electron, and the initial and final state angular momenta, whereas S and S_f denote the initial and final state spin, respectively. Restricting the discussion to the closed shell atom case, we have $L = 0$, and $S = J = 1/2$ for an Auger transition from the K-shell which yields

$$I_0 = \frac{1}{|\mathbf{k}|^2} \frac{1}{2} \sum_{\ell j} (2j+1) \begin{Bmatrix} 1/2 & j & J_f \\ 0 & \ell & L_f \\ 1/2 & 1/2 & S_f \end{Bmatrix}^2 \left| \langle L_f \| V \| (L=0\,\ell) L_f \rangle \right|^2. \tag{3.4}$$

The symmetry relations of the $9j$-symbol gives the selection rule $\ell = L_f$. Thus, the summation over ℓ can be omitted and (3.4) can be written as

$$I_0 = \frac{1}{|\mathbf{k}|^2} \frac{1}{4(2L_f+1)} \delta_{\ell,L_f} \sum_{j} (2j+1) \begin{Bmatrix} 1/2 & S_f & 1/2 \\ J_f & j & L_f \end{Bmatrix}^2$$
$$\times \left| \langle L_f \| V \| (L=0\,\ell=L_f) L_f \rangle \right|^2. \tag{3.5}$$

Eventually, with the symmetries of the $6j$-symbols the sum over j can be performed and we remain with

$$I_0 = \frac{1}{|\mathbf{k}|^2} \frac{1}{4(2L_f+1)(2S_f+1)} \left| \langle L_f \| V \| (L=0\,\ell=L_f) L_f \rangle \right|^2. \tag{3.6}$$

Equation (3.6) gives an interesting result. Due to the selection rule $L = 0$ for K-shell Auger transitions, we remain with the fact that only one partial wave is emitted. In order to explore (3.6), we investigate the KLL Auger decay in detail, which, within the X-ray notation, consists of the KL_1L_1, $KL_1L_{2,3}$ and $KL_{2,3}L_{2,3}$ Auger groups generated by the $1s^{-1} \to 2s^{-2}$, $1s^{-1} \to 2s^{-1}2p^{-1}$ and $1s^{-1} \to 2p^{-2}$ Auger transitions, resulting in ^1S and $^{1,3}P$ final states for the first and second Auger group, and in ^1S, ^1D and ^3P final state multiplets for the last Auger group, respectively. In closed shell atoms the initial K-shell hole state always has even parity. Taking parity conservation of the Coulomb interaction into account, we have an even parity partial wave emission for the KLL Auger transitions to the even parity final states and vice versa, shown in Table 3.1. As an important result, we find the $1s^{-1} \to 2p^{-2}$ Auger transitions to the final ^3P state as forbidden, since the angular momentum coupling

Table 3.1 Partial waves of the emitted Auger electrons for the different KLL Auger transitions

Auger tran.	Final state		Part. wave
KL_1L_1	$(2s^02p^6)$	^1S	ε s
$KL_1L_{2,3}$	$(2s^12p^5)$	$^1P^o$	ε p
		$^3P^o$	ε p
$KL_{2,3}L_{2,3}$	$(2s^22p^4)$	^1S	ε s
		^1D	ε d
		^3P	ε p†

The superscript "o" of the ^{2S+1}L states denotes an odd parity final state.
†: In closed shell atoms, the $KL_{2,3}L_{2,3}$ ^3P Auger transition is forbidden due to parity violation

rules allow for an emission of an odd partial wave, only, which contradicts parity conservation requiring an even partial wave.

Notc, that this is truc in cxtrcmc LS coupling, only. Though, Sicgbahn ct al. (1967) pointed out that there is evidence for a non-vanishing transition probability of the $KL_{2,3}L_{2,3}$ $^3P_2$ Auger line, which has been numerically confirmed by Lohmann and Fritzsche (1994). With increasing atomic number, e.g. considering rubidium or cesium, which also show strong lines in the optical spectrum, the $KL_{2,3}L_{2,3}$ ^3P channels open further which results in the well known fact of an observable $KL_{2,3}L_{2,3}$ ^3P_J multiplet (e.g., see Lohmann 2008, for related information).

KLL Auger transitions have been investigated in detail, both, experimentally and theoretically by several authors. A good compendium has been collected by Cleff and Mehlhorn (1969), shown in Fig. 3.1. Compared are the relative group ratios for the $I(KL_{2,3}L_{2,3})/I(KL_1L_1)$ and $I(KL_1L_{2,3})/I(KL_1L_1)$ Auger transitions, showing both ratios decreasing with increasing atomic number Z after rapidly increasing and passing through a maximum between $10 < Z < 20$. Investigations into this direction have been further extended by Kostroun et al. (1971) to the $I(KLX)/I(KLL)$ and $I(KXY)/I(KLL)$ Auger transition probability ratios which are shown in Fig. 3.2. Here, both ratios are increasing with increasing atomic number Z.

3.2 Auger Matrix Elements in *jj* Coupling

Following or even parallel to investigations of KLL Auger transitions, an important theoretical progress has been made by Asaad (1963a,b) realizing that jj is the basic and general description for analyzing Auger transitions. This has been confirmed later by Grant (1970) showing the jj coupled state functions to form the natural basis for a full relativistic treatment which even may influence light atom Auger transitions; the latter has also been pointed out by Chattarji (1976). We will not give a detailed discussion here, e.g. see Chattarji (1976), or Lohmann (2008) discussing the numerical realization, but simply resume and apply the results.

In the following we are changing the picture describing the Auger process via the transition of hole states. In jj this is equivalent to the electronic picture (e.g., see

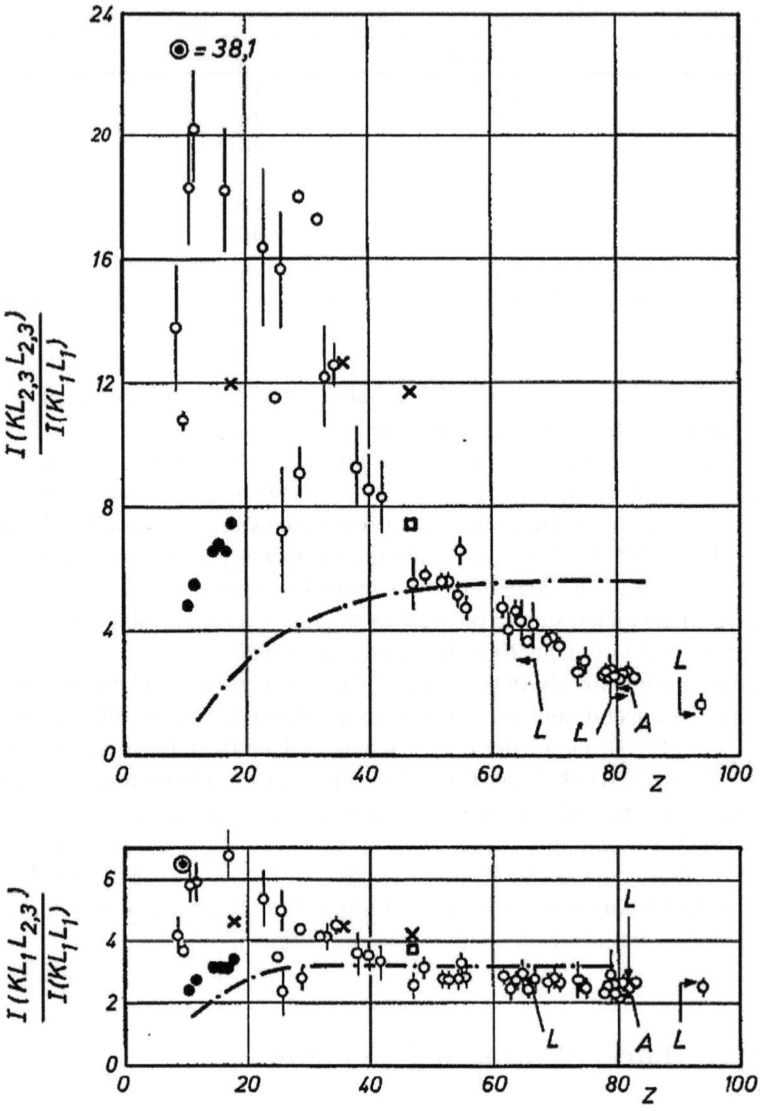

Fig. 3.1 Comparison of experimental and numerical relative group ratios $I(KL_{2,3}L_{2,3})$ $/I(KL_1L_1)$ *(top)* and $I(KL_1L_{2,3})/I(KL_1L_1)$ *(bottom)*; after Cleff and Mehlhorn (1969). (○) experimental values. Non-relativistic theory: (⊙) Krause et al. (1964) $Z = 10$; (●) Archard (1963) $Z = 11, 12, 15–18$; (×) Rubenstein (1955) $Z = 18, 36, 47$; (□) Burhop (1935) $Z = 47$, recalculated by Callan (1961); (— · —) Callan (1961). Calculated values for $Z \leq 36$ taking configuration interaction into account after Mehlhorn and Asaad (1966), and for $Z > 36$ after Asaad (1965). Relativistic theory: The arrows mark calculated values by Asaad (1959) for $Z = 80$, and Listengarten (1962) for $Z = 65, 81$, and 92

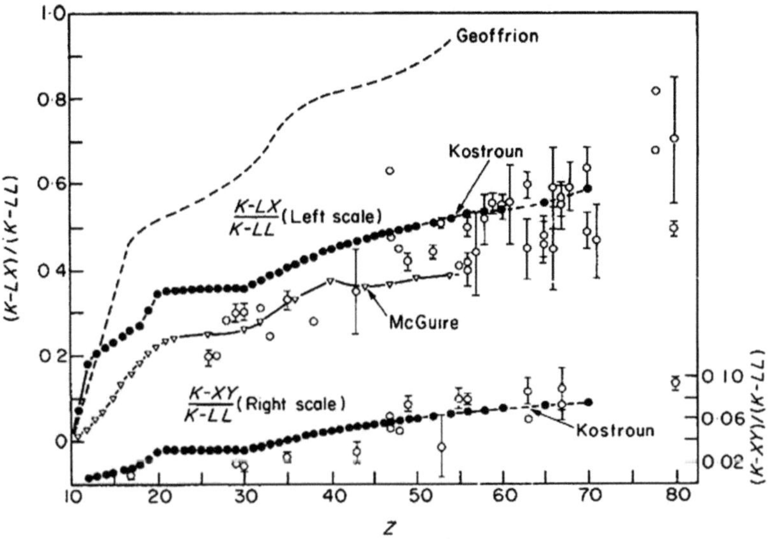

Fig. 3.2 Theoretical and experimental KLX/KLL and KXY/KLL Auger transition probability ratios as a function of the atomic number (after Kostroun et al. 1971). Calculated ratios: Geffrion and Nadeau (1959); Kostroun et al. (1971); McGuire (1970). Measured relative intensities of the K Auger electron groups assembled by Erman et al. (1964); except for the following ratios: Zn (Bellicard et al. 1956), Co (Bellicard et al. 1957), Te (Casey and Albridge 1969), and Ce and Nd after D'yakov and Rogachev (1962)

Grant 1970). Taking the Fermi character of the electronic or hole states into account, we have two indistinguishable processes prominent which participate in the Auger emission; see Fig. 3.3.

The *direct* process:

$$|N, L, J\rangle \longrightarrow |n_1, \ell_1, j_1\rangle, \tag{3.7a}$$

$$|\varepsilon, \ell, j\rangle \longrightarrow |n_2, \ell_2, j_2\rangle, \tag{3.7b}$$

and the *exchange* process:

$$|N, L, J\rangle \longrightarrow |n_2, \ell_2, j_2\rangle\,, \tag{3.8a}$$

$$|\varepsilon, \ell, j\rangle \longrightarrow |n_1, \ell_1, j_1\rangle\,. \tag{3.8b}$$

Here, J, j_1 and j_2 characterize the total angular momenta of the initial and the two final hole states of the singly and doubly ionized atom, respectively, while j refers to the Auger electron. The orbital angular momenta and the main quantum numbers are indicated analogously denoted by L, ℓ_1, ℓ_2, ℓ and N, n_1, n_2, whereas the Auger energy in the continuum is denoted by ε.

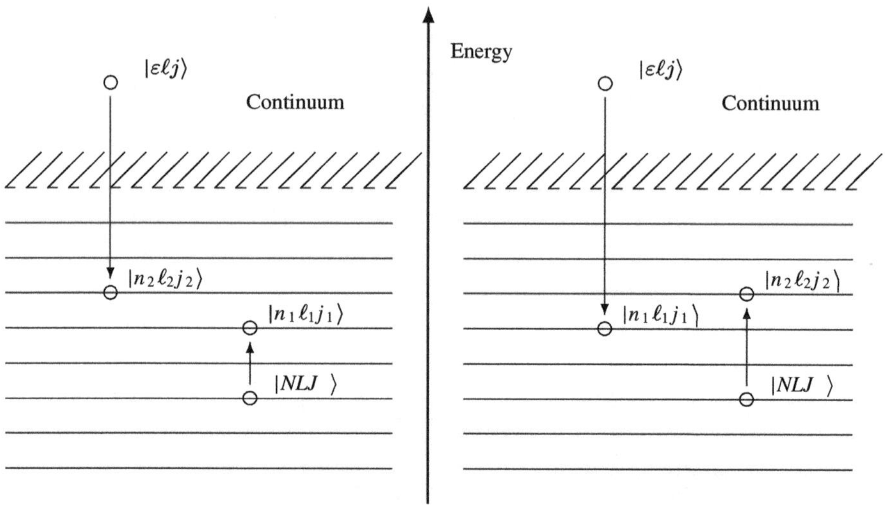

Fig. 3.3 (*Left*) direct Auger emission; the continuum hole state is transferred into an outer shell, the inner shell hole of the primary ion is shifted into an inner shell of the doubly ionized final state ($n_1 \leq n_2$). (*Right*) exchange Auger emission; the primary inner shell hole is shifted into an outer, the continuum hole is transferred into an inner shell of the doubly ionized final state, respectively

Due to their indistinguishability, both processes cannot be resolved within an experiment, and must be anti-symmetrized for satisfying the Fermi statistics of the electrons. Taking a usual multi-configuration approach into account, the reduced matrix elements can be written as

$$\begin{aligned}\langle (j_1 j_2) J_f \| V \| (J j) J_f \rangle = & \sum_{r=1}^{n_c} c_r(\alpha) \\ & \times \tau \Big(\langle (\phi(1)_{j_1}, \phi(2)_{j_2}) J_f(\alpha) \| V \| (\phi(1)_J, \phi(2)_j) J_f(\alpha) \rangle_D \\ & - \langle (\phi(1)_{j_1}, \phi(2)_{j_2}) J_f(\alpha) \| V \| (\phi(2)_J, \phi(1)_j) J_f(\alpha) \rangle_E \Big),\end{aligned} \tag{3.9}$$

where the indices D and E refer to the direct (3.7) and exchange (3.8) transitions, respectively. The functions, e.g. $\phi(1)_J \equiv \phi(1)_{NLJ}$, denote the occurring configuration state functions, with the notation 1, 2 abbreviating the coordinates $(r_1, \theta_1, \varphi_1)$ and $(r_2, \theta_2, \varphi_2)$ of the electronic and/or hole states, respectively. The $c_r(\alpha)$ are the configuration mixing coefficients for the state α, where n_c is the number of the contributing configuration state functions to the atomic state function (ASF), while τ has been introduced to account for the case of two equivalent electrons in the final state.

$$\tau = \begin{cases} 1 & \text{for non-equivalent electrons,} \\ \dfrac{1}{\sqrt{2}} & \text{for equivalent electrons.} \end{cases} \tag{3.10}$$

Equation (3.9) can be further performed by expanding the Coulomb operator into a series of Legendre polynomials,

$$\frac{1}{r_{12}} = \sum_{\lambda=0}^{\infty} \gamma_\lambda(r_1, r_2)\, P_\lambda(\cos\theta_{12}), \tag{3.11}$$

where γ_λ is defined as

$$\gamma_\lambda(r_1, r_2) = \frac{r_<^\lambda}{r_>^{\lambda+1}} \quad \text{with} \quad r_< = \begin{cases} r_1 & \text{for } r_1 < r_2 , \\ r_2 & \text{for } r_2 < r_1 , \end{cases} \tag{3.12}$$

and vice versa for $r_>$, while θ_{12} denotes the angle between (ϑ_1, φ_1) and (ϑ_2, φ_2) of the two vectors $\mathbf{r}_1$ and $\mathbf{r}_2$. The Legendre polynomials, on the other hand, can be expressed as a product of irreducible Racah tensors, which are proportional to the spherical harmonics,

$$\begin{aligned} P_\lambda(\cos\theta_{12}) &= \sum_\mu (-1)^\mu\, C_\mu^\lambda(\vartheta_1, \varphi_1)\, C_{-\mu}^\lambda(\vartheta_2, \varphi_2) \\ &= \frac{4\pi}{2\lambda+1} \sum_\mu (-1)^\mu\, Y_{\lambda\mu}(\vartheta_1, \varphi_1) Y_{\lambda-\mu}(\vartheta_2, \varphi_2) . \end{aligned} \tag{3.13}$$

For the further reduction of direct and exchange terms we refer to the derivations by Asaad (1963a, b) and Lohmann (1988, 1990). Eventually, we obtain the following expression for the reduced transition matrix elements

$$\begin{aligned} \langle (j_1 j_2) J_f \| V \| (Jj) J_f \rangle = \tau \Big(& \sum_\lambda (-1)^{J_f + j_2 + j} \mathcal{D}(\lambda, j) Q_s(\lambda, j) \\ & - \sum_\mu (-1)^{j_2 - j} \mathcal{E}(\mu, j) R_s(\mu, j) \Big) . \end{aligned} \tag{3.14}$$

Here, $\mathcal{D}$ and $\mathcal{E}$ denote the direct and exchange Slater integrals, respectively. They are defined as

$$\mathcal{D}(\lambda, j) = \mathcal{R}^{(\lambda)}(n_1\ell_1 j_1; n_2\ell_2 j_2; NLJ; \varepsilon\ell j) , \tag{3.15}$$

and

$$\mathcal{E}(\mu, j) = \mathcal{R}^{(\mu)}(n_1\ell_1 j_1; n_2\ell_2 j_2; \varepsilon\ell j; NLJ) , \tag{3.16}$$

where the Slater integrals are given as

$$\mathcal{R}^{(\lambda)}\left(a,b,c,d\right)=\iint R_a\left(r\right)R_b\left(s\right)\gamma\lambda\left(r,s\right)R_c\left(r\right)R_d\left(s\right)r^2s^2\,\mathrm{d}r\mathrm{d}s. \tag{3.17}$$

Here, (a, b, c, d) abbreviate the quantum numbers of the radial wavefunctions and γ_λ has been defined in (3.12). The Slater integrals show some useful symmetries, which can be easily derived by interchanging r and s. In general we have

$$\mathcal{R}^{(\lambda)}(a,b,c,d)=\mathcal{R}^{(\lambda)}(b,a,d,c) \tag{3.18}$$

and if the two first or the two second radial wavefunctions are identical (e.g. $a = b$) we have

$$\mathcal{R}^{(\lambda)}(a,a,c,d)=\mathcal{R}^{(\lambda)}(a,a,d,c)\,. \tag{3.19}$$

The coefficients Q_s and R_s basically consist of $3nj$-symbols. We have

$$Q_s(\lambda,j)=\Gamma(\hat{J})\sqrt{2J_f+1}\begin{pmatrix}J & j_1 & \lambda\\ 1/2 & -1/2 & 0\end{pmatrix}\begin{pmatrix}j & j_2 & \lambda\\ 1/2 & -1/2 & 0\end{pmatrix}\begin{Bmatrix}j_2 & j & \lambda\\ J & j_1 & J_f\end{Bmatrix}, \tag{3.20}$$

and

$$R_s(\mu,j)=\Gamma(\hat{J})\sqrt{2J_f+1}\begin{pmatrix}J & j_2 & \mu\\ 1/2 & -1/2 & 0\end{pmatrix}\begin{pmatrix}j & j_1 & \mu\\ 1/2 & -1/2 & 0\end{pmatrix}\begin{Bmatrix}j_1 & j & \mu\\ J & j_2 & J_f\end{Bmatrix}, \tag{3.21}$$

where the abbreviation

$$\Gamma(\hat{J})=\sqrt{(2J+1)(2j+1)(2j_1+1)(2j_2+1)} \tag{3.22}$$

is used. The coefficients Q_s and R_s have a useful symmetry. If we interchange j_1 and j_2 we find that

$$Q_s(j_1,j_2,\lambda,j)=R_s(j_2,j_1,\lambda,j)\,. \tag{3.23}$$

An analogous expression can be found for interchanging J and j. Thus, if either $j_1 = j_2$ or $j = J$ is fulfilled, the coefficients Q_s and R_s are identical.

Applying parity conservation and angular momentum coupling, we obtain specific selection rules. Generally, the *direct* term is only non-zero if

$$(-1)^{\ell_1+\lambda+L}=1 \quad\text{and}\quad (-1)^{\ell_2+\lambda+\ell}=1 \tag{3.24}$$

are simultaneously fulfilled. Further, the triangular conditions of angular momenta must be fulfilled, that is

$$\left|L-\ell_1\right|\le\lambda\le L+\ell_1 \quad\text{and}\quad \left|\ell_2-\ell\right|\le\lambda\le\ell_2+\ell\,. \tag{3.25}$$

Analogously,

$$(-1)^{\ell_1+\mu+\ell}=1 \quad\text{and}\quad (-1)^{\ell_2+\mu+L}=1 \tag{3.26}$$

and

$$|\ell_1 - \ell| \leq \mu \leq \ell_1 + \ell \quad \text{and} \quad |L - \ell_2| \leq \mu \leq L + \ell_2 \tag{3.27}$$

must be fulfilled for the *exchange* term. Thus, the summations over λ and μ in (3.14) are restricted to a finite number of elements, depending on the specific type of Auger transition.

If we are dealing with equivalent electrons, i.e.

$$n_1 = n_2, \quad \ell_1 = \ell_2, \quad \text{and} \quad j_1 = j_2 \tag{3.28}$$

for the final target state, the symmetries (3.23) and (3.19) apply, and from (3.10) we have $\tau = 1/\sqrt{2}$. Thus, (3.14) can be simplified which yields

$$\begin{aligned}\langle (j_1 j_1) J_f \| V \| J_f (Jj) \rangle &= \frac{1}{\sqrt{2}} \left(1 + (-1)^{J_f} \right) \\ &\quad \times \left(\sum_\lambda (-1)^{j_1 - j + 1} \mathcal{D}(\lambda, j) \mathcal{Q}_s(\lambda, j) \right).\end{aligned} \tag{3.29}$$

In this case the reduced matrix elements are only different form zero for an even total angular momentum J_f. This is however not a special result of our calculation but a general restriction in the jj coupling scheme based on the *Pauli principle.*

Applying the jj coupling results, we are able to re-write (3.2), expressing the relative intensities in short as

$$I_0 = \frac{1}{|\mathbf{k}|^2} \frac{1}{2J + 1} \sum_{\ell j} \tau^2 \left| D - E \right|^2, \tag{3.30}$$

abbreviating the anti-symmetrized direct and exchange terms by their capital letters.[1] Expanding (3.30) by inserting the results of (3.14), we obtain

$$\begin{aligned}I_0 = \frac{1}{|\mathbf{k}|^2} \frac{1}{2J + 1} \sum_{\ell j} \tau^2 \Bigg| &\sum_\lambda (-1)^{J_f + j_2 + j} \mathcal{D}(\lambda, j) \mathcal{Q}_s(\lambda, j) \\ &- \sum_\mu (-1)^{j_2 - j} \mathcal{E}(\mu, j) R_s(\mu, j) \Bigg|^2 .\end{aligned} \tag{3.31}$$

Equation (3.31) shows an interesting behaviour. While there is an incoherent sum of the angular and total angular momenta ℓ and j, the summation over the order λ and μ of the direct and exchange Slater integrals must be performed coherently. General expressions for specific Auger transition probabilities are somewhat tedious but can be calculated straightforward from (3.31). On the other hand, the calculation of KLL

[1] A similar anti-symmetrized expression is obtained in any coupling scheme due to the Fermionic character of the electrons.

Auger transitions is very simple for closed shell atoms. Inserting $L = 0$ and $J = 1/2$ in (3.31), and applying the selection rules and triangular conditions (3.24)–(3.27), we obtain the restrictions

$$\lambda = \ell_1 \quad \text{and} \quad \mu = \ell_2 \ , \tag{3.32}$$

for any Auger transition with an initial $ns_{1/2}$ hole state, corresponding to the K, L_1, M_1, ..., etc. series. Thus, in (3.31) the summations over λ and μ, both, collapse to a single term, and, after rearranging the phase factor, we may write

$$\begin{aligned} I_0 = \frac{1}{|\mathbf{k}|^2} \frac{1}{2} \sum_{\ell j} \tau^2 \big| &\mathcal{D}(\lambda = \ell_1, j) Q_s(\lambda = \ell_1, j) \\ &+ (-1)^{J_f} \mathcal{E}(\mu = \ell_2, j) R_s(\mu = \ell_2, j) \big|^2 \ . \end{aligned} \tag{3.33}$$

From (3.33) the short-hand notation of (3.30) becomes obvious. Applying the results of (3.29) in (3.30), the case of equivalent electrons can be further reduced

$$I_0 = \frac{1}{|\mathbf{k}|^2} \delta_{\ell_1, \ell_2} \delta_{\ell, L_f} \sum_j \big| \mathcal{D}(\lambda = \ell_1, j) Q_s(\lambda = \ell_1, j) \big|^2 \ , \tag{3.34}$$

with the intensity being non-zero for J_f even, only. Generally, the sum over j is solely over two terms, and consist of a single term for the emission of $\varepsilon s_{1/2}$ partial waves. The selection rule $\ell = L_f$ somehow contradicts the jj coupling description, however, we have chosen the Kronecker symbol in order to point out the collapse of the sum over ℓ to a single term identical to the angular momentum of the final state. Inserting the relevant quantum numbers in (3.33) and (3.34), we are able to identify the contributing partial waves and Slater integrals, shown in Table 3.2. The ${}^{2S+1}L_J$ notation for the final states has been chosen to outline the relation to the LS coupling states, though, besides parity, the final state total angular momentum J_f is the only good quantum number in jj coupling.

Comparison with Table 3.1 shows that jj coupling generally allows for a larger variety of partial waves in contrast to extreme LS coupling. However, for most of the KLL Auger transitions we still remain with a single partial wave emission.

In jj coupling, Auger transitions to the $KL_{2,3}L_{2,3}\,{}^3P_J$ final states are no longer forbidden by angular momentum coupling rules, and do occur in heavy atoms; e.g. see Sect. 3.1 or Graham et al. (1961). The relative line intensities for the KLL Auger final fine structure states are shown in Table 3.3 as a function of the participating Slater integrals.

Of particular interest is the $KL_{2,3}L_{2,3}\,{}^3P_1$ Auger transition, as the selection rules allow for the emission of two partial waves with different angular momentum ℓ in this transition, only, whereas it is forbidden in extreme LS coupling. As can be seen from Table 3.3 the relative intensity for the ${}^3P_1$ must be almost zero for comparatively light atoms having a small fine structure splitting. Hence, the contributing $np_{1/2}$ and

Table 3.2 Partial waves and Slater integrals of the emitted Auger electrons in the jj coupling scheme for the different KLL Auger transitions

Auger tran.	Final state		Part. waves	Slater int.
KL_1L_1	$(2s^0_{1/2}2p^2_{1/2}2p^4_{3/2})$	$^1S_0$	$\varepsilon s_{1/2}$	$\mathcal{D}(0)=\mathcal{E}(0)$
KL_1L_2	$(2s^1_{1/2}2p^1_{1/2}2p^4_{3/2})$	$^1P^o_1$	$\varepsilon p_{1/2}$	$\mathcal{D}(0)$
			$\varepsilon p_{1/2,3/2}$	$\mathcal{E}(1)$
		$^3P^o_0$	$\varepsilon p_{1/2}$	$\mathcal{D}(0), \mathcal{E}(1)$
KL_1L_3	$(2s^1_{1/2}2p^2_{1/2}2p^3_{3/2})$	$^3P^o_1$	$\varepsilon p_{3/2}$	$\mathcal{D}(0)$
			$\varepsilon p_{1/2,3/2}$	$\mathcal{E}(1)$
		$^3P^o_2$	$\varepsilon p_{3/2}$	$\mathcal{D}(0), \mathcal{E}(1)$
KL_2L_2	$(2s^2_{1/2}2p^0_{1/2}2p^4_{3/2})$	$^1S_0$	$\varepsilon s_{1/2}$	$\mathcal{D}(1)=\mathcal{E}(1)$
KL_2L_3	$(2s^2_{1/2}2p^1_{1/2}2p^3_{3/2})$	$^1D_2$	$\varepsilon d_{3/2,5/2}$	$\mathcal{D}(1)$
			$\varepsilon d_{3/2}$	$\mathcal{E}(1)$
		$^3P_1$	$\varepsilon s_{1/2}$, $\varepsilon d_{3/2}$	$\mathcal{D}(1), \mathcal{E}(1)$
KL_3L_3	$(2s^2_{1/2}2p^2_{1/2}2p^2_{3/2})$	$^3P_0$	$\varepsilon s_{1/2}$	$\mathcal{D}(1)=\mathcal{E}(1)$
		$^3P_2$	$\varepsilon d_{3/2,5/2}$	$\mathcal{D}(1)=\mathcal{E}(1)$

The superscript "o" of the $^{2S+1}L_J$ states denotes an odd parity final state. Note, that the Slater integrals are different for the different partial waves; e.g. $\mathcal{E}(1)$ of $\varepsilon s_{1/2}$ differs from $\mathcal{E}(1)$ of $\varepsilon d_{3/2}$, etc

Table 3.3 The jj coupling relative intensities of the emitted Auger electrons for the different KLL Auger transitions as a function of the Slater integrals

Final state		$I_0\lvert\mathbf{k}\rvert^2$
$(2s^0_{1/2}2p^2_{1/2}2p^4_{3/2})$	$^1S_0$	$\lvert\mathcal{D}(0,1/2)\rvert^2$
$(2s^1_{1/2}2p^1_{1/2}2p^4_{3/2})$	$^1P^o_1$	$\frac{3}{2}\left(\lvert\mathcal{D}(0,1/2)-\frac{1}{9}\mathcal{E}(1,1/2)\rvert^2+\frac{8}{81}\lvert\mathcal{E}(1,3/2)\rvert^2\right)$
	$^3P^o_0$	$\frac{1}{2}\lvert\mathcal{D}(0,1/2)-\frac{1}{3}\mathcal{E}(1,1/2)\rvert^2$
$(2s^1_{1/2}2p^2_{1/2}2p^3_{3/2})$	$^3P^o_1$	$\frac{3}{2}\left(\lvert\mathcal{D}(0,3/2)+\frac{1}{9}\mathcal{E}(1,3/2)\rvert^2+\frac{8}{81}\lvert\mathcal{E}(1,1/2)\rvert^2\right)$
	$^3P^o_2$	$\frac{5}{2}\lvert\mathcal{D}(0,3/2)-\frac{1}{3}\mathcal{E}(1,3/2)\rvert^2$
$(2s^2_{1/2}2p^0_{1/2}2p^4_{3/2})$	$^1S_0$	$\frac{1}{9}\lvert\mathcal{D}(1,1/2)\rvert^2$
$(2s^2_{1/2}2p^1_{1/2}2p^3_{3/2})$	$^1D_2$	$\frac{4}{15}\left(\frac{1}{24}\lvert\mathcal{D}(1,3/2)-5\mathcal{E}(1,3/2)\rvert^2+\lvert\mathcal{D}(1,5/2)\rvert^2\right)$
	$^3P_1$	$\frac{4}{27}\left(\lvert\mathcal{D}(1,1/2)-\mathcal{E}(1,1/2)\rvert^2+\frac{1}{4}\lvert\mathcal{D}(1,3/2)-\mathcal{E}(1,3/2)\rvert^2\right)$
$(2s^2_{1/2}2p^2_{1/2}2p^2_{3/2})$	$^3P_0$	$\frac{2}{9}\lvert\mathcal{D}(1,1/2)\rvert^2$
	$^3P_2$	$\frac{2}{15}\left(\frac{2}{3}\lvert\mathcal{D}(1,3/2)\rvert^2+\lvert\mathcal{D}(1,5/2)\rvert^2\right)$

The superscript "o" of the $^{2S+1}L_J$ states denotes an odd parity final state

$np_{3/2}$ wavefunctions to the Slater integrals can be interpreted as almost equivalent electrons, and thus, $\mathcal{D}(1)\approx\mathcal{E}(1)$, which results in $I_0\sim 0$.

Equations (3.33) and (3.34) are not restricted to Auger emission from the K-shell, only. Instead, the results of Tables 3.2 and 3.3 can be applied to any Auger transition with an initial $ns_{1/2}$ state. Auger transition probabilities as a function of the direct and exchange Slater integrals have been tabulated by Asaad (1963a, b) for initial s and p hole states and for final state vacancies ranging from s to d holes. A review on Auger and Coster-Kronig transition probabilities, as well as fluorescence yields, including extended tables, has been given by Bambynek et al. (1972).

3.3 Excited Sodium KLL Auger Transitions

We have outlined the general features of the KLL Auger spectrum for closed shell atoms in the previous Sects. 3.1 and 3.2. However, even in such a simple system as the KLL Auger decay, everything is different for the general case. Let us consider open shell atoms, and, particularly focus on sodium. Whereas in closed-shell systems, like the rare gases, the KLL Auger spectrum has to be isotropic with respect to the angle of Auger emission, this is no longer true for open shell systems. This is caused by the angular momentum coupling of the valence shell electrons with the inner shell 1s-hole. Na KLL Auger transitions have been investigated by Lohmann and Fritzsche (1994) for photoionization of the $^2S_{1/2}$ ground state, focussing on the angular distribution of the singly ionized $^3S_1$ state; —a detailed discussion of the alkali KLL Auger transitions can be found in the book by Lohmann (2008). While large angular distribution anisotropy parameters have been found for most of the Auger transitions, their experimental determination is feasible. On one hand, this is caused by the small fine structure splitting, on the other, the resonant intermediate state has to be aligned. As has been investigated (Lohmann and Fritzsche 1994; Lohmann 2008), for the non-excited KLL Auger alkali spectra an alignment can only be generated by relativistic effects, not prominent in comparatively light sodium.

A possible way of generating a sufficient alignment is to excite the outer valence electron to the p-shell. Due to the unequally populated sub-levels such an excited state can show an alignment. The excitation can usually be done by laser pumping, and the excited system can be investigated further, e.g. by a subsequent photoionization. Experiments of this type have been performed by several groups (e.g. Bizau et al. 1985; Carre et al. 1990; Dorn et al. 1995; Baier et al. 1994a), and have been reviewed by Wuilleumier and Meyer (2006), while Lohmann et al. (1996) provided a detailed numerical analysis of the anisotroic KLL Auger decay of laser excited sodium.

Bizau et al. (1993) performed an experiment measuring the Auger spectra of laser pumped sodium, exciting the 3s electron to the 3p shell, while Dorn et al. (1995), supported by the corresponding theoretical analysis (Grum- Grzhimailo and Dorn 1995; Zatsarinny 1995), investigated the laser excited $Na^*2s2p^6n\ell$ Auger states following inner shell ionization by electron impact, applying high-resolution electron spectroscopy. By coupling of an outer p-electron with an inner shell s-hole a non-zero alignment is possible even in a non-relativistic treatment which, on the other hand, simplifies the experimental determination of an angular anisotropy.

For the following discussion, let us consider photoexcited sodium

$$h\nu + \mathrm{Na(3s)}\,^2S_{1/2} \longrightarrow \mathrm{Na^*(3p)}\,^2P_{1/2,3/2}\,. \tag{3.35}$$

While the sodium ground state is isotropic, since its overall s-shell governed character, this is different for exciting the valence electron to the 3p-shell, resulting in an alignment of the photoexcited state (e.g. see Kronast et al. 1986). Subsequently, an inner shell K-hole can be created, for instance by a synchrotron beam, and, eventually, the singly ionized, excited intermediate state decays via Auger emission leaving the excited 3p valence electron as a spectator,

$$\gamma_{syn} + \mathrm{Na}^{*}(3\mathrm{p})\,^{2}\mathrm{P} \longrightarrow \mathrm{Na}^{+*}(1\mathrm{s}^{-1}3\mathrm{p})\,^{1,3}\mathrm{P} + e_{e} \tag{3.36a}$$

$$\hookrightarrow \mathrm{Na}^{++*} + e_{Auger}\,. \tag{3.36b}$$

Equation (3.36) illustrates the description of the Auger emission as a two-step process, which can be applied most useful in the numerical approaches. The two-step model of Auger emission has been experimentally well-observed (e.g. see Mehlhorn 1990).

Alternatively to the primary photoexcitation (3.35) with a subsequent photoionization (3.36) followed by Auger emission, and theoretically more easier to deal with, let us interchange the photoionization and excitation processes. For this, we consider a primary photoionization of the outer 3s electron,

$$h\nu + \mathrm{Na}(3\mathrm{s})\,^{2}\mathrm{S}_{1/2} \longrightarrow \mathrm{Na}^{+}(2\mathrm{p}^{6})\,^{1}\mathrm{S}_{0} + e_{e}\,, \tag{3.37}$$

which can be achieved via laser beam techniques. Due to its closed shell structure the remaining Na^{+} ion must be in a $^{1}\mathrm{S}_{0}$ state. In a second step, the 1s electron of the sodium cation is excited by a synchrotron beam to the 3p Rydberg level,

$$\gamma_{syn} + \mathrm{Na}^{+}(2\mathrm{p}^{6})\,^{1}\mathrm{S} \longrightarrow \mathrm{Na}^{+*}(1\mathrm{s}^{-1}3\mathrm{p})\,^{1,3}\mathrm{P} \tag{3.38a}$$

$$\hookrightarrow \mathrm{Na}^{++*} + e_{Auger}\,, \tag{3.38b}$$

while the intermediate excited sodium Rydberg state eventually decays via Auger emission.

The photoionization/excitation and the photoexcitation/ionization processes, respectively, can be distinguished from each other. Either by detecting the intermediate atomic or ionic state after the primary photoexcitation or ionization, or by energy analysis of the triggering apparatus. From a theoretical point of view the angular distribution analysis of the photoionization (3.37) can be dealt with straightforward since the residual ion must be in a $^{1}\mathrm{S}_{0}$ state. Thus, no alignment can be obtained as we are remaining with a closed shell structure. The subsequent photoexcitation and Auger decay (3.38) are, from its theoretical analysis, identical to a resonant Auger emission which has been extensively investigated by several groups, particularly for the rare gases (von Raven et al. 1990; Hergenhahan et al. 1991, 1999; Chen 1993; Tulkki et al. 1993; de Gouw et al. 1995; Aksela and Mursu 1996; Aksela et al. 1997; Farhat et al. 1997; Meyer et al. 2001, and many others); we also refer to the discussions in Sects. 3.4 and 4.3.3. For a theoretical analysis and specific examples see the book by Lohmann (2008, Sects. 2.6, 4.6 and 4.7). For a photoexcitation it has been shown (Kronast et al. 1986) that the alignment of the excited state takes a value of $\mathcal{A}_{20} = 1/\sqrt{2}$ independent of the specific Auger emission (see Table 3.13 and the discussion in Sect. 3.5.1). The only disadvantage of the photoexcitation method is that in such a decay scheme the $Na^{+*}(1s^{-1}3p)\,^3P_0$ and $^3P_2$ Rydberg levels cannot be populated due to the dipole selection rules. This can only be achieved via the alternative photoexcitation/ionization process; see (3.35) and (3.36).

3.3.1 Relative and Absolute Line Intensities

Applying the LS coupling scheme, the relative line intensities can be obtained from (3.3). Considering the possible initial states of the singly ionized laser excited sodium atom, we have a single p-electron in the open shell which can couple with the K-hole in two ways, i.e. to form a $^3P_{0,1,2}$ and a $^1P_1$ state, respectively. Inspecting the more general case of any singlet state by inserting $S = 0$ into (3.3), the $9j$-symbol reduces to a $6j$-symbol, and the triangular relations of the $9j$-symbols yield $S_f = 1/2$. Thus, as general result we obtain

$$I_0^{(LS)}(^1L_J \to {}^4L_{f\,J_f}) = 0 \tag{3.39}$$

for any singlet Auger transition to a final quartet state. Only the possible doublet states are allowed in the extreme LS coupling limit. Inserting the relevant quantum numbers for the singlet to doublet Auger transitions into (3.3) yields

$$\begin{aligned} I_0^{(LS)}(^1L_J \to {}^2L_{f\,J_f}) &= \frac{1}{|\mathbf{k}|^2}\,\frac{1}{2(2J+1)^2}\,\delta_{J,L}\sum_{\ell j}(2j+1) \\ &\quad\times \begin{Bmatrix} J_f & J & j \\ \ell & 1/2 & L_f \end{Bmatrix}^2 \left|\langle L_f \| V \| (L\,\ell)L_f\rangle\right|^2 , \end{aligned} \tag{3.40}$$

and by performing the sum over j, we remain with

$$\begin{aligned} I_0^{(LS)}(^1L_J \to {}^2L_{f\,J_f}) &= \frac{1}{|\mathbf{k}|^2}\,\frac{1}{2(2J+1)^2(2L_f+1)} \\ &\quad\times \delta_{J,L}\sum_{\ell}\left|\langle L_f \| V \| (L\,\ell)L_f\rangle\right|^2 , \end{aligned} \tag{3.41}$$

while for the specific case of an initial $^1P_1$ state, we get

$$I_0^{(LS)}(^1P_1 \to {}^2L_{f\,J_f}) = \frac{1}{|\mathbf{k}|^2}\,\frac{1}{18(2L_f+1)}\sum_{\ell}\left|\langle L_f \| V \| (L=1\,\ell)L_f\rangle\right|^2 . \tag{3.42}$$

Due to the triangular condition

$$\left|L_f - 1\right| \le \ell \le L_f + 1, \tag{3.43}$$

the summation over ℓ in (3.42) is over one or two terms, only, depending on the parity of the Auger transition.

The unresolved multiplet intensities can be expressed as

$$I_0^{M \to J_f} = \sum_J I_0^{(LS)}(J \to J_f) \,, \tag{3.44}$$

and by using (3.3) we obtain

$$I_0^{M \to J_f} = \frac{1}{|\mathbf{k}|^2} \sum_{\ell j J} \frac{2j+1}{2J+1} \begin{Bmatrix} J & j & J_f \\ L & \ell & L_f \\ S & 1/2 & S_f \end{Bmatrix}^2 \left| \langle L_f \| V \| (L\,\ell) L_f \rangle \right|^2 . \tag{3.45}$$

Inserting $J = L = 1$ and $S = 0$ in (3.45), we end up again with (3.42).

3.3.2 Interpretation and Analysis of Spectra

As pointed out (Lohmann and Fritzsche 1994) Auger spectra of open shell systems are often complex and cause a great deal of difficulties in their interpretation compared to the closed shell atoms (i.e. mainly to the spectra of the rare gases). Even for atoms with a simple shell structure the number of possible decay lines increases enormously often showing a rather small fine structure splitting. This complicates the experimental identification of the Auger lines. Therefore, the fine structure splitting has not been properly resolved in most of such experiments. Experimental data for laser pumped sodium have been given by Bizau et al. (1993) who applied high resolution techniques. However, a clear resolution of the fine structure has not been obtained yet for most of the sodium Auger lines. The data by Lohmann et al. (1996) therefore provide a set of consistent data which might trigger further investigations in this field.

3.3.3 Designation of States

Though Lohmann et al. (1996) performed their calculations in an intermediate coupling scheme we will follow them and designate the individual Auger transitions in the LSJ notation. This scheme has been used to classify most of the available data in the literature. Furthermore, sodium is a relatively light atom, i.e. LS coupling can still be assumed as a good approximation. To denote the various final states beyond their total angular momentum and parity we are using their main contribution in the configuration expansion of the wavefunctions and their sequence of total energies. The KLL Auger lines following an excitation of the four possible

$[1s^{-1}3p]\,^1P_1$ and $[1s^{-1}3p]\,^3P_{0,1,2}$ initial states result in a total of 41 allowed final states with 18 of them having even parity. They all belong to the $KL_1L_{2,3}$ part of the KLL spectrum. Within the chosen classification scheme we denote them as $(^1P)3p\,^2S_{1/2}$, $^2P_{1/2,3/2}$, $^2D_{3/2,5/2}$, $(^3P)3p\,^2S_{1/2}$, $^2P_{1/2,3/2}$, $^2D_{3/2,5/2}$, and $(^3P)3p$ $^4S_{3/2}$, $^4P_{1/2,3/2,5/2}$, $^4D_{3/2,5/2,7/2}$ transitions. The ^2P and ^2D states occur twice due to the possible coupling. The remaining 23 states are of odd parity which is characterized by an 'o' superscript. They can be identified as KL_1L_1 $(^1S)3p$ $^2P^o_{1/2,3/2}$ transitions, whereas the $KL_{2,3}L_{2,3}$ Auger series lead to $(^1S)3p$ $^2P^o_{1/2,3/2}$, $(^1D)3p$ $^2P^o_{1/2,3/2}$, $^2D^o_{3/2,5/2}$, $^2F^o_{5/2,7/2}$, and $(^3P)3p$ $^2S^o_{1/2}$, $^2P^o_{1/2,3/2}$, $^2D^o_{3/2,5/2}$ final doublet states. Again, some of them occur several times due to the possible different coupling. Eventually, we get the $(^3P)3p\,^4S^o_{3/2}$, $^4P^o_{1/2,3/2,5/2}$, and $^4D^o_{1/2,3/2,5/2,7/2}$ final quartet states. Note, that transitions of the $KL_{2,3}L_{2,3}$ $(^3P)3p$ type are forbidden in extreme LS coupling, because the selection rule $\Delta L = 0$ cannot be fulfilled without parity violation (e.g. see the book by Chattarji 1976). However, the comparison with experimental data is not without problems since most experiments do not fully resolved the fine structure of the final states.

3.3.4 Auger Transition Energies

The initial and final ionic bound states have been generated by using the GRASP atomic structure package (Dyall et al. 1989; Grant and Parpia 1992). Beyond the Coulomb interaction in the self-consistent-field procedure, these calculations include the transverse Breit interaction as well as estimates of the main QED corrections as perturbations. In this approximation, Lohmann et al. (1996) obtained an energy splitting of the excited $[1s^{-1}3p]\,^3P_{0,1,2}$ triplet states of $\Delta E(^3P_0 - {}^3P_1) = 0.83 \times 10^{-4}$ au and $\Delta E(^3P_0 - {}^3P_2) = 3.73 \times 10^{-4}$ au which can hardly be resolved within an experiment. On the other hand, the $[1s^{-1}3p]$ $^1P_1$ singlet state with $\Delta E(^3P_0 - {}^1P_1) = 60.6 \times 10^{-4}$ au is comparatively well separated from the triplet levels. The total decay rate of all four levels shows an almost constant value of $\Gamma \sim 94.9 \times 10^{-4}$ au (see also the next section). Thus, it is almost impossible to resolve the fine structure levels of the triplet state. Even the separation of the singlet from the triplet state requires highly advanced experimental effort.

From the independent computations for the initial and final state configurations the Auger energies are obtained as differences of the total energies

$$E_{Auger} = E_f - E_i \ . \tag{3.46}$$

The Auger transitions energies are given in atomic units [au] and have been used for the calculation of excited Na KLL spectra (see Figs. 3.4–3.6 later on). In sodium, unfortunately, experimental Auger energies are rare to compare with theory.

Comparing the Auger energies of the laser excited sodium KLL transitions with the ground state transitions obtained by Lohmann and Fritzsche (1994) shows a generally small shift towards lower energies, usually not exceeding 1 eV.

3.3.5 Auger Rates and Relative Intensities

Structure of the Spectrum

The absolute Auger rates and relative intensities have been calculated applying (3.2) for all transitions between the four initial and the allowed final states (Lohmann et al. 1996). The total decay rates for the initial $^1P_1$ and ^3P_J states, where each correspond to 100 % intensity, are approximately equal $\Gamma \sim 94.92 \times 10^{-4}$ au, confirming the small contribution of correlation and relativistic effects to the total decay widths. Correlation effects mainly concern the re-distribution of the partial decay rates. They might become further important if the number of configuration state functions (CSF) is increased, e.g. if CSF of excited states like $2s^2 2p^3 nd(ns)$ are included.

The initial singlet and triplet terms are still too close to each other to allow for a separation. Assuming a Lorentz profile the relative Auger line intensities can be expressed as

$$I_0^{rel}(E) = \frac{\Gamma}{2\pi} \frac{1}{(E - E_0)^2 + \Gamma^2/4}\,, \tag{3.47}$$

where E_0 denotes the Auger energy and Γ is the decay width (Frauenfelder and Henley 1979). Normalizing the total spectrum to 100 and using (3.47), the absolute line profiles can be obtained (Mayer-Kuckuk 1977) as

$$I_0^{abs}(E) = \frac{\pi \Gamma}{2} \frac{\Gamma_{tot}}{100} I_0^{rel}(E) = \frac{\Gamma_{tot}}{100} \frac{\Gamma^2/4}{(E - E_0)^2 + \Gamma^2/4}\,, \tag{3.48}$$

where Γ_{tot} is the total Auger decay rate.

A numerical $\mathrm{FWHM_{Na}}$ of $\Gamma = 0.45$ eV has been assumed for generating the Lorentz profiles of the Auger lines. The excited line spectra of both, the singlet and the triplet states, are shown in Fig. 3.4. The LS coupling character of the spectrum is clearly seen, showing unresolved groups of fine structure multiplets, only. The manifold of the contributing Auger lines have been indicated by vertical bars. As can be seen, the KLL behaviour of the spectrum is pronounced, showing the well-known Na D-doublet which, for the excited state transitions, is exhibited via a ^2F, ^2D and ^2P structure.

Concerning the validity of the data it is worth to consider the special case of extreme LS coupling. Since sodium is a relatively light atom predictions in the LS coupling limit should be close to the calculations by Lohmann et al. (1996). The $KL_{2,3}L_{2,3}(^3P)3p\ ^2S_{1/2}$, $^2P_{1/2,3/2}$, and $^2D_{3/2,5/2}$ and the $^4S_{3/2}$, $^4P_{1/2,3/2,5/2}$, and $^4D_{1/2,3/2,5/2,7/2}$ Auger transitions are strongly forbidden in LS coupling;

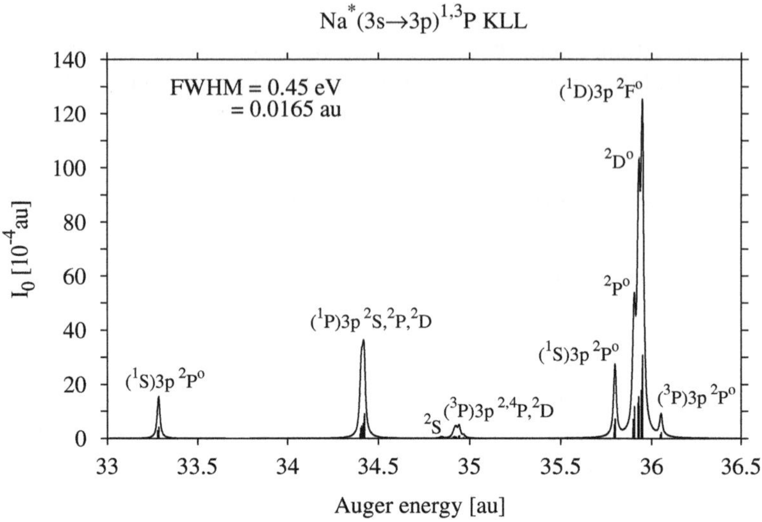

Fig. 3.4 The 3s → 3p laser excited KLL Auger spectrum of sodium after ionization of the 1s shell. The Auger decay propagates from the resonantly excited unresolved $^1P_1$ and $^3P_{0,1,2}$ fine structure states

see Sect. 3.1 or Chattarji (1976). We therefore expect an Auger rate which is almost zero in sodium. This is confirmed by the numerical data, except for the $KL_{2,3}L_{2,3}(^3P)3p\ ^2P_{1/2,3/2}$ transitions which we explain by a relatively strong mixing with the final $KL_{2,3}L_{2,3}(^1S)3p$ and $KL_{2,3}L_{2,3}(^1D)3p$ $J = 1/2, 3/2$ states.

Magnifying Fig. 3.4, the impressive three-peak structure of the sodium $KL_{2,3}L_{2,3}(^1D)3p\ ^2L_J$ lines dominates (see Fig. 3.5), while the low intensity $KL_{2,3}L_{2,3}$ $(^3P)3p\ ^2P_{1/2,3/2}$ *LS*-forbidden Auger lines are emerging around 36.05 au. From the prediction of (3.39) in the *LS* coupling limit we also expect vanishing Auger rates for the $KL_1L_{2,3}(^3P)3p$ $(^1P_1 \rightarrow\ ^4L_{fJ_f})$ quartet state transitions. This is confirmed by our calculations as can be seen from Fig. 3.4. The $KL_1L_{2,3}(^3P)3p$ $(^3P_J \rightarrow\ ^4L_{fJ_f})$ Auger lines show a small but non-zero intensity, which can be identified as part of the two-peak structure between 34.9 and 35.0 au of the low intensity peaks shown in Fig. 3.6 corresponding to the $KL_1L_{2,3}(^3P)3p$ Auger lines.

Approximately 65 % of the intensity of the spectrum is covered by the $KL_{2,3}L_{2,3}(^1D)3p$ $(^{1,3}P^o_J \rightarrow\ ^2L_{fJ_f}^o)$ doublet Auger lines which can be identified to stem from the well-known *sodium* D-*doublet* (see Fig. 3.5). Another 7.5 and 4.5 % is contributed from the $KL_{2,3}L_{2,3}(^1S)3p$ and $KL_1L_1(^1S)3p\ ^2P^o_{1/2,3/2}$ doublet lines, respectively. I.e. ∼77 % of the total intensity of the spectrum is caused by Auger decay to final states with odd parity (see Fig. 3.4). ∼17 % arises from Auger decay to the even parity $KL_1L_{2,3}(^1P)3p^2L_{fJ_f}$ doublet Auger lines. All together these transitions cover ∼94 % of the total Auger rate. The few remaining percents stem from the weaker allowed $KL_1L_{2,3}(^3P)3p$ and $KL_{2,3}L_{2,3}(^3P)3p$ Auger lines.

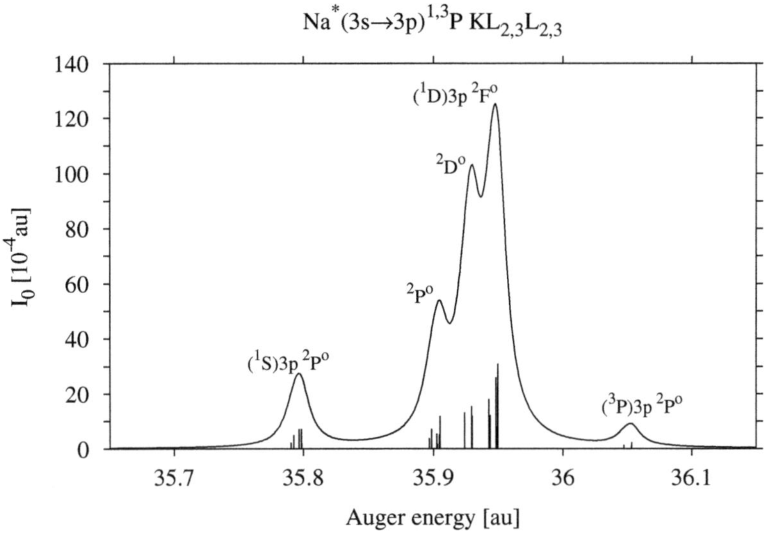

Fig. 3.5 The multiplet splitting of the 3s → 3p laser excited Na $KL_{2,3}L_{2,3}$ Auger lines (Na D-doublet) following 1s shell ionization. The Auger decay is from the resonantly excited unresolved $^1P_1$ and $^3P_{0,1,2}$ fine structure states

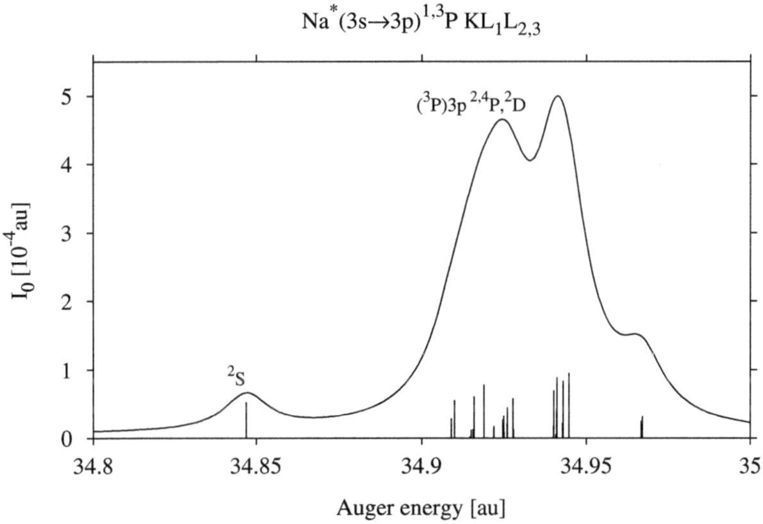

Fig. 3.6 The multiple splitting of the low intensity 2s, $2p^5(^3P)\,^{2,4}L_J$ Auger lines, stemming from the laser excited 3s → 3p Na $KL_1L_{2,3}(^3P)3p$ Auger transitions

Here, the 2 % contribution from the LS-forbidden $KL_{2,3}L_{2,3}(^3P)3p^2P_{1/2,3/2}$ Auger transitions should be noted.

Eventually, we point out that the absolute total rates of the spectrum turn out to be independent of the initial fine structure state. Further, considering extreme LS coupling and averaging over the possible initial triplet states, the ratio R of the intensities should be $R = {}^1P_1 : \sum_J^3 P_J = 1 : 3$. This result is well reproduced by the calculations.

Comparison with the $1s^{-1}3s$ KLL Diagram Lines

Due to the fact that Lohmann and Fritzsche (1994) only included Auger transitions from the initial $^3S_1$ state in their calculations and not considered the $^1S_0$ state transitions we are not able to directly compare the excited Na KLL results with the diagram ground state data. Further difficulties occur due to the fact that the number of final states is enormously increased for laser excited sodium transitions. We are however able to compare the group intensities. In Table 3.4 the group intensities for the KL_1L_1, $KL_1L_{2,3}$, and $KL_{2,3}L_{2,3}$ Auger groups of the laser excited sodium are shown in contrast to the diagram line results by Lohmann and Fritzsche (1994) and to the experimental data by Hillig et al. (1974), respectively.

The comparison with Lohmann and Fritzsche (1994) shows the group intensity for the KL_1L_1 Auger transitions reduced by a factor of two, while the intensity for the $KL_1L_{2,3}$ group is only 2/3 of the diagram lines. This is mainly due to a decrease in the Auger intensities of the $2s2p^5(^1P)3p$ final state transitions. On the other hand, the intensity for the $KL_{2,3}L_{2,3}$ group is increased by approximately 5/4 which is mainly caused by an increasing intensity for the $2s^22p^4(^1D)3p$ final state transitions. This comparison is however not without problems since Auger transitions from the $^1S_0$ state have not been included.

For comparison with experimental data obtained for the diagram Na KLL Auger lines we only discuss the data by Hillig et al. (1974), which are the most reliable ones (see Lohmann and Fritzsche 1994, for further references). We find the intensities for the KL_1L_1 and $KL_1L_{2,3}$ group reduced by ~20 % whereas the group intensity for the $KL_{2,3}L_{2,3}$ Auger transitions has increased by the same amount.

Table 3.4 Relative Auger KLL group intensities for the laser excited Na^{+*} ion

Na	KLL group intensities		
Transition	(a)	(b)	(c)
KL_1L_1	4.39	8.67	5.1 ± 0.5
$KL_1L_{2,3}$	19.85	30.02	24.0 ± 1.6
$KL_{2,3}L_{2,3}$	75.77	61.31	70.9 ± 3.1

a Lohmann et al. (1996) for the $^1P_1$ state, in comparison with data for the Na^+ ion;
b Lohmann and Fritzsche (1994),
c Hillig et al. (1974). See Sect. 3.3.5 for further explanations

3.3.6 Angular Distribution

Generating the Auger spectrum by photoionization and taking the dipole approximation into account, the angular distribution of the Auger electrons may be written as

$$I(\theta) = \frac{I_0}{4\pi}\left(1 + \alpha_2 \mathcal{A}_{20}(J) P_2(\cos\theta)\right) . \tag{3.49}$$

Due to the two-step-model the anisotropy parameter can be factorized in the alignment tensor $\mathcal{A}_{20}(J)$ describing the ion A^+ with total angular momentum J, and the anisotropy coefficient α_2 which depends on the matrix elements of Auger decay.

A general expression for the coefficient α_2 in terms of matrix elements may be found in the book by Lohmann (2008) and can be written as

$$\begin{aligned}\alpha_2 = &\left(\sum_{\ell j}\left|\langle J_f \| V \| (Jj)J_f\rangle\right|^2\right)^{-1} \\ &\times \sqrt{5(2J+1)} \sum_{\ell\ell' jj'} (-1)^{J_f+J-1/2+j+j'+\ell'} i^{\ell+\ell'} \\ &\times \cos(\sigma_{\ell'}^{j'} - \sigma_{\ell}^{j}) \langle J_f \| V \| (Jj)J_f\rangle \langle J_f \| V \| (Jj')J_f\rangle \\ &\times \sqrt{(2j+1)(2j'+1)} \begin{pmatrix} j' & j & 2 \\ 1/2 & -1/2 & 0 \end{pmatrix} \begin{Bmatrix} j' & j & 2 \\ J & J & J_f \end{Bmatrix} .\end{aligned} \tag{3.50}$$

Here, J and J_f denote the total angular momenta of the intermediate and final ionic states, respectively. Whereas j and ℓ are total angular momentum and angular momentum of the emitted Auger electron. The total scattering phase is denoted as σ_ℓ^j and V is the Coulomb operator.

Angular Anisotropy of KLL Lines in the LS Coupling

In studying the Auger dynamics it is often useful to consider the LS coupling limit which may allow for further insights and predictions without applying large computer codes. Furthermore, for light elements, like sodium, LS coupling can be expected as a relatively good approximation. Thus, predictions within the LS coupling scheme can be assumed to be close to our more sophisticated multiconfigurational Dirac–Fock (MCDF) results. We will further discuss this point in Sect. 3.3.8.

General expressions for the parameter α_2^{LS} may be found in the literature (Klar 1980b; Kabachnik et al. 1988). In both articles, however, the discussion has been restricted to the case of the rare gases. Using the same formalism as in (3.50) we obtain

$$
\begin{aligned}
\alpha_2^{(LS)} &= \left(\sum_{\ell j}(2j+1)\begin{Bmatrix} J & j & J_f \\ L & \ell & L_f \\ S & 1/2 & S_f \end{Bmatrix}^2 \left|\langle L_f \| V \| (L\ell)L_f\rangle\right|^2\right)^{-1} \\
&\times \sqrt{5(2J+1)} \sum_{\ell\ell' jj'} (-1)^{J_f+J-1/2+j+j'+\ell'} i^{\ell+\ell'} \\
&\times \cos(\sigma_{\ell'}-\sigma_\ell)\langle L_f \| V \| (L\ell)L_f\rangle\langle L_f \| V \| (L\ell')L_f\rangle \\
&\times \sqrt{(2j+1)(2j'+1)}\begin{pmatrix} j' & j & 2 \\ 1/2 & -1/2 & 0 \end{pmatrix}\begin{Bmatrix} j' & j & 2 \\ J & J & J_f \end{Bmatrix} \\
&\times \sqrt{(2j+1)(2j'+1)}\begin{Bmatrix} J & j & J_f \\ L & \ell & L_f \\ S & 1/2 & S_f \end{Bmatrix}\begin{Bmatrix} J & j' & J_f \\ L & \ell' & L_f \\ S & 1/2 & S_f \end{Bmatrix}, \qquad (3.51)
\end{aligned}
$$

where L, S and L_f, S_f denote orbital and spin angular momentum of the initial and final state, respectively. Note that the scattering phase depends only on the angular momentum ℓ in the above equation and not on the total angular momentum j as in (3.50).

Inserting $S = 0$ into (3.51), reducing the $9j$-symbols and applying some lengthy but straightforward angular momentum algebra the anisotropy parameter may be expressed as

$$
\begin{aligned}
\alpha_2^{(LS)}({}^1L_J) &= \delta_{1/2,S_f}\delta_{J,L}\left(\sum_{\ell}\left|\langle L_f \| V \| (L\ell)L_f\rangle\right|^2\right)^{-1} \\
&\times \sqrt{5(2L+1)}\sum_{\ell\ell'}(-1)^{-J-L_f+\ell'} i^{\ell+\ell'} \\
&\times \cos(\sigma_{\ell'}-\sigma_\ell)\langle L_f \| V \| (L\ell)L_f\rangle\langle L_f \| V \| (L\ell')L_f\rangle \\
&\times \sqrt{(2\ell+1)(2\ell'+1)}\begin{pmatrix} \ell' & \ell & 2 \\ 0 & 0 & 0 \end{pmatrix}\begin{Bmatrix} \ell' & \ell & 2 \\ L & L & L_f \end{Bmatrix}. \qquad (3.52)
\end{aligned}
$$

Equation (3.52) gives two important results for Auger transitions from an initial singlet state. First, only anisotropy parameters to a final doublet state can have a non-vanishing angular anisotropy. This is in accordance with the result of (3.39), which forbids transitions to quartet states. Secondly, (3.52) is independent of the total angular momentum J_f of the final state, i.e. the anisotropy parameter $\alpha_2^{(LS)}$ has a constant value for any fine structure state of a given final state multiplet (see the discussion in Sect. 3.3.8).

We now consider the initial ${}^1P_1$ state, and insert $L = J = 1$ into the (3.52). Applying parity conservation and the usual angular momentum coupling rules to the partial waves there are three possibilities (the parity is denoted by π):

1. $\ell = \ell' = L_f \neq 0$, and $\pi_{Auger} = (-1)^{L_f}$, or
2. $\ell = \ell' = 1, L_f = 0$, and $\pi_{Auger} = -1$, or
3. $\ell, \ell' = L_f \pm 1$, and $\pi_{Auger} = (-1)^{L_f \pm 1}$.

Table 3.5 The anisotropy coefficients $\alpha_2^{(LS)}$ for the ${}^1\mathrm{P}_1 \longrightarrow {}^2L_{f\,J_f}$ KLL Auger transitions in extreme LS coupling

KLL Auger transitions	
Final state	$\alpha_2^{(LS)}$
${}^2\mathrm{S}_{1/2}$	$-\sqrt{2} \sim -1.4142$
${}^2\mathrm{P}_{1/2,3/2}$	$\frac{1}{\sqrt{2}} \sim 0.7071$
${}^2\mathrm{D}^{\mathrm{o}}_{3/2,5/2}$	$\frac{1}{\sqrt{2}} \sim 0.7071$

The data are independent of the Auger energy, the nuclear charge of the atom, and of the fine structure of the final states. An odd parity of the states is denoted by an "o" superscript; after Lohmann et al. (1996)

Only in the latter case is the anisotropy coefficient still dependent on its matrix elements. This is directly caused by parity conservation and can be seen in analogy to the so-called *normal* and *anomalous* parity transitions introduced by Kabachnik and Sazhina (1984).

Inserting the quantum numbers for the first case we obtain

$$\alpha_2^{(LS)}({}^1\mathrm{P}_1) = \delta_{1/2,S_f}(-1)^{L_f-1}\sqrt{15}\,(2L_f+1) \times \begin{pmatrix} L_f & L_f & 2 \\ 0 & 0 & 0 \end{pmatrix} \begin{Bmatrix} L_f & L_f & 2 \\ 1 & 1 & L_f \end{Bmatrix} \tag{3.53}$$

for the anisotropy parameter. The anisotropy coefficient for the second case can be easily obtained from (3.52) by inserting $L_f = 0$. The anisotropy parameters for the possible Auger transitions for the two cases are listed in Table 3.5.

Eventually, if we sum over the final state fine structure in (3.51) we obtain

$$\begin{aligned} \alpha_2^{\Sigma(LS)} &= \Big(\sum_{\ell} |\langle L_f \| V \| (L\ell)L_f\rangle|^2\Big)^{-1} \\ &\times \sqrt{5(2J+1)}\,(2L+1)\sum_{\ell\ell'}(-1)^{J+S-L_f+\ell'}i^{\ell+\ell'} \\ &\times \cos(\sigma_{\ell'}-\sigma_\ell)\,\langle L_f\|V\|(L\ell)L_f\rangle\langle L_f\|V\|(L\ell')L_f\rangle \\ &\times \sqrt{(2\ell+1)(2\ell'+1)}\begin{pmatrix} \ell' & \ell & 2 \\ 0 & 0 & 0 \end{pmatrix}\begin{Bmatrix} \ell' & \ell & 2 \\ L & L & L_f \end{Bmatrix}\begin{Bmatrix} J & J & 2 \\ L & L & S \end{Bmatrix}. \end{aligned} \tag{3.54}$$

Inserting $S = 0$ in the above equation again yields (3.52). This equation has been found already to be independent of the final state fine structure.

3.3.7 Angular Distribution for an Unresolved Resonance

The fine structure of the resonantly excited, singly ionized intermediate state can often be hardly resolved. For some Auger transitions this is due to the fact that the fine structure splitting of the states is smaller than their natural line widths. Therefore a coherent excitation takes place. Such effects have been first investigated by Mehlhorn and Taulbjerg (1980) and Kabachnik et al. (1994).

Supposing a spin independent primary ionization process and the target atoms randomly oriented then, for an unresolved multiplet $M(LS)$ of the singly ionized intermediate state, the angular distribution of the Auger electrons may be written as (Kabachnik et al. 1994)

$$I_{M\to J_f}(\theta) = \frac{I_0^{M\to J_f}}{4\pi}\left(1 + \alpha_2^{M\to J_f}\mathcal{A}_{20}(L)P_2(\cos\theta)\right). \tag{3.55}$$

Though the anisotropy parameter can still be factorized, the alignment tensor $\mathcal{A}_{20}(L)$ now depends on the angular momentum L only, and the anisotropy coefficient $\alpha_2^{M\to J_f}$ can be written as a coherent sum over the total angular momentum J

$$\begin{aligned}\alpha_2^{M\to J_f} &= \frac{1}{N}\sum_{JJ'} A_2(J,J')\,(-1)^{J+L+S}\\ &\quad\times \frac{\sqrt{(2J+1)(2J'+1)(2L+1)}}{2S+1}\begin{Bmatrix} J & L & S\\ L & J' & 2\end{Bmatrix}\frac{\Gamma_{JJ'}}{\omega_{JJ'}^2+\Gamma_{JJ'}^2}\end{aligned} \tag{3.56}$$

where $\omega_{JJ'} = E_J - E'_J$ is the energy splitting, $\Gamma_{JJ'} = (\Gamma_J + \Gamma'_J)/2$, and Γ_J is the total width of the level J, and

$$\begin{aligned}A_2(J,J') &= (-1)^{J_f+J-1/2}\sqrt{5(2J+1)(2J'+1)}\\ &\quad\times\sum_{\ell\ell'jj'}(-1)^{\ell-\ell'}\sqrt{(2j+1)(2j'+1)}\sqrt{(2\ell+1)(2\ell'+1)}\\ &\quad\times\langle J_f\|V\|(Jj)J_f\rangle\langle J_f\|V\|(J'j')J_f\rangle^*\\ &\quad\times\begin{pmatrix}\ell & \ell' & 2\\ 0 & 0 & 0\end{pmatrix}\begin{Bmatrix} j' & j & 2\\ J & J' & J_f\end{Bmatrix}\begin{Bmatrix} j & \ell & 1/2\\ \ell' & j' & 2\end{Bmatrix}.\end{aligned} \tag{3.57}$$

The relative intensity is proportional to N

$$N = \sum_J \frac{2J+1}{(2L+1)(2S+1)}\frac{\Gamma_{J\to J_f}}{\Gamma_J}, \tag{3.58}$$

where

$$\Gamma_{J\to J_f} = 2\pi\sum_{\ell j}|\langle J_f\|V\|(Jj)J_f\rangle|^2 \tag{3.59}$$

is the partial width.

Note, that our expressions (3.56) and (3.58) differ from those by Kabachnik et al. (1994) by a factor of $\sqrt{2L+1}$ due to different normalization conventions. Otherwise one would obtain different relative intensities for a singlet state applying either (3.2) or (3.58).

Note, that the initial $^1P_1$ state of the sodium spectra can be used as a test case. Inserting $J = L = 1$ and $S = 0$ into (3.56)–(3.58) one ends up with (3.50) again.

3.3.8 Angular Distribution: Anisotropy Coefficients α_2

Generally, in KLL Auger transitions, a non-vanishing anisotropy parameter α_2 can only occur due to the coupling of the angular momenta of the inner shell vacancy and the outer open shell electrons. In the considered case of excited singly ionized sodium we have one p-electron in the outer shell which couples with the inner shell 1s-hole in two different ways, i.e. by generating either a $^3P_{0,1,2}$ or a $^1P_1$ state. The $^3P_0$ state is of no interest because no alignment can be produced during the ionization process and thus, no anisotropic angular distribution can occur. As can be seen from (3.50) the anisotropy parameter α_2 vanishes in this case. However, if the initial Auger state has a total angular momentum $J = 1$ or $J = 2$ both, a non-zero alignment $\mathcal{A}_{20}$ and an anisotropy coefficient $\alpha_2 \neq 0$ is allowed. Lohmann and Fritzsche (1994) found that a non-zero alignment $\mathcal{A}_{20} \neq 0$ for the initial state of KLL Auger transitions is only possible in a full-relativistic treatment. For a primary photoionization, in contrast, by neglecting the spin-orbit interaction in the generation of the photoelectron spectrum, and following the work of Bußert and Klar (1983), one can expect a non-zero alignment even in a non-relativistic description for the $^3P_{1,2}$ and $^1P_1$ states.

The numerical results of Lohmann et al. (1996) for the anisotropy parameter α_2 are shown in Table 3.6 for the KL_1L_1 and $KL_1L_{2,3}$ Auger transitions and in Table 3.7 for the $KL_{2,3}L_{2,3}$ Auger transitions for the $^3P_{1,2}$ and $^1P_1$ initial states, respectively.

While Lohmann et al. (1996) calculated large anisotropy coefficients for most of the low intensity $KL_{2,3}L_{2,3}(^3P)3p$ Auger transitions their discussion is useless since they can hardly be measured in any experiment. Therefore, we have only listed those anisotropy parameters where a non-zero (i.e. at least $> 10^{-5}$ au) intensity has been calculated.

LS Coupling Predictions

For the interpretation of the data it is useful to investigate the α_2 parameters in the LS coupling limit, too. As has been discussed in Sect. 3.3.6, in pure LS coupling, the anisotropy parameters of angular distribution for some of the $^1P_1$ KLL Auger transitions are independent of the Auger matrix elements. In more detail, these are the Auger transitions to the even parity $^2S_{1/2}$ and $^2P_{1/2,3/2}$, and to the odd parity

Table 3.6 The angular anisotropy coefficients α_2 for the KL_1L_1 and $KL_1L_{2,3}$ Auger transitions of 3s → 3p excited sodium

Na KLL Auger transitions I			
Final state	α_2 parameters		
	$^3P_1$	$^3P_2$	$^1P_1$
$\|2p^6(^1S)3p\ ^2P^o_{3/2}\rangle$	~0	~0	~0
$\|2p^6(^1S)3p\ ^2P^o_{1/2}\rangle$	~0	−0.836	~0
$\|2s2p^5(^1P)3p\ ^2S_{1/2}\rangle$	0.706	−0.837	−1.411
$\|2s2p^5(^1P)3p\ ^2P_{3/2}\rangle$	~0	0.673	0. 705
$\|2s2p^5(^1P)3p\ ^2P_{1/2}\rangle$	−0.665	−0.837	0.705
$\|2s2p^5(^1p)3p\ ^2D_{5/2}\rangle$	−0.141	−0.170	−0.141
$\|2s2p^5(^1p)3p\ ^2D_{3/2}\rangle$	0.138	0.673	−0.141
$\|2s2p^5(^3p)3p\ ^2S_{1/2}\rangle$	−1.414	−0.837	−1.413
$\|2s2p^5(^3p)3p\ ^2p_{1/2}\rangle$	−0.701	−0.837	0.701
$\|2s2p^5(^3p)3p\ ^2p_{3/2}\rangle$	0.281	0.811	0.621
$\|2s2p^5(^3p)3p\ ^2D_{3/2}\rangle$	−0.076	0.797	−0.061
$\|2s2p^5(^3p)3p\ ^2D_{5/2}\rangle$	−0.141	−0.035	−0.141
$\|2s2p^5(^3p)3p\ ^4p_{1/2}\rangle$	−1.149	−0.837	−0.351
$\|2s2p^5(^3p)3p\ ^4p_{3/2}\rangle$	0.652	−0.513	0.273
$\|2s2p^5(^3p)3p\ ^4p_{5/2}\rangle$	−0.141	0.820	−0.141
$\|2s2p^5(^3p)3p\ ^4D_{1/2}\rangle$	0.684	−0.837	0.698
$\|2s2p^5(^3p)3p\ ^4D_{3/2}\rangle$	0.145	0.800	0.260
$\|2s2p^5(^3p)3p\ ^4D_{5/2}\rangle$	−0.141	0.202	−0.141
$\|2s2p^5(^3p)3p\ ^4D_{7/2}\rangle$	–	−0.239	–
$\|2s2p^5(^3p)3p\ ^4S_{3/2}\rangle$	0.706	−0.834	0.707

At the top of columns 2–4 the initial state of the Auger transition is shown. No anisotropy can occur for the initial $^3P_0$ state. The symbol ~0 denotes an anisotropy coefficient $< 10^{-4}$. For details see Sect. 3.3.8 (after Lohmann et al. 1996)

Table 3.7 Anisotropy coefficients α_2 for the $KL_{2,3}L_{2,3}$ Auger transitions of excited sodium

Na KLL Auger transitions II			
Final state	α_2 parameters		
	$^3P_1$	$^3P_2$	$^1P_1$
$\|2s^22p^4(^1S)3p\ ^2P^o_{1/2}\rangle$	~0	−0.841	~0
$\|2s^22p^4(^1S)3p\ ^2P^o_{3/2}\rangle$	~0	~0	~0
$\|2s^22p^4(^1D)3p\ ^2P^o_{1/2}\rangle$	−0.702	−0.837	−0.707
$\|2s^22p^4(^1D)3p\ ^2P^o_{3/2}\rangle$	0.517	~0	−0.707
$\|2s^22p^4(^1D)3p\ ^2D^o_{5/2}\rangle$	−0.144	0.598	0.705
$\|2s^22p^4(^1D)3p\ ^2D^o_{3/2}\rangle$	−0.573	~0	0.699
$\|2s^22p^4(^1D)3p\ ^2F^o_{7/2}\rangle$	−0.202	−0.240	−0.202
$\|2s^22p^4(^1D)3p\ ^2F^o_{5/2}\rangle$	0.196	0.598	−0.202
$\|2s^22p^4(^3P)3p\ ^2P^o_{1/2}\rangle$	−0.635	−0.836	−0.689
$\|2s^22p^4(^3P)3p\ ^2P^o_{3/2}\rangle$	0.501	0.005	−0.691

See Table 3.6 for further information (after Lohmann et al. 1996)

$^2D^o_{3/2,5/2}$ final states.[2] Comparing the numerical data for the anisotropy coefficients of Tables 3.6 and 3.7 with the extreme *LS* coupling results of Table 3.5 shows good agreement.

Furthermore, as a result of (3.52) any anisotropy coefficients for Auger transitions from an initial singlet state should be independent of the final state fine structure, e.g. the α_2 parameter should be the same for the $KL_1L^1_{2,3}P_1 \rightarrow {}^2D_{3/2}$ and ${}^1P_1 \rightarrow {}^2D_{5/2}$ Auger transitions, respectively. This is confirmed for almost all possible final doublet states of the laser excited sodium spectra. Deviations occurring for the $KL_1L_{2,3}(^3P)3p^2P_{1/2,3/2}$ and $^2D_{3/2,5/2}$ fine structure doublets, however, demonstrate that, though the *LS* coupling limit allows for good predictions for most of the relevant data of the sodium Auger spectra (see also Lohmann and Fritzsche 1994), a relativistic MCDF approach with intermediate coupling can yield totally different results, mainly for the low intensity lines, and where coupling to other fine structure states becomes important.

Equation (3.52) also shows that all anisotropy parameters for ${}^1P_1$ to final quartet state transitions should vanish in the *LS* coupling limit. Though Lohmann et al. (1996) calculated large anisotropy coefficients for these Auger transitions their further discussion is useless since they have been found as strongly forbidden in the *LS* coupling what has been confirmed by the calculations.

However, one should keep in mind that where the $KL_{2,3}L_{2,3}(^3P)3p$ quartet final state Auger transitions are generally forbidden in the *LS* coupling by parity violation even for Auger transitions from the initial triplet state, the $KL_1L_{2,3}(^3P)3p\ {}^3P$ to final quartet state transitions are allowed. Their numbers have been therefore included in the tables.

Anisotropy Parameters for an Unresolved Resonance

As has been discussed in Sect. 3.3.4, the fine structure splitting of the initial ${}^3P_{0,1,2}$ states is of the same order or even less than their partial widths. Thus, the ${}^3P$ fine structure states can be assumed as coherently excited by the primary photoionization process. Therefore, considering a possible experiment measuring the anisotropy coefficients of laser excited sodium, the equations of Sect. 3.3.7 apply. Lohmann et al. (1996) calculated the anisotropy parameter $\alpha_2^{M\rightarrow J_f}$ for the unresolved initial ${}^1P$ and ${}^3P$ state multiplets for some of the Auger lines covering most of the total intensity of the excited sodium KLL spectra. These data are shown in Table 3.8. The α_2 parameters of the Auger transitions of the ${}^1P$ initial state can be used as a test, since they have to be identical to the numbers obtained for the resolved fine structure calculation for the ${}^1P_1$ state which is confirmed by the results of Lohmann et al. (1996).

The relative intensities of the ${}^3P$ to final state Auger transitions have been also given in Table 3.8. Even though the total decay rate is approximately constant for

[2] Note, that e.g. the final $^2P^o_{1/2,3/2}$ odd or the $^2D_{3/2,5/2}$ even parity states do depend on the Auger matrix elements.

Table 3.8 The anisotropy coefficients $\alpha_2^{M\to J_f}$ for the coherently excited KLL Auger transitions of sodium

Na (1s^{-1}3p) KLL Auger transitions						
Final states	$\alpha_2^{M\to J_f}$ parameters					
	I_0	^{3}P$_{av}$		^{3}P$_{co}$		^{1}P
$\|2s2p^5(^1p)3p\ ^2s_{1/2}\rangle$	1.93	−0.230		0.562		−1.378
$\|2s2p^5(^1p)3p\ ^2p_{3/2}\rangle$	3.74	0.374	0.017	−0.111	−0.273	0.704
$\|2s2p^5(^1p)3p\ ^2p_{1/2}\rangle$	1.90	−0.687		−0.591		0.720
$\|2s2p^5(^1p)3p\ ^2D_{5/2}\rangle$	5.59	−0.141	0.088	−0.011	0.052	−0.141
$\|2s2p^5(^1p)3p\ ^2D_{3/2}\rangle$	3.85	0.420		0.144		−0.140
$\|2s^22p^4(^1D)3p\ ^2p^o_{1/2}\rangle$	3.76	−0.699	−0.112	−0.062	0.275	−0.705
$\|2s^22p^4(^1D)3p\ ^2p^o_{3/2}\rangle$	7.76	0.172		0.438		−0.706
$\|2s^22p^4(^1D)3p\ ^2D^o_{5/2}\rangle$	13.55	0.284	0.093	−0.167	−0.274	0.707
$\|2s^22p^4(^1D)3p\ ^2D^o_{3/2}\rangle$	9.14	−0.191		−0.433		0.697
$\|2s^22p^4(^1D)3p\ ^2F^o_{7/2}\rangle$	17.92	−0.201	0.061	−0.012	0.077	−0.202
$\|2s^22p^4(^1D)3p\ ^2F^o_{5/2}\rangle$	13.88	0.398		0.191		−0.203

Data are listed for the strongest Auger lines only. The whole multiplet has been normalized to 100 %. The displayed Auger lines in the table cover ~83 % of the total intensity. (I_0): The relative intensities for the unresolved initial ^{3}P state transitions. (^{3}P$_{av}$): The anisotropy parameters averaged over the initial state fine structure. ($^3P_{co}$): The anisotropy parameters for a coherently excited ^{3}P state; after Lohmann et al. (1996)

the initial states, strong deviations occur between the individual decay channels. In Table 3.8 we show the anisotropy coefficients α_2 in different schemes of excitation. To demonstrate the effect of a coherent excitation, we also include the values for the anisotropy parameters averaged over the initial ^{3}P state fine structure. Large deviations between both numbers indicate a strong modification of the anisotropy of the individual contributions due to a coherent excitation. Comparing the data of both calculations for the initial triplet state large differences have been found. Changes by factors of 2 up to 10 and even reversal of sign occur. This coincides with the investigations by Kabachnik et al. (1994) who studied the angular anisotropy of the Ne KL-LLL satellite Auger decay. They also found their data strongly influenced by coherence effects in the initial state.

As an important result of this section we point out that though the fine structure of the initial state has not been resolved one still remains with large anisotropy parameters even for Auger lines with a relatively high, i.e. measurable, intensity. Finally, even by summing over the fine structure of the final state multiplet one still remains with a measurable anisotropy parameter. These numbers have been also included in Table 3.8. For the averaged initial state data summing over the final state fine structure leads to almost vanishing anisotropy parameters except for the final $2s^22p^4(^1D)3p\ ^2P$ state. To the contrary, for the coherently excited ^{3}P state we still remain with an anisotropy coefficient of $\sim\pm 0.27$ for the $2s2p^5(^1P)3p\ ^2P$, $2s^22p^4(^1D)3p\ ^2P$, and $2s^22p^4(^1D)3p\ ^2D$ Auger lines.

The large anisotropy parameters for the initial ^{1}P state Auger transitions remain undisturbed since their numbers show the same sign and magnitude for all of the final fine structure states. Thus, mainly the $^1\mathrm{P} \rightarrow {}^2\mathrm{L_f}$ Auger lines of the laser excited singly ionized sodium can be seen as good candidates for observing large anisotropy parameters in an angle resolved Auger experiment. In particular, the $^1\mathrm{P} \rightarrow 2\mathrm{s}^2 2\mathrm{p}^4(^1\mathrm{D})3\mathrm{p}\ ^2\mathrm{D}$ Auger line should combine a high intensity with a large anisotropy parameter. The large anisotropy parameter for the $^1\mathrm{P} \rightarrow 2\mathrm{s}2\mathrm{p}^5(^1\mathrm{P})3\mathrm{p}\ ^2\mathrm{S}$ Auger transition should also be noted.

Following the photoionization/excitation decay scheme of (3.37) and (3.38) we are able to assume an alignment value of $\mathcal{A}_{20} = 1/\sqrt{2}$ common for the subsequent Auger decay of the photoexcited $^{1,3}\mathrm{P}_1$ states. Using the alignment value and applying the anisotropy data of Table 3.8, we obtain the angular distribution of the emitted Auger electrons for the coherently excited Na KLL Auger transitions. Figure 3.7 shows the anisotropic angular distribution for selected Auger transitions. As can be seen, the intensity can either show a pronounced or less pronounced maximum or minimum at $\theta = 90°$, depending on the sign and magnitude of the anisotropy parameters. Note, that all graphs coincide at the magic angle $\theta_m \simeq 54.7°$ as the second Legendre polynomial equals zero.

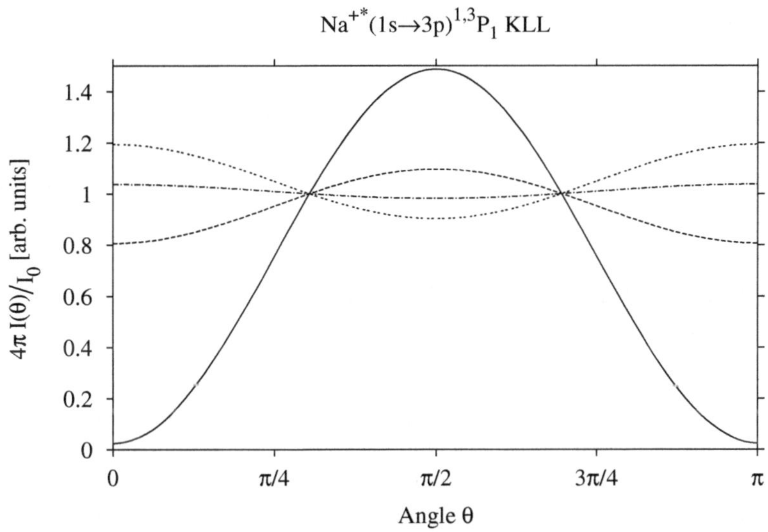

Fig. 3.7 The angular distribution of the emitted Auger electrons for selected Auger transitions of the excited $\mathrm{Na}^{+*}(1\mathrm{s}^{-1} \rightarrow 3\mathrm{p})\ ^{1,3}\mathrm{P}_1$ KLL Auger decay. Utilizing the photoionization/excitation process as per (3.37) and (3.38) an alignment of $\mathcal{A}_{20} = 1/\sqrt{2}$ has been assumed. The angular distribution data for the anisotropy coefficient have been taken from Table 3.8. The Auger transitions are depicted as follows; (——): $^1\mathrm{P} \longrightarrow {}^2\mathrm{S}$; (- - -): $^3\mathrm{P} \longrightarrow {}^2\mathrm{P}$; (— · — · —): $^3\mathrm{P} \longrightarrow {}^2\mathrm{D}$; (- - - - - -): $^3\mathrm{P} \longrightarrow {}^2\mathrm{P^o}$

Comparison with the Diagram Lines

As already pointed out in Sect. 3.3.2 a comparison with the Na KLL Auger transition data from the $^3S_1$ state by Lohmann and Fritzsche (1994) is not without problems (also see Sect. 4.8 of Lohmann 2008). Particularly, we find a totally different behaviour of the anisotropy parameter for the laser excited and ground state Na KLL Auger transitions, respectively. Whereas for the KLL $^3S_1$ state transitions the most intense Auger lines, i.e. Auger transitions to the $2s2p^5(^1P)3s$ and $2s^22p^4(^1D)3s$ final states, show an almost vanishing anisotropy we find large anisotropy parameters (i.e. at least a magnitude larger) for their counterparts in the laser excited sodium case. This effect can be explained within the following picture.

Remembering that sodium is a light atom, LS coupling can be seen as a relatively good approximation. Using the LS coupling results for the anisotropy parameter (see (9) of Lohmann and Fritzsche 1994), we get $\alpha_2 = 0$ for all Auger transitions to final doublet states. Thus, Lohmann and Fritzsche (1994) obtained the anisotropy parameters for Auger transitions to the final $2s2p^5(^1P)3s$ and $2s^22p^4(^1D)3s$ states of almost vanishing magnitude, even within an MCDF approach.

In contrast, for the laser excited case, even in the non-relativistic limit, the anisotropy parameter α_2 can be generally non-zero for Auger transitions to final doublet states, e.g. see (3.51) and (3.53). Due to this behaviour, large anisotropy coefficients have been found in the calculations for the $2s2p^5(^1P)3p$ and $2s^22p^4(^1D)3p$ final state transitions, respectively.

This is directly caused by the fact that for the laser excited sodium KLL spectrum non-zero anisotropy parameters can occur for certain Auger transitions in the non-relativistic limit, which is in contrast to the discussion of Sect. 3.1, and has been found as forbidden for the case of the sodium KLL ground state Auger transitions.

3.4 Advanced Experiments: Resonant $Ar^*(2p^{-1}_{1/2}4s_{1/2} + 3d_{3/2})_{J=1}$ Auger Emission

While the modulus of the Auger transition matrix elements yields information about the magnitude of the Auger decay width, the angular distribution of Auger electrons allows for accessing the relative scattering phases; see (3.49)–(3.50). The next step beyond an angle resolved analysis in order of gaining a more detailed insight into the Auger emission dynamics is by investigating not only angle but spin resolved Auger electrons. During the last decades a number of experiments and theoretical investigations have been performed (see Lohmann 2008, for further references). A recent state of the art experiment has been performed by Turri et al. (2007). Applying the experimentally well-observed two-step model (e.g. see Mehlhorn 1990), the angle and spin resolved, resonantly excited Auger decay of argon is considered, where the primary excitation is either to the $4s_{1/2}$ or $3d_{3/2}$ Rydberg levels

$$\gamma + Ar(^1S_0) \longrightarrow \begin{cases} Ar^*(2p_{1/2}^{-1}, 4s_{1/2})_{J=1}, \\ Ar^*(2p_{1/2}^{-1}, 3d_{3/2})_{J=1}, \end{cases}$$
$$\hookrightarrow Ar^{+*}(3p_{1/2,3/2}^{-2}, 4s_{1/2} + 3d_{3/2,5/2})L_{J=1/2,\ldots,9/2}^{2S+1=2,4} + e_{Auger}. \tag{3.60}$$

For the independent treatment of the two decay processes we refer to the discussion in Sect. 3.4.4.

3.4.1 Theoretical Considerations

The general equations for the angular distribution and the cartesian components of the spin polarization vector have been derived by Kleiman et al. (1999a, b) in the helicity frame of the emitted Auger electrons with the axis of Auger emission as quantization axis; see Fig. 3.8. For such geometry, both, the angular distribution and the spin polarization vector can be related to an arbitrarily oriented coordinate frame, e.g. the laboratory frame, via two angles θ and ϕ defined with respect to the quantization axis. The two-step model allows for a factorization of the angular distribution and spin polarization parameters. In addition, the dipole approximation restricts the number of parameters, i.e. only parameters with rank $K \leq 2$ contribute. Thus, we obtain the angular distribution for an arbitrarily polarized photon beam as

$$I(\theta,\phi)^{\gamma} = \frac{I_0}{4\pi}\left\{1 + \alpha_2\left(\mathcal{A}_{20}P_2(\cos\theta) + \sqrt{\frac{3}{2}}\left[Re\mathcal{A}_{22}\cos 2\phi - Im\mathcal{A}_{22}\sin 2\phi\right]\sin^2\theta\right)\right\}, \tag{3.61}$$

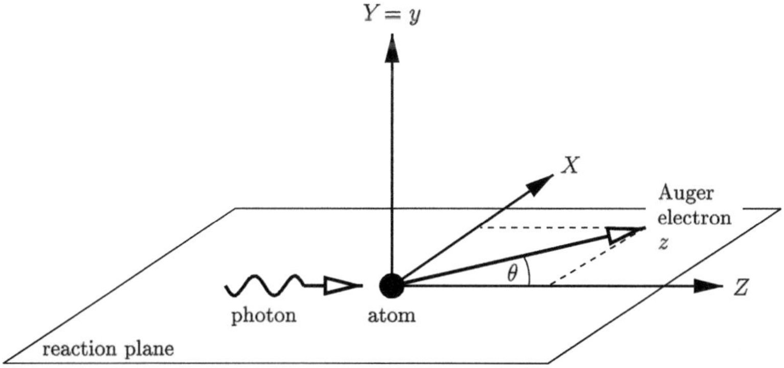

Fig. 3.8 The reaction plane and coordinate frame defined for synchrotron beam excitation (Lohmann et al. 2003b)

where I_0 denotes the total intensity, the superscript γ indicates the photonic excitation, and $P_2(\cos\theta)$ is the second Legendre polynomial.

The cartesian components of the spin polarization vector can be expressed as

$$P_x(\theta,\phi)^\gamma = \frac{1}{N^\Gamma}\Big(\xi_1\mathcal{A}_{10} + \sqrt{6}\,\xi_2\big[Re\mathcal{A}_{22}\sin 2\phi + Im\mathcal{A}_{22}\cos 2\phi\big]\Big)\sin\theta, \tag{3.62}$$

$$P_y(\theta,\phi)^\gamma = \frac{-3}{2N^\Gamma}\,\xi_2\Big(\mathcal{A}_{20} - \sqrt{\frac{2}{3}}\big[Re\mathcal{A}_{22}\cos 2\phi - Im\mathcal{A}_{22}\sin 2\phi\big]\Big)\sin 2\theta, \tag{3.63}$$

and

$$P_z(\theta,\phi)^\gamma = \frac{1}{N^\Gamma}\,\delta_1\mathcal{A}_{10}\cos\theta, \tag{3.64}$$

where we introduced the abbreviation $N^\Gamma = 4\pi I(\theta,\phi)^\gamma/I_0$. Note, that the numerator of (3.64) remains independent of the azimuth ϕ. This can be explained by the fact that, irrespective of the specific choice of the X- and Y-axes of the laboratory frame, the z-axis of the helicity frame is always contained in the reaction plane. The dynamics of the photoexcitation into the intermediate excited Ar* state is described by a set of state multipoles $\mathcal{A}_{KQ}$. Particularly, $\mathcal{A}_{10}$ and $\mathcal{A}_{20}$ are known as orientation and alignment parameters. For our case of a primary photoexcitation they become constant numbers, and take their maximum values for specific geometries, e.g. see Kronast et al. (1986). For the definition of the state multipoles $\mathcal{A}_{K0}$ in terms of the magnetic substate population see (4.65).

The Auger decay dynamics is described by the (intrinsic) angular distribution parameter α_2 and spin polarization parameters δ_1 and ξ_1, referring to the transferred (TSP) and ξ_2 related to the dynamic (DSP) spin polarization, respectively. These parameters are combinations of bilinear products of the matrix elements and phases of the second step Auger decay, only.

Inspecting (3.61)–(3.64), we find that only the component p_x of the spin polarization vector, i.e. (3.62), allows for observing either the TSP or the DSP within the same experimental set-up (see Sect. 3.4.2). Here, the first depends on the orientation parameter $\mathcal{A}_{10}$, while the latter depends on the real and imaginary components $\mathcal{A}_{22}$ of the alignment tensor.

3.4.2 Experimental Set-Up

The measurements of Turri et al. (2007) have been performed at the beamline 4.0.2 of the advanced light source (ALS) at Lawrence Berkeley National Laboratory in Berkeley, CA. The elliptically polarized undulator (EPU) has been set to deliver either

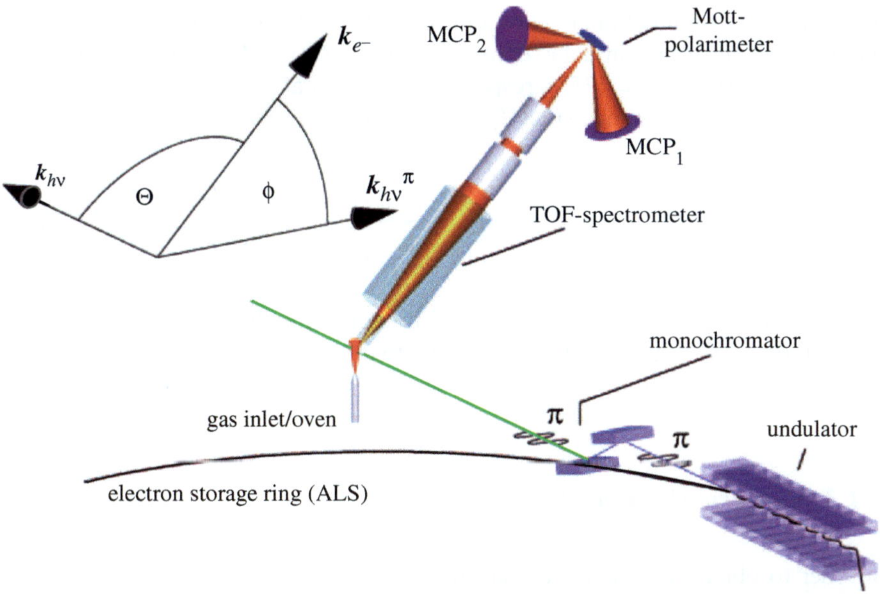

Fig. 3.9 Experimental set-up with undulator beamline, TOF electron analyzer and Mott detector (Lohmann et al. 2005; Lohmann 2008)

linearly or circularly polarized light (polarization 100 % in both cases), with energy of 246.4 or 249.0 eV to excite the $2p_{1/2}^{-1}4s$ and $2p_{1/2}^{-1}3d$ resonances, respectively. The experimental geometry is illustrated in Fig. 3.9. The emitted Auger electrons have been collected in a plane perpendicular to the photon propagation direction and at 45° with respect to the plane of the storage ring, and their kinetic energy has been measured by time-of-flight (TOF) detectors (Snell et al. 2002). A Mott detector of the Rice type (Burnett et al. 1994; Snell et al. 2000a), operated at 25 kV, with a Sherman function $S_{\mathrm{eff}} = 0.13 \pm 0.02$, mounted after the drift tube of the TOF detector, measured the spin polarization along the photon propagation axis. Geometrical asymmetries of the apparatus have been accounted for, by the standard technique of reversing the helicity of the photons when using circularly polarized light and rotating by 90° the polarization of linearly polarized light. Within the described geometry, the equations for the TSP and DSP can be derived from 3.62 and take the simple forms (Snell et al. 2002; Lohmann et al. 2005, and refs. therein)

$$P_{trans}(\theta) = \frac{2\sqrt{3}\xi_1}{2\sqrt{2} - \alpha_2}, \tag{3.65}$$

and

$$P_{dyn}(\theta, \phi) = \frac{6\xi_2}{2\sqrt{2} - \alpha_2}. \tag{3.66}$$

Note, that the azimuthal angle ϕ becomes redundant for the TSP. The transferred and dynamic electron spin polarizations, corresponding to $P_{trans}(\theta_{exp} = 90°)$ and $P_{dyn}(\theta_{exp} = 90°, \phi_{exp} = 135°)$, respectively, can be calculated from the four measured spin-up and down intensities I_1^+, I_1^-, I_2^+, and I_2^- as follows

$$P_{trans,dyn}(\theta_{exp}, \phi_{exp}) = \frac{1}{S_{\rm eff}} \frac{\sqrt{I_1^+ I_2^-} - \sqrt{I_1^- I_2^+}}{\sqrt{I_1^+ I_2^-} + \sqrt{I_1^- I_2^+}} \,. \tag{3.67}$$

Note, that the experimental angle $\Phi = 45°$ for collecting the electrons with respect to the storage ring plane (see Fig. 3.9) refers to $\phi = 135°$ in our chosen coordinate frame.

3.4.3 Numerical Methods

In order to obtain the numerical data from (3.65) and (3.66), Turri et al. (2007) employed a relativistic distorted wave approximation (RDWA).[3] Here, the bound state wavefunctions of the excited intermediate and the ionized final atomic states are constructed using the multiconfigurational Dirac-Fock (MCDF) computer code of Grant et al. (1980). Intermediate coupling has been taken into account obtaining the mixing coefficients by applying the average level calculation mode (see Grant et al. 1980). The calculation of the Auger transition matrix elements has been performed applying a relaxed orbital method (Lohmann 1999a, b; Lohmann and Kleiman 2001), utilizing the ANISO program package (Lohmann 2008), calculating the bound electron wavefunctions of the intermediate state in the field of the excited atom, whereas obtaining the bound electron wavefunctions of the final state in the field of the singly ionized atom.

The atomic state function (ASF) of the intermediate excited and the singly ionized final state have been constructed as linear combinations of configuration state functions (CSF)

$$\left|\psi_\alpha(PJM)\right\rangle = \sum_{r=1}^{n_c} c_r(\alpha) \left|\gamma_r PJM\right\rangle . \tag{3.68}$$

The configuration states $\left|\gamma_r PJM\right\rangle$ are constructed from antisymmetrized products of Dirac orbitals which are eigenstates of the total (one-electron) angular momentum and parity. The label γ_r distinguish the occupation of the different subshells and angular coupling schemes (see Grant 1970, for further details). n_c is the number of CSF included in the expansion and $c_r(\alpha)$, $r = 1, \ldots, n_c$, are the configuration mixing coefficients for the state α.

---

[3] This method has been outlined in detail in the book by Lohmann (2008).

Turri et al. (2007) generated the intermediate excited state as a linear combination of the five possible jj-coupled $Ar^*(2p_{1/2,3/2}^{-1}4s_{1/2})_{J=1}$ and $Ar^*(2p_{1/2,3/2}^{-1}3d_{3/2,5/2})_{J=1}$ CSF. Their calculation yields the $Ar^*(2p_{3/2}^{-1}4s_{1/2})_{J=1}$ and $Ar^*(2p_{1/2}^{-1}4s_{1/2})_{J=1}$, as well as the $Ar^*(2p_{1/2}^{-1}3d_{3/2})_{J=1}$ ASF as almost pure states. This allows for separately discussing our $L_2M_{2,3}M_{2,3}$ Auger decay data obtained from the $2p_{1/2}^{-1} \rightarrow 4s$ and $2p_{1/2}^{-1} \rightarrow 3d$ excitation processes, respectively. However, strong intermediate coupling has been found between the $Ar^*(2p_{3/2}^{-1}3d_{3/2})_{J=1}$ and the $Ar^*(2p_{3/2}^{-1}3d_{5/2})_{J=1}$ ASF.

A configuration interaction (CI) calculation has been performed for the final ionic state. I.e., all possible linear combinations of the $Ar^+(3p_{1/2,3/2}^{-2}4s_{1/2})$ and $Ar^+(3p_{1/2,3/2}^{-2}3d_{3/2,5/2})$ jj-coupled states, forming a basis set of 36 CSF, have been included in generating the final state ASF (36 CSF-CI).[4]

It is noticeable, that the collapse of the 3d orbital has no effect on the resonant Auger decay of the exited 2p states, as it causes a strong satellite line in the inner valence region of Ar, only (Becker and Shirley 1996).

Eventually, the continuum wavefunction of the Auger electron has been evaluated by solving the Dirac equation with an intermediate coupling potential where electron exchange with the continuum has been taken into account. The intermediate coupling potential is constructed from the mixed CSF of the final ionic state. Thereby it has been taken into account, that the ejected electron moves within the field of the residual ion. With this, the Auger transition matrix elements have been obtained for calculating the relevant angular anisotropy and spin polarization parameters, respectively. Note, that both are not functions of the transition matrix elements, only, but explicitly depend on the scattering phases. Further information may be found in Lohmann (1999a,b); Lohmann and Kleiman (2001) or in Lohmann (2008).

3.4.4 Numerical Versus Experimental Results

Turri et al. (2007) obtained numerical data for the relative intensities, angular distribution and spin polarization parameters from their 36 CSF-CI for the $Ar^*(2p_{1/2}^{-1} \rightarrow 4s_{1/2})_{J=1}$ and $Ar^*(2p_{1/2}^{-1} \rightarrow 3d_{3/2})_{J=1}$ $L_2M_{2,3}M_{2,3}$ resonant Auger decay. They find the high J part for certain multiplets suppressed due to internal J-dependent selection rules (Lohmann et al. 2005). Particularly, Auger transitions to the $J_f = 7/2$ and $9/2$ final states are suppressed for the $Ar^*(2p_{1/2}^{-1}4s_{1/2})_{J=1}$, and to the $J_f = 9/2$ final fine structure states for the $Ar^*(2p_{1/2}^{-1}3d_{3/2})_{J=1}$ intermediate excited states, respectively.

[4] The $Ar(3p^4nl)$ photoelectron satellites have been measured varying the photoexcitation energy across the $Ar^*(2p_{1/2}^{-1}4s_{1/2})_{J=1}$ and $Ar*(2p_{3/2}^{-1}3d_{3/2,5/2})_{J=1}$ resonances (Fritzsche et al. 2007), finding correlation patterns between the initial and the resulting final fine structure states as a function of the photoelectron energy. While the calculation of Turri et al. (2007) confirms strong correlations for the final state configuration interaction (CI), they obtained the initial states of the Auger transition as almost pure.

As has been discussed in Lohmann et al. (2005), this results in surviving of the low J fine structure components of a multiplet, only, with no partner for polarization cancellation of the DSP.

The relevant spin-up and down intensities have been obtained by performing a weighted sum over the fine structure states, and where appropriate even summing over two or more unresolved lines, in order to allow for a comparison with the experimental low resolution line spectra. Note, that the numerical spectrum usually needs to be shifted by an energy offset, and re-normalized to the experimental total intensities. Since the 36 CSF-CI does not cover the full energy range of the observed spectrum, a comparison with the experimental data for the DSP and TSP has been possible for lines 1–7 of the measured Auger spectra, only. The assignment between the line and peak numbers may be found in Table 3.11 later on. The numerical results of lines 1–9 for the $Ar^*(2p_{1/2}^{-1} \rightarrow 3d_{3/2})_{J=1}$ $L_2M_{2,3}M_{2,3}$ spectrum are positioned outside the observed energy range. However, they have been used in an earlier spin-resolved analysis of the unresolved peak structure of the $Ar^*(2p_{1/2,3/2}^{-1} \rightarrow 4s_{1/2})_{J=1}$ $L_{2,3}M_{2,3}M_{2,3}$ spectrum (Lohmann et al. 2005) and have been included in the calculation for a consistent comparison of the data.

In Fig. 3.10, the spin-unresolved spectrum (average of spin-up and spin-down spectra) of the $2p_{1/2}^{-1}3d$ Auger decay measured with circularly polarized light is displayed. It consists of some 15 peaks, where every peak is a manifold of many overlapping components, corresponding to the transitions to the different final states

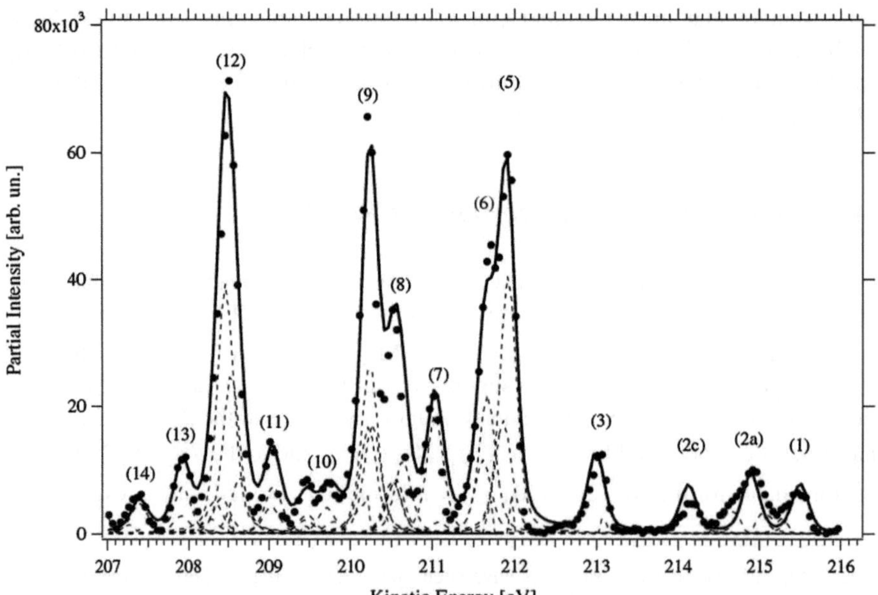

Fig. 3.10 The $Ar^*2p_{1/2}^{-1}3d_{3/2_{J=1}}L_2M_{2,3}M_{2,3}$ Auger spectrum averaged over the electronic spin. Peak numbers as assigned in Table 3.9. (●) experimental data. (——) manifold. (- - - - - -) manifold components (for further details see text); after Turri et al. (2007)

of the singly ionized argon atom. The correct assignment of the peaks is not straightforward, especially for time-of-flight measurements, where small errors in the kinetic energy scale may occur due to the time to energy conversion. The number of the Ar^+ states, their binding energy and the relative intensities of the Auger transitions from the $Ar2p_{1/2}^{-1}3d$ state are known (Mursu et al. 1996; Langer et al. 1997; de Gouw et al. 1995; Chen 1993). The angular distribution has also been measured for many transitions (Farhat et al. 1997; Langer et al. 1997; Chen 1993) whereas Turri et al. (2007) have been first in measuring the transferred as well as the dynamic spin polarization. From the values of Mursu et al. (1996); Farhat et al. (1997); Langer et al. (1997), Turri et al. (2007) have been able to identify the components of all the peaks. Their results are shown in Table 3.9.

Very weak transitions have been neglected and only the final states that bring significant contribution to the decay of the $2p_{1/2}^{-1}3d$ excited state are reported in column 2 of Table 3.9. Labelling of the peaks is consistently with Lohmann et al. (2005), starting from the most strongly bound one around 33.5 eV binding energy. The identification of line 2c of the Ar*$(2p_{1/2}^{-1}3d_{3/2})_{J=1}$ Auger spectrum as $^2G_{9/2}$ is not straightforward, as experimentally the fine structure splitting of the ^{2}G doublet of ~3 meV is hard to resolve. Thus, it cannot be stated whether the $J_f = 7/2$ or $9/2$ state has been observed. On the other hand, the numerical 36 CSF-CI RDWA approach has not been able to generate the $J_f = 9/2$ final state due to internal selection rules which are suppressing the emission of an $\varepsilon f_{7/2}$ partial wave for the considered transition (see Lohmann et al. 2005, and the discussion above). These two findings allows for an interesting hypothesis. Taking the assignment by Mursu et al. (1996), the particular state under discussion would originate from a "three-body interaction". In the usual picture of multi-step two-body interaction, such a final state would require, first, the generation of the $^2G_{7/2}$ state by Auger decay and, subsequently, a shake-up of the Rydberg $3d_{3/2}$ into the $3d_{5/2}$ electron. Such an explanation is however unlikely to happen and overexceeds the interpretation of the Auger decay as caused by two-body interaction. Eventually, this leaves us with the possibility of at least qualitatively explaining line 2c by an interaction between the two electrons involved in the Auger decay and the shaking Rydberg electron, resulting in a simultaneous three-electron recombination in the final state. However, neither the resolution of the experiment by Turri et al. (2007) nor the line intensity is good enough to prove the line designation of the former experiment. Hence, better resolved measurements and calculations with more extended basis sets are necessary to corroborate this interesting aspect.

For a more quantitative comparison, Turri et al. (2007) used a least-squares fit method, where Gaussian functions were used for the manifold components. The relative energy and the relative intensity of the components of the same manifold have been forced to be equal to the values by Mursu et al. (1996) whereas the fitting procedure has been allowed to adjust the position of the different manifolds. The width of the Gaussian functions was estimated from the few peaks consisting of one or almost one component only, and it was fixed in the fitting procedure. Finally, it has been further assumed that all the components of the same manifold have similar anisotropy parameters and simply each manifold's overall intensity has been rescaled

Table 3.9 Peaks assignment and their measured spin polarization

Peak	Final state	Binding energy [eV]	$Ar^*2p_{1/2}^{-1}3d$ TSP	$Ar^*2p_{1/2}^{-1}3d$ DSP	$Ar^*2p_{1/2}^{-1}4s$ TSP	$Ar^*2p_{1/2}^{-1}4s$ DSP
1	$3p^4(^3p)3d\,^4F_{3/2,5/2}$	33.50	−0.14 (0.2)	0.2 (0.3)	−0.05 (0.06)	0.03 (0.03)
2a–b	$3p^4(^3p)3d\,^4P$	34.05	0.25 (0.15)	−0.1 (0.2)	−0.36 (0.06)	0.03 (0.03)
2c	$3p^4(^1D)3d\,^2G_{7/2,9/2}$	34.88	0.13 (0.25)	0.7 (0.4)	–	–
3	$3p^4(^1D)3d\,^2F_{5/2}$	36.00	0.33 (0.2)	−0.15 (0.2)	–	–
4	$3p^4(^1S)4s\,^2S_{1/2}$	36.50	–	–	0.80 (0.2)	−0.06 (0.07)
5	$3p^4(^1D)3d\,^2D_{5/2}$	37.13	0.1 (0.05)	0.06 (0.09)	−0.36 (0.2)	−0.35 (0.1)
	$3p^4(^1D)3d\,^2D_{3/2}$	37.19				
6	$3p^4(^1D)3d\,^2P_{3/2}$	37.38				
	$3p^4(^1D)3d\,^2P_{1/2}$	37.44				
7	$3p^4(^1S)3d\,^2D_{5/2}$	38.03	0.3 (0.1)	−0.03 (0.2)	–	–
	$3p^4(^1S)3d\,^2D_{3/2}$	38.07				
8	$3p^4(^3p)5s\,^2P_{1/2}$	38.46	0.01 (0.05)	0.11 (0.09)	0.11 (0.2)	0.19 (0.13)
	$3p^4(^3p)4d\,^4D_{5/2}$	38.55				
	$3p^4(^3p)4d\,^4D_{3/2}$	38.57				
9	$3p^4(^1D)3d\,^2S_{1/2}$	38.59				
	$3p^4(^3p)4d\,^4F_{5/2}$	38.83				
	$3p^4(^3p)4d\,^4F_{3/2}$	38.86				
	$3p^4(^3p)4d\,^4P_{3/2}$	38.88				
10	–	39.24	−0.18 (0.15)	0.05 (0.2)	–	–
	$3p^4(^3p)4d\,^2P_{1/2,3/2}$	39.31				
	$3p^4(^3p)4d\,^2D_{5/2}$	39.63				
	$3p^4(^3p)4d\,^2D_{3/2}$	39.65				
11	$3p^4(^1D)5s\,^2D_{5/2}$	40.04	−0.3 (0.2)	0.45 (0.2)	0.45 (0.4)	0.56 (0.3)
	$3p^4(^3p)4f\ J_f = 3/2$	40.07				
12	$3p^4(^3p)6s\,^4P_{5/2}$	40.41	0.18 (0.07)	0.05 (0.1)	–	–
	$3p^4(^1D)4d\,^2D_{5/2}$	40.52				
	$3p^4(^1D)4d\,^2F_{5/2}$	40.59				
	$3p^4(^3p)6s\,^2P_{1/2}$	40.63				
	$3p^4(^3p)5d\,^4P_{3/2}$	40.72				
	$3p^4(^3p)5d\,^4P_{5/2}$	40.78				
13	$3p^4(^3p)5d\,^2D_{3/2}$	41.10	0.21 (0.2)	0.4 (0.2)	–	–
	$3p^4(^3p)5d\,^2P_{1/2}$	41.12				
14	$3p^4(^3p)6d\,^4P_{1/2}$	41.61	−0.12 (0.2)	0 (0.02)	–	–

TSP Transferred spin polarization. *DSP* Dynamic spin polarization. Peak numbers as assigned in Fig. 3.10; after Turri et al. (2007)

to fit the experimental data. The last approximation, though not strictly correct (Farhat et al. 1997), does not seem to introduce too much error. The results are the dashed and continuous curves in Fig. 3.10.

Figure 3.11 depicts the spin-resolved spectra for the TSP (top) and DSP (bottom), respectively. Because of the complex structure of the manifolds, no efforts have been taken to separate their components in the spin-resolved experimental spectra nor to fit each manifold with an analytical function. Rather, the manifold areas have been taken for obtaining the spin polarization, assuming that the peak broadening due to

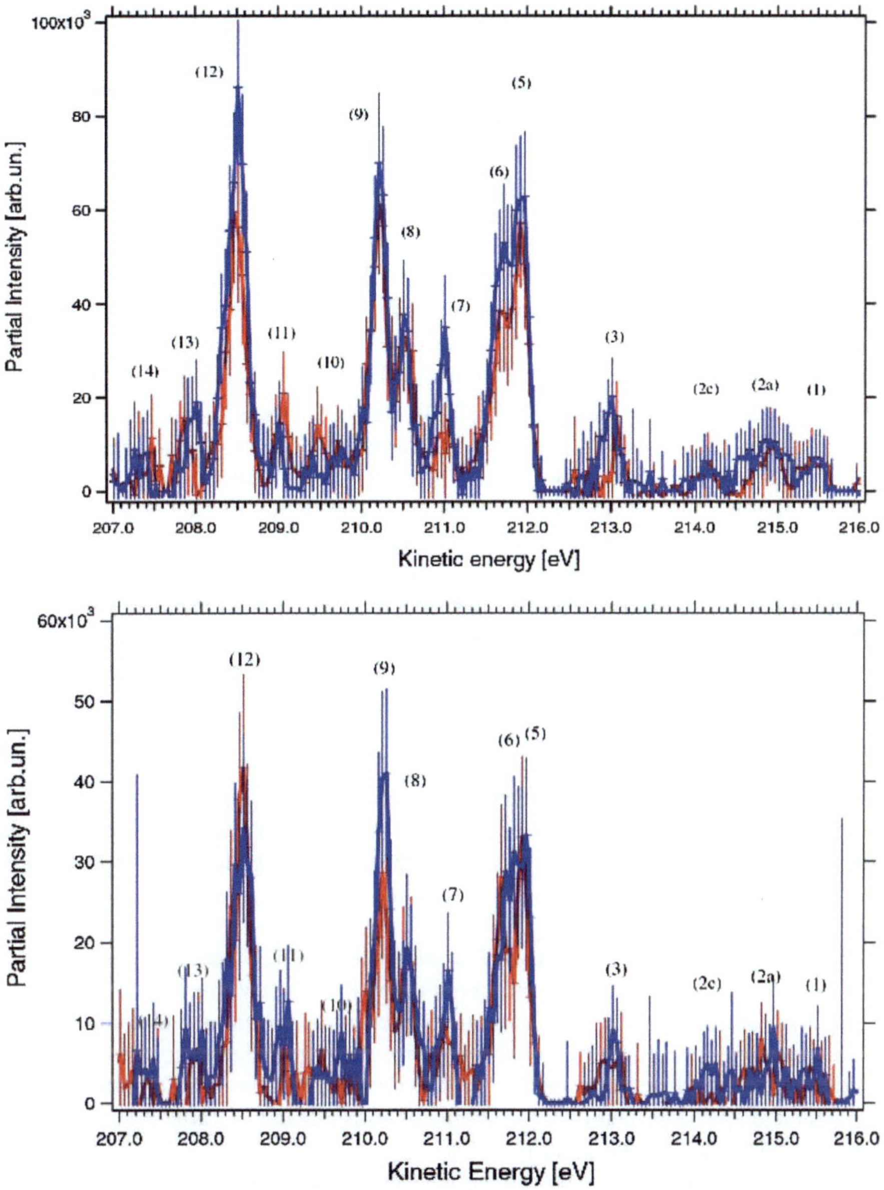

Fig. 3.11 The spin resolved spectrum of the $Ar^*2p^{-1}_{1/2}3d_{3/2_{J=1}}$ Auger decay. Peak numbers as assigned in Table 3.9. Experimental results measured with (*top*) circularly and (*bottom*) linearly polarized light. (*Blue*) spin parallel, (*red*) spin anti-parallel to the direction of photon propagation (after Turri et al. 2007)

finite instrumental resolution does not significantly contribute to the overall area of the manifold. The results are reported in columns 4 and 5 of Table 3.9. For the sign of the polarization, the same notation as in Lohmann et al. (2005) have been adopted, where a positive spin polarization indicates that the electron is emitted preferentially with the spin parallel, rather than anti-parallel, to the photon propagation direction. The errors in Table 3.9 account for both the statistical error (evaluated from the manifold areas) and the uncertainty in the Sherman function.

The analysis for the collected spectra of the Auger decay of the $Ar^*2p_{1/2}^{-1}4s$ state has been performed and the results are shown in columns 6 and 7 of Table 3.9. The latter show reasonable agreement with previous measurements for peaks 1–4 (Lohmann et al. 2005), they have larger errors due to a lower statistic of the measurements. In general, the TSP is stronger for the decay of the $2p_{1/2}^{-1}4s$ state than for the $2p_{1/2}^{-1}3d$ state. In contrast, the $2p_{1/2}^{-1}3d$ state shows significant amount of the DSP for peaks 2c, 11 and 13. The measured large DSP of ~70 % for line 2c leaves us with more ambiguities. Unfortunately, the 36 CSF-CI RDWA is able to generate the line, though, with a vanishing DSP, which might be interpreted as in favour of the discussed three-body interaction hypothesis (see above). Peak 12 in Fig. 3.11 (bottom) suggests that its components have strong DSP, though the total polarization vanishes when the manifold is not resolved. Also, peaks 11 and the unresolved peaks 8–9 show similar values of the DSP in the $2p_{1/2}^{-1}4s$ and $2p_{1/2}^{-1}3d$ decay.

The numerical data obtained for the TSP and DSP from the related spin-up and spin-down intensities for the experimental angles $\theta_{exp} = 90°$ and $\phi_{exp} = 135°$ are shown in Table 3.10 together with the unresolved group intensities forming the peaks 1–7. Note, that the $2p_{1/2}^{-1}3d \rightarrow (3p^2[^1S]4s)\,^2S_{1/2}$ transition, i.e. line 30 (peak 4) should have a vanishing intensity due to its single channel character, while it is prominent for the $2p_{1/2}^{-1}4s \rightarrow (3p^2[^1S]4s)\,^2S_{1/2}$ Auger decay (Lohmann et al. 2005). This is corroborated by the numerical results yielding almost no intensity for peak 4, and coincides with the experimental data (see Figs. 3.10 and 3.11) showing no evidence for this line.

The results of the 36 CSF-CI calculations for the $2p_{1/2}^{-1}3d$ state are compared to the TSP and DSP which has been measured in the 33.5–37 eV binding energy range in Fig. 3.12, and in Table 3.11. The relative intensities, positions and widths of the peaks, as obtained by the fitting of the spin-unresolved spectrum using the procedure described above and depicted in Fig. 3.10, were used. These were combined with the calculated spin polarization reported in columns 6 and 8 of Table 3.11, resulting in the full bold lines depicted in Fig. 3.12.

The calculations correctly reproduce the DSP, with the only exception of peak 2c. In contrast, they strongly overestimate the TSP of peak 5, and they find the wrong sign of the polarization of peaks 3 and 7. The remaining agreement between calculation and theory would suggest that for the $2p_{1/2}^{-1}3d$ state it is the TSP, rather than the DSP as has been shown for the $2p_{1/2}^{-1}4s$ state (Lohmann et al. 2005), which is more sensitive to the calculational details. To test this hypothesis, an extended approach, including a larger basis set than the 36 CSF-CI, should be tempted. Such an extended-basis calculation is also expected to give more accurate results for the

Table 3.10 Our tentative *LSJ* coupling assignment of the observed and calculated peaks (36 CSF-CI RDWA)

$Ar^*(3d_{3/2})\, L_2M_{2,3}M_{2,3}$					
Final states	no.	peak	int.	pol. %	
(a)	(b)	(c)	I_0†	p_y	p_x
$(3P^2[^3P]\,4s)\,^4P_{1/2,...,5/2}$	1–3	1a	6.48	0.24	−8.11
$(3P^2[^3P]\,3d)\,^4D_{1/2,...,7/2}$	4–7				
$(3P^2[^3P]\,4s)\,^2P_{1/2,3/2}$	8–9	1b	3.90	−1.24	46.03
$(3P^2[^3P]\,3d)\,^4F_{3/2,...,9/2}$	10–13	1c	11.29	0.21	21.26
$(3P^2[^3P]\,3d)\,^2P_{1/2,3/2}$	14–15	2a	37.72	0.60	−16.71
$(3P^2[^3P]\,3d)\,^4P_{1/2,...,5/2}$	16–18				
$(3P^2[^1D]\,4s)\,^2D_{3/2,5/2}$	19–20				
$(3P^2[^3P]\,3d)\,^2F_{5/2,7/2}$	21–22				
$(3P^2[^3P]\,3d)\,^2D_{3/2,5/2}$	23–24	2b	0.69	−13.79	−15.93
$(3P^2[^1D]\,3d)\,^2G_{7/2,9/2}$	25–26	2c	1.56	0.00	60.01
$(3P^2[^1D]\,3d)\,^2F_{5/2,7/2}$	27–28	3	6.68	−14.83	−21.25
$(3P^2[^1D]\,3d)\,^2S_{1/2}$	29				
$(3P^2[^1S]\,4s)\,^2S_{1/2}$	30	4	0.02	0.00	79.95
$(3P^2[^1D]\,3d)\,^2D_{3/2,5/2}$	31–32	5	11.10	9.56	85.95
$(3P^2[^1D]\,3d)\,^2P_{1/2,3/2}$	33–34	6	8.52	−0.16	77.56
$(3P^2[^1S]\,3d)\,^2D_{3/2,5/2}$	35–36	7	12.05	0.80	−10.36

The calculated relative peak intensities are given together with the degree of dynamic, $P_y(\theta_{exp}, \phi_{exp})$, and transferred, $P_x(\theta_{exp})$, spin polarization, respectively. Note, that transitions to fine structure terms with $J > 7/2$ are suppressed due to J-dependent selection rules. †: The total intensity has been normalized to 100. *a* Our tentative assignment of the unresolved *LSJ* fine structure terms. *b* Line numbers according to our numerical 36 CSF *jj* coupled approach. *c* Peak numbers as assigned in the calculated spectrum

relative line intensities, which are not well reproduced by the 36 CSF-CI calculation, in particular the large intensities of peaks 5–7 as compared to peaks 1–3. An attempt into that direction has been recently performed by Fritzsche et al. (2007).

3.5 Generation and Detection of Inner Shell Holes

The generation and decay of inner shell holes is best described in the scheme of density matrix theory (e.g. Balashov et al. 2000; Blum 1996). A variety of theoretical investigations have been performed, and we are found to glance a few of them, which already sounds to be an almost impossible task. In the following, we will give a short compendium of the current state of the art research.

3.5.1 Polarization States of Photons

The polarization of an exciting or ionizing photon beam is usually described in terms of *Stokes parameters* (see Born and Wolf 1970). There notation is not unique in the

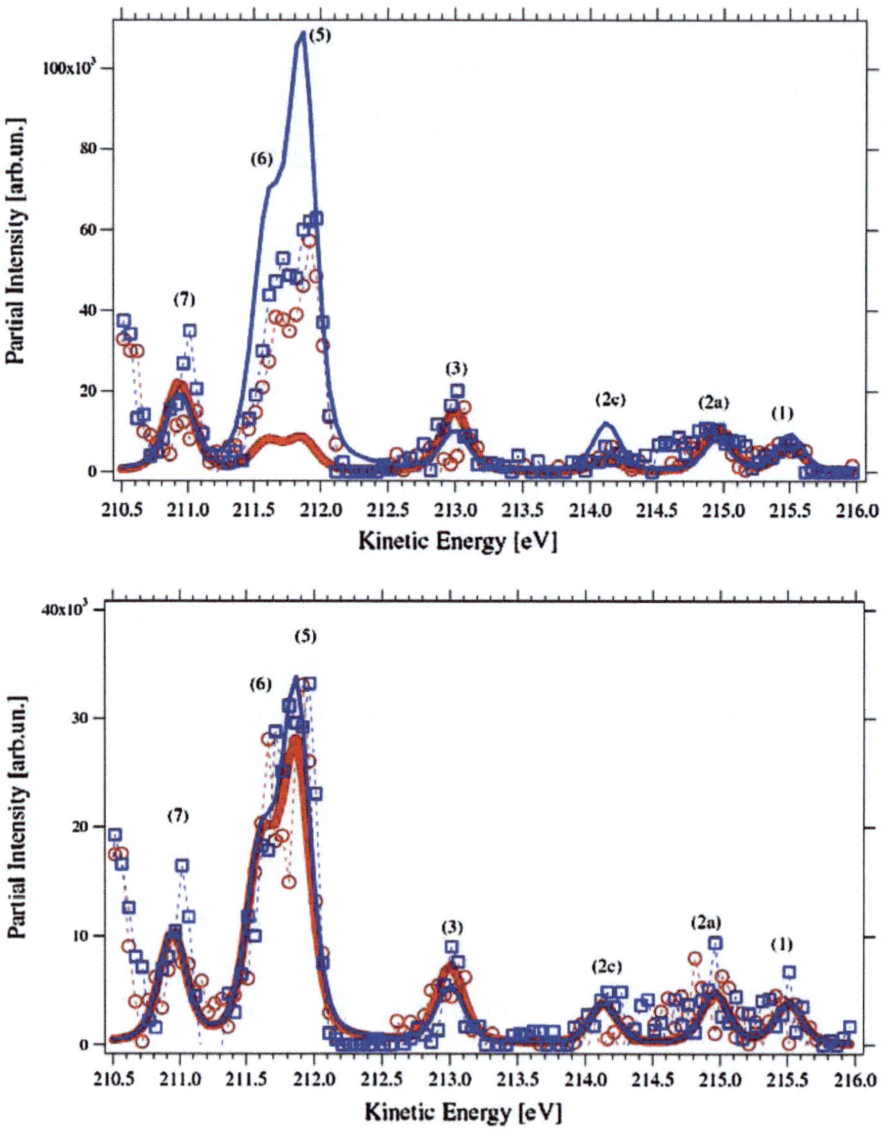

Fig. 3.12 (*top*) TSP and (*bottom*) DSP of the $Ar^{*}2p_{1/2}^{-1}3d_{3/2}{}_{J=1}$ Auger decay. Peak numbers as assigned in Tables 3.9 and 3.10. Experimental data: (□) spin parallel; (○) spin anti-parallel to the direction of photon propagation. Numerical data: (*blue*) spin parallel, (*red*) spin anti-parallel. (*dashed lines*) to guide the eye, only (after Turri et al. 2007)

literature, though the two sets of parameters $\eta_i, i = 1, 2, 3$ and $P_i, i = 1, 2, 3$ have been used most frequently. Following Blum (1996), and supposing a coordinate frame with the z-axis parallel to the photon beam, the circular polarization of light can be described as

Table 3.11 Measured (Exp.) and calculated (Theo.) transferred (TSP) and dynamic (DSP) spin polarization for the $Ar^{*}2p_{1/2}^{-1}3d$ Auger decay; after Turri et al. (2007)

Peak	Final state	State number[a]	Binding energy [eV]	TSP		DSP	
				Exp.	Theo.	Exp.	Theo.
1	$3p^4(^3P)3d\,^4F_{3/2,5/2}$	10–13	33.50	−0.14 (0.2)	0.21	0.2 (0.3)	~0
2a–b	$3p^4(^3p)3d\,^4P$	14–24	34.05	0.25 (0.15)	−0.17	−0.1 (0.2)	~0
2c	$3p^4(^1D)3d\,^2G_{7/2,9/2}$	25–26	34.88	0.13 (0.25)	0.60	0.7 (0.4)	0
3	$3p^4(^1D)3d\,^2F_{5/2}$	27–29	36.00	0.33 (0.2)	−0.21	−0.15 (0.2)	−0.15
4	$3p^4(^1S)4s\,^2S_{1/2}$	30	36.50	–	0.80[b]	–	0
5	$3p^4(^1D)3d\,^2D_{5/2}$	31–32	37.13	0.1 (0.05)	0.86[c]	0.06 (0.09)	0.10[c]
	$3p^4(^1D)3d\,^2D_{3/2}$		37.19				
6	$3p^4(^1D)3d\,^2P_{3/2}$	33–34	37.38		0.78[d]		~0[d]
	$3p^4(^1D)3d\,^2P_{1/2}$		37.44				
7	$3p^4(^1S)3d\,^2D_{5/2}$	35–36	38.03	0.3 (0.1)	−0.10	−0.03 (0.2)	0.01
	$3p^4(^1S)3d\,^2D_{3/2}$	35–36	38.07				

For further explanations see text
[a]The state numbers refer to the numerical 36 jj-coupled CSF-CI RDWA calculation.
[b]Calculations state a high TSP but a vanishing intensity for this line, which in fact could not be measured experimentally
[c] Data refer to Peak 5 of Fig. 3.12
[d] Data refer to Peak 6 of Fig. 3.12

$$\eta_2 = P_3 = \frac{I_+ - I_-}{I_{tot}}, \tag{3.69}$$

where $I_\pm$ denotes the intensity of right or left handed circularly polarized light, respectively. I_{tot} denotes the total photon beam intensity.

Describing the degree of linear polarization usually requires a so-called *reaction plane* which settles the x-axis in this plane. With this, we are able to express the linear photon polarization via the two parameters

$$\eta_3 = P_1 = \frac{I(0°) - I(90°)}{I_{tot}}, \tag{3.70}$$

and

$$\eta_1 = P_2 = \frac{I(45°) - I(135°)}{I_{tot}}. \tag{3.71}$$

For our choice of coordinate frame $I(\alpha)$ denotes the intensity transmitted by a Nicol prism oriented at an angle α with respect to the x-axis.

The Stokes parameters can be related to the density matrix of photon polarization. In particular, assuming an arbitrarily polarized photon beam and choosing the quantization axis along the incoming beam direction, we write

$$\hat{\rho}(\gamma_{Syn}) = \frac{I}{2}\begin{pmatrix} 1+\eta_2 & 0 & -\eta_3+i\eta_1 \\ 0 & 0 & 0 \\ -\eta_3-i\eta_1 & 0 & 1-\eta_2 \end{pmatrix}, \tag{3.72}$$

where the notation of Kleiman (2001) has been adopted. Due to the transverse character of light, the photons cannot populate the magnetic substate parallel to their direction of propagation, which results in the zeros in (3.72).

Alternatively, for describing the photon polarization, we may expand the density matrix into an irreducible tensorial set, so-called state multipoles $\mathcal{T}^{+}_{\Gamma\gamma}$. We write

$$\hat{\rho}(\gamma_{Syn}) = \sum_{\Gamma\gamma}\left\langle \mathcal{T}^{+}_{\Gamma\gamma}\right\rangle \mathcal{T}_{\Gamma\gamma} \tag{3.73}$$

for the photonic polarization state. Applying (3.72) and (3.73) yields general relations between both representations shown in Table 3.12. In particular, we note that for any arbitrarily polarized photon beam only the tensors $\langle\mathcal{T}^{+}_{\Gamma 0}\rangle$, with $\Gamma \leq 2$ and $\langle\mathcal{T}^{+}_{2\pm 2}\rangle$ can be non-zero. However, this depends on the choice of quantization axis. As can be seen from Table 3.12 the photon alignment $\langle\mathcal{T}^{+}_{20}\rangle$ is directly proportional to the monopole. From a physical point of view, this demonstrates the fact that a photon beam is always aligned due to the transverse character of the electromagnetic field.

3.5.2 Photoexcitation State Multipoles

The case of photoexcitation has been already discussed in detail in Sect. 3.4 reviewing the Auger decay of resonantly excited argon via synchrotron radiation. Here, we briefly consider the primary excitation process

Table 3.12 The state multipoles of an arbitrarily polarized photon beam and their connection to the Stokes parameters

State multipoles Stokes parameters	Stokes parameters State multipoles
$\left\langle \mathcal{T}^{+}_{00}\right\rangle = \frac{I}{\sqrt{3}}$	$I = \sqrt{3}\left\langle \mathcal{T}^{+}_{00}\right\rangle$
$\left\langle \mathcal{T}^{+}_{10}\right\rangle = \frac{I}{\sqrt{2}}\eta_2$	$I\eta_2 = \sqrt{2}\left\langle \mathcal{T}^{+}_{10}\right\rangle$
$\left\langle \mathcal{T}^{+}_{1\pm 1}\right\rangle = 0$	
$\left\langle \mathcal{T}^{+}_{20}\right\rangle = \frac{1}{\sqrt{2}}\left\langle \mathcal{T}^{+}_{00}\right\rangle = \frac{I}{\sqrt{6}}$	$I = \sqrt{6}\left\langle \mathcal{T}^{+}_{20}\right\rangle$
$\left\langle \mathcal{T}^{+}_{2\pm 1}\right\rangle = 0$	
$\left\langle \mathcal{T}^{+}_{22}\right\rangle = \frac{I}{2}\left(-\eta_3+i\eta_1\right)$	$I\eta_3 = -2Re\left\langle \mathcal{T}^{+}_{22}\right\rangle$
$\left\langle \mathcal{T}^{+}_{2-2}\right\rangle = \left\langle \mathcal{T}^{+}_{22}\right\rangle^{*} = \frac{I}{2}\left(-\eta_3-i\eta_1\right)$	$I\eta_1 = 2Im\left\langle \mathcal{T}^{+}_{22}\right\rangle$

The photon beam axis has been chosen as quantization axis. Multipoles of rank $K > 2$ must be zero due to dipole selection rules; after Lohmann (2008)

$$\gamma + A(J_0) \longrightarrow A^*(J)\,. \tag{3.74}$$

Such processes have been extensively investigated with respect to angle and spin resolved Auger emission, first by Kronast et al. (1986), and in detail in the book by Lohmann (2008). Using the photon beam axis as quantization axis, applying the usual methods of state multipoles (e.g. see Blum 1996) and tensor algebra (e.g. see Brink and Satchler 1962), and using the results of Sect. 3.5.1, we obtain a relation between the state multipoles $\mathcal{A}_{KQ}$ describing the excited atomic state and the degree of polarization $\langle \mathcal{T}^+_{\Gamma\gamma} \rangle$ of the exciting photons,

$$\mathcal{A}_{KQ} = \delta_{K,\Gamma}\,\delta_{Q,\gamma}\,3\sqrt{2J+1}\,(-1)^{K+J_0+J+1} \begin{Bmatrix} 1 & 1 & K \\ J & J & J_0 \end{Bmatrix} \left\langle \mathcal{T}^+_{\Gamma\gamma} \right\rangle. \tag{3.75}$$

In particular, $\mathcal{A}_{20}$ and $\mathcal{A}_{10}$ refer to the parameters of alignment and orientation, respectively. Due to the dipole approximation and the application of selection rules, in (3.75), the parameters $\mathcal{A}_{KQ}$ are independent of the excitation dipole matrix elements and become a simple function of geometric factors, only. Moreover, (3.75) implies that the alignment tensor $\mathcal{A}_{20}$ and its component $\mathcal{A}_{22}$ are not independent but are related to each other by a common factor,

$$\mathcal{A}_{2\pm2} = -\sqrt{\frac{3}{2}}\,\mathcal{A}_{20}\,(\eta_3 \mp \mathrm{i}\eta_1)\,. \tag{3.76}$$

For the case of excitation from a $J_0 = 0$ state, e.g. for photoexcitation of the rare gases which requires an excitation into $J = 1$ states, 3.75 can be further reduced which simply yields

$$\mathcal{A}_{KQ}(J_0 = 0 \to J = 1) = \delta_{K,\Gamma}\,\delta_{Q,\gamma}\,\delta_{J,1}\,\sqrt{3}\left\langle \mathcal{T}^+_{\Gamma\gamma} \right\rangle. \tag{3.77}$$

Table 3.13 shows particular results obtained for the alignment and orientation parameters for specific cases of interest. For example, the data of the second row are related to the rare gases or the earth alkalis, whereas rows 2–3 yield alignment and orientation data for photoexcitation from the $^2S_{1/2}$ ground state of the alkalis, while rows 4–5 apply to photoexcitation of the $^2P_{3/2}$ ground state of the halogenes. The rare gas data have been used in the analysis of the resonantly excited $Ar^*(2p^{-1}_{1/2}3d + 4s)_{J=1}$ Auger decay in Sect. 3.4.

3.5.3 State Multipoles of Photoionization

A more complicated case in inner shell hole production is by photoionization. Instead of exciting an inner shell electron into a Rydberg state, the electron can be knocked out by a photon beam leaving a singly ionized target behind,

Table 3.13 Alignment and orientation parameters $\mathcal{A}_{KQ}$ and their relation to the Stokes parameters for photoexcitation from various ground states J_0 to the possible final states J

$J_0 \to J$	$\mathcal{A}_{10}$	$\mathcal{A}_{1\pm1}$	$\mathcal{A}_{20}$	$\mathcal{A}_{2\pm1}$	$\mathcal{A}_{2\pm2}$
$0 \to 1$	$\sqrt{\frac{3}{2}}\eta_2$	0	$\frac{1}{\sqrt{2}}$	0	$\frac{-\sqrt{3}}{2}(\eta_3 \mp i\eta_1)$
$1/2 \to 1/2$	η_2	0	0	0	0
$1/2 \to 3/2$	$\frac{\sqrt{5}}{2}\eta_2$	0	$\frac{1}{2}$	0	$\frac{-\sqrt{3}}{2\sqrt{2}}(\eta_3 \mp i\eta_1)$
$3/2 \to 1/2$	$\frac{-1}{2}\eta_2$	0	0	0	0
$3/2 \to 3/2$	$\frac{1}{\sqrt{5}}\eta_2$	0	$\frac{-2}{5}$	0	$\frac{\sqrt{6}}{5}(\eta_3 \mp i\eta_1)$

The photon beam axis has been chosen as quantization axis

$$\gamma + A(J_0) \longrightarrow A^+(J) + e. \tag{3.78}$$

Basic investigations concerning angular distribution and spin polarization of the emitted photoelectrons have been invented by Klar and Kleinpoppen (1982) and many others.[5] With respect to the subsequent Auger emission, here, the alignment and orientation are no though the general relations remain similar. The showcase of argon Auger emission and related data has been already discussed in Sect. 3.4.

Intensive investigations concerning the generation of inner shell holes via photoionization and their subsequent Auger decay have been performed by Kleiman et al. (1999a,b). Reducing the general equations by inserting the related nj-symbols and performing the possible summations, we are introducing the photoionization parameter $B_{Phot}(K)$ as[6]

$$\begin{aligned} B_{Phot}(K) = \delta_{K,\Gamma}\,\delta_{Q,\gamma}\,\frac{1}{2J_0+1} & \sum_{\ell j J_1 J_1'} \langle (Jj)J_1 \|\mathbf{d}\| J_0\rangle \langle (Jj)J_1' \|\mathbf{d}\| J_0\rangle^* \\ & \times (-1)^{J_0+j-J+J_1-J_1'} \sqrt{(2J_1+1)(2J_1'+1)} \\ & \times \begin{Bmatrix} 1 & 1 & K \\ J_1' & J_1 & J_0 \end{Bmatrix} \begin{Bmatrix} J_1' & J_1 & K \\ J & J & j \end{Bmatrix}, \end{aligned} \tag{3.79}$$

where J_1 denotes the total angular momentum of the combined final ion–photo electron system, and $\mathbf{d}$ abbreviates the dipole operator. Using (3.79), the orientation can expressed in terms of the photoionization parameters as

$$\mathcal{A}_{10} = \sqrt{\frac{3}{2}}\,\frac{B_{Phot}(1)}{B_{Phot}(0)}\,\eta_2\,, \tag{3.80}$$

[5] See the discussion in Sects. 2.3.3, 4.2.1, and 4.2.2.

[6] The definition of $B_{Phot}(K)$ differs by a phase factor to that of Lohmann (2008); it coincides with Kleiman et al. (1999a) or Lohmann and Kleiman (2006).

while the alignment parameter is obtained as

$$\mathcal{A}_{20} = \frac{1}{\sqrt{2}} \frac{B_{Phot}(2)}{B_{Phot}(0)} . \tag{3.81}$$

As has been outlined in Sect.3.5.2, see Table 3.13, the alignment component $\mathcal{A}_{22}$ can be directly related to the alignment tensor $\mathcal{A}_{20}$ via (3.76).

Inserting $K = 0$ into (3.79), the normalization factor can be reduced to

$$B_{Phot}(0) = \delta_{J_1,J_1'} \frac{1}{2J_0+1} \frac{1}{\sqrt{3(2J+1)}} \sum_{\ell j J_1} \left|\langle (Jj)J_1 \|\mathbf{d}\| J_0\rangle\right|^2 . \tag{3.82}$$

Inserting (3.79) and (3.82) into (3.80) and (3.81), the orientation and alignment tensors can be combined and expressed as

$$\begin{aligned}
\mathcal{A}_{K0} = & \left(\sum_{\ell j J_1} \left|\langle (Jj)J_1 \|\mathbf{d}\| J_0\rangle\right|^2\right)^{-1} \\
& \times \kappa_K \sqrt{\frac{3(2J+1)}{2}} \sum_{\ell j J_1 J_1'} \langle (Jj)J_1 \|\mathbf{d}\| J_0\rangle \langle (Jj)J_1' \|\mathbf{d}\| J_0\rangle^* \\
& \times (-1)^{J_0+j-J+J_1-J_1'} \sqrt{(2J_1+1)(2J_1'+1)} \\
& \times \begin{Bmatrix} 1 & 1 & K \\ J_1' & J_1 & J_0 \end{Bmatrix} \begin{Bmatrix} J_1' & J_1 & K \\ J & J & j \end{Bmatrix} ,
\end{aligned} \tag{3.83}$$

where

$$\kappa_K = \begin{cases} \sqrt{3}\eta_2 & \text{if } K = 1, \\ 1 & \text{if } K = 0, 2. \end{cases} \tag{3.84}$$

Considering a closed shell electronic structure, i.e. $J_0 = 0$, which occurs for the rare gases or the earth alkalis, (3.83) can be further reduced since the selection rule $J_1 = J_1' = 1$ applies which yields

$$\begin{aligned}
\mathcal{A}_{K0}(J_0 = 0 \rightarrow J_1 = 1) = & \left(\sum_{\ell j} \left|\langle (Jj)1 \|\mathbf{d}\| 0\rangle\right|^2\right)^{-1} \\
& \times \kappa_K \sqrt{\frac{3(2J+1)}{2}} \sum_{\ell j} (-1)^{K+j-J} \begin{Bmatrix} 1 & 1 & K \\ J & J & j \end{Bmatrix} \left|\langle (Jj)1 \|\mathbf{d}\| 0\rangle\right|^2 .
\end{aligned} \tag{3.85}$$

Inserting explicit numbers for the geometrical factors in (3.85), upper and lower bounds for the alignment and orientation parameters of the $J_0 = 0 \rightarrow J_1 = 1$ photoionization can be derived (Kleiman et al. 1999a), which are shown in Table 3.14.

Table 3.14 Upper and lower bounds for the alignment and orientation parameters $\mathcal{A}_{20}$ and $\mathcal{A}_{10}$ for selected total angular momenta J of the final ionic state

	$\mathcal{A}_{10}$		$\mathcal{A}_{20}$	
J	lb	ub	lb	ub
1/2	−0.500	1.000	–	–
3/2	−0.671	1.118	−0.400	0.500
5/2	−0.732	1.025	−0.428	0.374
7/2	−0.764	0.982	−0.436	0.327
9/2	−0.783	0.957	−0.440	0.303
11/2	−0.797	0.941	−0.442	0.288

For the orientation a 100 % circularly polarized photon beam has been assumed ($\eta_2 = 1$). *lb* lower bound; *ub* upper bound; after Kleiman et al. (1999a)

Equation (3.83) can be analyzed in more detail within the LS coupling scheme. Applying the well-known jj–LS recoupling formalism (e.g. see Rotenberg et al. 1959) the reduced dipole matrix elements can be expressed as

$$\begin{aligned}\langle (Jj)J_1\|\mathbf{d}\|J_0\rangle &= \langle ([LS]J,[\ell\ 1/2]j)J_1\|\mathbf{d}\|(L_0S_0)J_0\rangle \\ &= \delta_{S_0S_1}\sqrt{(2J+1)(2j+1)(2J_1+1)(2S_0+1)(2J_0+1)} \\ &\quad\times \sum_{L_1}(-1)^{S_0+J_0+L_1+1}\sqrt{2L_1+1}\,\langle (L\ell)L_1\|\mathbf{d}\|L_0\rangle \\ &\quad\times \begin{Bmatrix} J_1 & J_0 & 1 \\ L_0 & L_1 & S_0 \end{Bmatrix}\begin{Bmatrix} L & \ell & L_1 \\ S & 1/2 & S_0 \\ J & j & J_1 \end{Bmatrix}. \end{aligned} \tag{3.86}$$

Due to the fact, that the dipole operator acts in the Hilbert subspace of the angular momenta, only, the important dipole selection rule $S_0 = S_1$ applies. Besides the general selection rule for the total angular momenta J_0 and J_1 for electric (E1) dipole transitions, the LS coupling scheme yields an additional selection rule for the angular momenta

$$|L_0 - 1| \le L_1 \le L_0 + 1. \tag{3.87}$$

Hence, the angular momentum L_1 of the combined photoelectron–ion final state system is restricted to a maximum of three values. Considering photoionization from an atomic $^1S_0$ ground state, (3.87) allows for a value of $L_1 = 1$, only. Thus, the sum over L_1 in (3.86) collapses into a single term leaving a direct proportionality between the jj- and LS-coupled dipole matrix elements

$$\begin{aligned}&\langle ([LS=1/2]J,[\ell\ 1/2]j)J_1 = 1\|\mathbf{d}\|(L_0=S_0=0)J_0=0\rangle \\ &= \delta_{S_0S_1}\,(-1)^{L+j+3/2}\langle (L\ell)L_1 = 1\|\mathbf{d}\|L_0=0\rangle \\ &\quad\times\sqrt{\frac{(2J+1)(2j+1)}{2}}\begin{Bmatrix} L & \ell & 1 \\ j & J & 1/2 \end{Bmatrix}. \end{aligned} \tag{3.88}$$

Employing (3.88), we are able to re-write (3.85) in the LS coupling approximation which, resulting in a final $^1\mathrm{P}_1$ state of the combined photoion–electron system, gives

$$\mathcal{A}_{K0}(^1\mathrm{S}_0 \to {}^1\mathrm{P}_1) = \left(\sum_{\ell j}(2j+1)\begin{Bmatrix} L & \ell & 1 \\ j & J & 1/2 \end{Bmatrix}^2 |\langle (L\ell)1\|\mathbf{d}\|0\rangle|^2\right)^{-1}$$
$$\times\, \kappa_K\sqrt{\frac{3(2J+1)}{2}}\sum_{\ell j}(-1)^{K+j-J}(2j+1)$$
$$\times \begin{Bmatrix} 1 & 1 & K \\ J & J & j \end{Bmatrix}\begin{Bmatrix} L & \ell & 1 \\ j & J & 1/2 \end{Bmatrix}^2 |\langle (L\ell)1\|\mathbf{d}\|0\rangle|^2 . \qquad (3.89)$$

In (3.89) the dipole matrix elements are independent of the partial wave total angular momenta. Performing the sum over j in the numerator and in the denominator, (3.89) can be further reduced

$$\mathcal{A}_{K0}(^1\mathrm{S}_0 \to {}^1\mathrm{P}_1) = \left(\sum_{\ell}|\langle (L\ell)1\|\mathbf{d}\|0\rangle|^2\right)^{-1}$$
$$\times\, \kappa_K(2L+1)\sqrt{\frac{3(2J+1)}{2}}\begin{Bmatrix} L & L & K \\ J & J & 1/2 \end{Bmatrix}$$
$$\times \sum_{\ell}(-1)^{\ell+1/2-J}\begin{Bmatrix} L & L & K \\ 1 & 1 & \ell \end{Bmatrix} |\langle (L\ell)1\|\mathbf{d}\|0\rangle|^2 . \qquad (3.90)$$

Due to the LS coupling, (3.90) yields additional selection rules which further influence the upper and lower bounds of the alignment and orientation parameters shown in Table 3.15. As can be seen, the LS coupling results in boundary limits more tight than the jj coupling case. Particularly, the lower bound of the alignmentparameter $\mathcal{A}_{20}$ is restricted to positive values in this approximation. Considering the fact that LS coupling is best applied to light atoms, a negative alignment can be expected preferably for heavy atoms. This has been confirmed in an experiment by Karvonen et al. (1999) measuring an alignment of $\mathcal{A}_{20} = -0.13$ close to the ionization threshold after photoionization of the xenon $3\mathrm{d}_{5/2}$ shell. As can be seen from Table 3.15, for large angular momenta L the upper and lower bounds converge to the same value for the related different total angular momenta $J = L \pm 1/2$.

The photoionization alignment and orientationparameters are energy dependent. Applying (3.90), we are able to plot the orientation $\mathcal{A}_{10}$ and the alignment $\mathcal{A}_{20}$ as a function of the energy. Photoionization data have been considered first by Berezhko and Kabachnik (1977) applying a rather crude model. Short after Berezhko et al. (1978a) published numerical alignment data applying a Hartree–Slater (HS) approximation. A state of the art survey through most of the periodic table applying a relaxed orbital method within a single-configurational Hartree–Fock approach (HF) utilizing the Cowan code (Cowan 1981) has been given by Kleiman and Lohmann (2003).

Table 3.15 Upper and lower bounds for the alignment and orientation parameters $\mathcal{A}_{20}$ and $\mathcal{A}_{10}$ in the LS coupling scheme for selected angular L and total angular momenta J of the final ionic state

		$\mathcal{A}_{10}$		$\mathcal{A}_{20}$	
L	J	lb	ub	lb	ub
0	1/2	–	–	–	–
1	1/2	−0.500	1.000	–	–
1	3/2	−0.559	1.118	0.050	0.500
2	3/2	−0.671	1.006	0.100	0.350
2	5/2	−0.683	1.025	0.107	0.374
3	5/2	−0.732	0.976	0.134	0.321
3	7/2	−0.736	0.982	0.136	0.327
4	7/2	−0.764	0.954	0.153	0.300
4	9/2	−0.766	0.957	0.154	0.303
5	9/2	−0.783	0.940	0.165	0.286
5	11/2	−0.784	0.941	0.166	0.288

For the orientation a 100 % circularly polarized photon beam has been assumed ($\eta_2 = 1$). *lb* lower bound; *ub* upper bound; after Kleiman et al. (1999a)

The Cowan code allows for the inclusion of the Darwin- and mass-velocity-term, both. Though, the semi-relativistic Hartree–Fock method (HFR) is not able to incorporate the spin-orbit interaction (Cowan 1981; Kleiman and Lohmann 2003). Within the HFR approach the dipole transition matrix elements $D_{\varepsilon\ell,nL}$ can be expressed in terms of the related radial dipole integrals $R_{\varepsilon\ell,nL}$ via

$$D_{\varepsilon\ell,nL} \equiv \langle (L\ell)1\|\mathbf{d}\|0\rangle = (-1)^{L+1}\sqrt{2\ell+1}\begin{pmatrix} L & \ell & 1 \\ 0 & 0 & 0 \end{pmatrix} R_{\varepsilon\ell,nL}, \tag{3.91}$$

whereas the radial dipole integrals can be written in their length-form as

$$R_{\varepsilon\ell,nL} = \int_0^\infty R_{\varepsilon\ell}(r) r^3 R_{nL}(r)\, dr = \int_0^\infty P_{\varepsilon\ell}(r) r P_{nL}(r)\, dr. \tag{3.92}$$

Recalling (3.80) and (3.81) and applying the LS coupling scheme, Kleiman (2002) found interrelations between orientation and alignment. Expressing the ratio of the modulus of the contributing dipole transition matrix elements in terms of the alignment parameters $\mathcal{A}_{20}$ interdependence to the orientation $\mathcal{A}_{10}$ can be obtained

$$\mathcal{A}_{10}(J = L + 1/2) = \eta_2\,(2L+3)\sqrt{\frac{5L}{3(L+2)}} \times \left(\mathcal{A}_{20}(J = L+1/2) - \sqrt{\frac{L(L+2)}{5(2L+1)(2L+3)}}\right), \tag{3.93a}$$

and

$$\mathcal{A}_{10}(J = L - 1/2) = \eta_2\,(2L - 1)\sqrt{\frac{5(L+1)}{3(L-1)}} \times \left(\mathcal{A}_{20}(J = L - 1/2) - \sqrt{\frac{(L-1)(L+1)}{5(2L-1)(2L+1)}}\right). \tag{3.93b}$$

Equation (3.93) yields an interesting feature. Realizing that the orientation parameter can be zero for a specific photoelectron energy, we find the alignment as an analytic function in this case, independent of the ionization dynamics.

$$\mathcal{A}_{10}(J = L + 1/2) = 0 \implies \mathcal{A}_{20}(J = L + 1/2) = \sqrt{\frac{L(L+2)}{5(2L+1)(2L+3)}}, \tag{3.94a}$$

and

$$\mathcal{A}_{10}(J = L - 1/2) = 0 \implies \mathcal{A}_{20}(J = L - 1/2) = \sqrt{\frac{(L-1)(L+1)}{5(2L-1)(2L+1)}}. \tag{3.94b}$$

Selected examples of interdependence are shown in Figs. 3.13 and 3.14 for the orientation and alignment parameters $\mathcal{A}_{10}(J = L + 1/2)$ and $\mathcal{A}_{20}(J = L + 1/2)$ of the singly photoionized atoms Kr, Sr, Pd, and Cd for the inner shell $4p_{3/2}$ hole states, and for the $4d_{5/2}$ inner shell hole states of Pd, Cd, Xe, and Ba. The photon beam is assumed as completely right-handed circularly polarized, i.e. $\eta_2 = 1$. The parameters $\mathcal{A}_{10}(J = L - 1/2)$ and $\mathcal{A}_{20}(J = L - 1/2)$ can be obtained from the figures by means of the proportionality relations (Berezhko et al. 1978a; Kabachnik and Lee 1989; Kleiman et al. 1999a),

$$\mathcal{A}_{10}(J' = L - 1/2) = \frac{L+1}{L}\sqrt{\frac{2L-1}{2L+3}}\,\mathcal{A}_{10}(J = L + 1/2), \tag{3.95a}$$

and

$$\mathcal{A}_{20}(J' = L - 1/2) = \sqrt{\frac{(L-1)(L+1)(2L+3)}{L(L+2)(2L-1)}}\,\mathcal{A}_{20}(J = L + 1/2). \tag{3.95b}$$

Discussing the orientation and alignment given in Figs. 3.13 and 3.14 is not ambiguous as the shape of the curves basically show only four characteristic features denoted as: *main transition, shape resonance, centrifugal barrier*, and *Cooper minimum*.

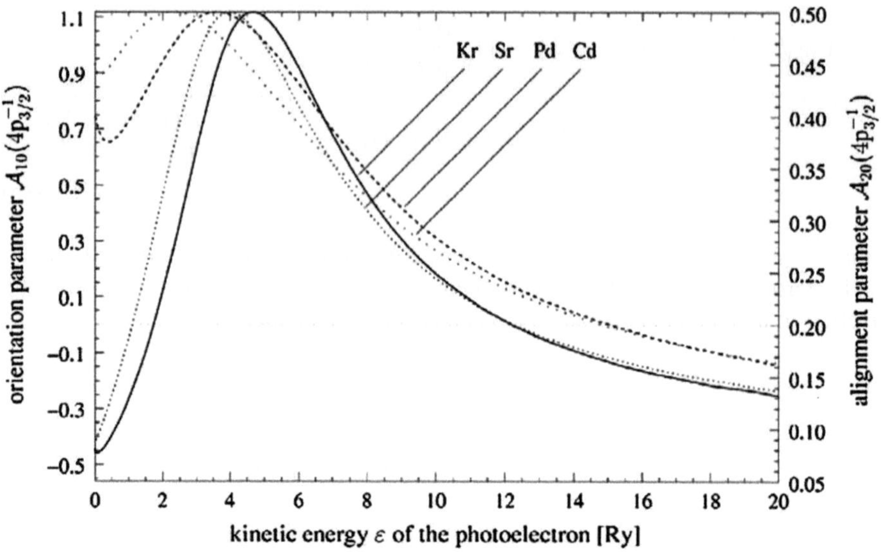

Fig. 3.13 Alignment $\mathcal{A}_{20}$ and orientation $\mathcal{A}_{10}$ data and their interdependence of the inner shell $4p_{3/2}$ hole state photoionization of krypton, strontium, palladium, and cadmium as a function of the kinetic energy ε of the photoelectron; after Kleiman and Lohmann (2003)

The so-called *main transition* $nL \rightarrow \varepsilon L+1$ usually dominates the $nL \rightarrow \varepsilon L-1$ transition, and therefore governs the dipole transition (e.g. Amusia, 1990; Cooper 1962; Manson and Cooper 1968; Schmidt 1992). In case the main transition dominates, both, the orientation and alignment parameters take values close to their lower limits given in Table 3.15; hence, the alignment is very small and therefore difficult to measure. On the other hand, the sign of the orientation can be used as an indicator which of the two transitions is the stronger one.

The main transition may be enhanced by the so-called *shape resonance* (Berezhko et al. 1978a; Schmidt 1992, 1997), which is connected with the shape of the effective potential determining the radial wavefunctions of the continuum electron.[7] Shape resonances above threshold can be seen in the $4p_{3/2}$ photoionization of krypton and palladium around 0.1 and 0.4 Ry, respectively (see Fig. 3.13), while for Sr and Pd the shape resonance is below threshold.

There are, however, energy regions in which the main transition is much weaker than the other one, i.e. $R^2_{\varepsilon L+1,nL} \ll R^2_{\varepsilon L-1,nL}$. Here, the orientation as well as the alignment parameters reach values close to their upper limits given in Table 3.15; i.e. measurable alignment values preferably in the near-threshold region. This is due to a *repulsive barrier* in the effective potential (see Fig. 2.29) used to evaluate the radial wavefunctions of the continuum electron. The barrier is caused by the *centrifugal potential* which is much more pronounced for the $\varepsilon L+1$ partial wave of the photoelectron than for $\varepsilon L-1$ (Berezhko et al. 1978a; Cooper 1964; Manson and Cooper 1968). This *centrifugal barrier* effect can be observed in the $4d_{5/2}$ photoi-

[7] We also refer to the discussion in Sect. 2.3.6; particularly, see Fig. 2.29.

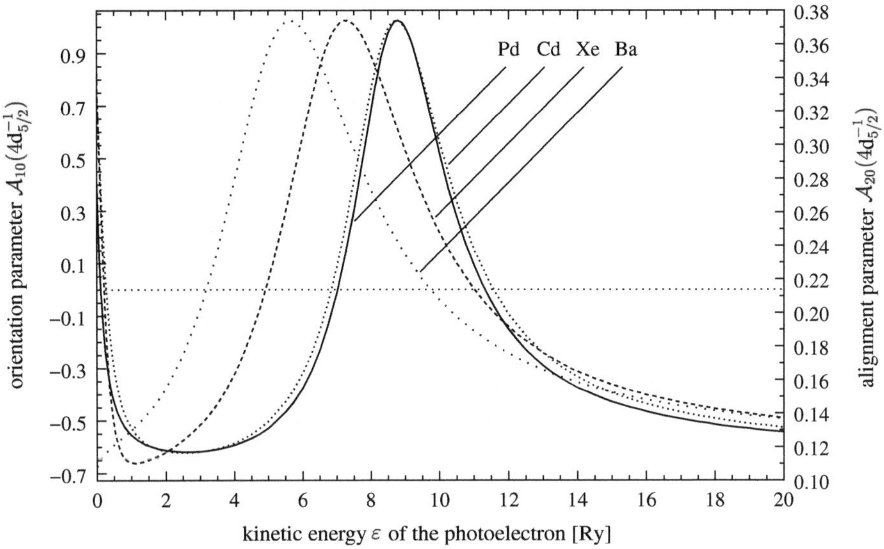

Fig. 3.14 Alignment $\mathcal{A}_{20}$ and orientation $\mathcal{A}_{10}$ data and their interdependence of the inner shell $4d_{5/2}$ hole state photoionization of palladium, cadmium, xenon, and barium as a function of the kinetic energy ε of the photoelectron; after Kleiman and Lohmann (2003)

onization of Pd, Cd, and Xe showing a strongly increasing alignment and orientation, respectively (see Fig. 3.14).

Large values of orientation and alignment at ionization threshold are not necessarily caused by a repulsive barrier in the effective potential, in particular, if the curve falls off slightly; this results in quite large values for the alignment parameter beyond the near-threshold region where a barrier is insignificant. Here, the dipole integral $R_{\varepsilon L+1,nL}$ of the main transition is nearly vanishing as the areas under the curve $P_{\varepsilon L+1}(r) r P_{nL}(r)$ with different signs almost cancel each other at small kinetic energies.

The orientation and alignment parameters can even reach their respective upper limits if the radial dipole integral $R_{\varepsilon L+1,nL}$ of the main transition changes sign and thus passes through a zero; the minimum in the absolute photoionization cross section, which is caused by this zero crossing, is the well-known *Cooper minimum* (Cooper 1962, 1964; Manson and Cooper 1968). Note, that according to Amusia (1990), the radial dipole integral $R_{\varepsilon L-1,nL}$ is different from zero for all elements with a $^1S_0$ atomic ground state; the HFR calculation by Kleiman and Lohmann (2003) does not reveal any exception to this finding. Cooper minima can occur at every kinetic energy ε, i.e. at ionization threshold, in the near-threshold region and far beyond the near-threshold region; however, no evidence for a Cooper minimum below the ionization threshold has been found (Berezhko et al. 1978a). Cooper minima may be found in the $4p_{3/2}$ photoionization of Kr, Sr, Pd, and Cd (see Fig. 3.13), as well as in the $4d_{5/2}$ photoionization of Pd, Cd, Xe, and Ba illustrated in Fig. 3.14.

For photoionization from the atomic $^1S_0$ ground state, there are two empirical rules for the occurrence of Cooper minima caused by the main transition.

1. A Cooper minimum does not occur if the bound state wavefunction $P_{nL}(r)$ is nodeless for any distance $r > 0$ (Cooper 1962, 1964). Radial wavefunctions of that kind belong to photoionization of the 1s, 2p, 3d and 4f shells. The HFR calculation by Kleiman and Lohmann (2003) does not reveal any exception to this comparatively old rule.
2. A Cooper minimum caused by a zero in the radial dipole integral $R_{\varepsilon L+1,nL}$ of the main transition cannot turn up if the $nL + 1$ shell does not exist or if this shell is not vacant (Amusia 1990). However, this more restrictive rule is not strictly met. Exceptions to this rule, besides others, have been reported, e.g for the photoionization of Zn, Cd, Yb and Hg with a vacancy in the 3p, 4p, 4d and 5p shell, respectively (Berezhko et al. 1978a; Kleiman and Lohmann 2003).

Eventually, it is worth noting that reasons for a large alignment at ionization threshold or in the near-threshold region can be different in the HS approach (Berezhko et al. 1978a) compared to the more advanced HFR approach (Kleiman and Lohmann 2003).

3.5.4 Orientation and Alignment of Inner Shell Photoionization

Selected alignment and orientation data are illustrated in Figs. 3.15, 3.16, 3.17, 3.18, 3.19, and 3.20 compared to the sparse available experimental results. Figure 3.15 displays experimental data for the alignment parameter $\mathcal{A}_{20}$ of singly photoionized magnesium with a $2p_{3/2}$ subshell vacancy (Kämmerling et al. 1992b). Apart from the two values directly at ionization threshold, the measured data in the near-threshold region are much larger than the HFR results owing to core-excited resonances (see Kämmerling et al. 1992b, for details) which cannot be reproduced within a single-configurational HF approach. At energies ε larger than about 0.5 Ry, the agreement between the measured values (including error bars) and the HFR results becomes better, and from about 3.5 Ry upwards the agreement is very good. The latter is due to the dominance of the radial dipole integral $R_{\varepsilon \mathrm{d},2\mathrm{p}}$ of the main transition; see (3.91) and (3.92). As a result the alignment parameter is not very sensitive to a deviation from the exact numerical ratio $R_{\varepsilon \mathrm{d},2\mathrm{p}}/\mathrm{R}_{\varepsilon \mathrm{s},2\mathrm{p}}$ (Kämmerling et al. 1989). Because of the resonances, which occur in the near-threshold region and shift the alignment to larger values, the HS approach (Berezhko et al. 1978a) incidentally yields better values than the HFR approach. However, directly at ionization threshold the HS values are much too large and an almost linear decrease of the plot with increasing ε is not corroborated experimentally. The energy dependence of the alignment parameter in the near-threshold region is very well reproduced by threshold-adapted calculations which base on relativistic random-phase approximation (RRPA) or many-body perturbation theory (MBPT); see Kämmerling et al. (1992b).

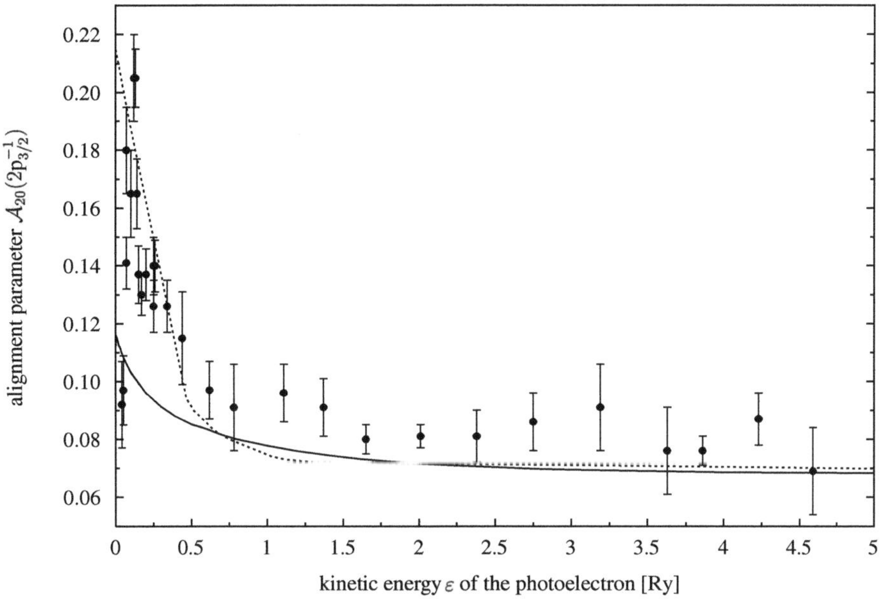

Fig. 3.15 Alignment parameter $\mathcal{A}_{20}$ of the photoionized inner shell $2p_{3/2}$ hole state of magnesium as a function of the kinetic energy ε of the photoelectron; numerical results: (——) HFR (Kleiman and Lohmann 2003), (· · · · · ·) HS (Berezhko et al. 1978a), (●) experimental data (Käammerling et al. 1994); after Kleiman and Lohmann (2003)

Figure 3.16 illustrates measured values for the alignment parameter $\mathcal{A}_{20}$ of calcium with a vacancy in the $3p_{3/2}$ subshell (Kämmerling et al. 1994; de Fanis et al. 1999). The four values in the near-threshold region, which have been obtained by de Fanis et al. (1999), fit well into the earlier measurement of Kämmerling et al. (1994). However, the agreement between the measured values and the numerical results of the HFR calculation is not that good. The HFR calculation places the position of the Cooper minimum at about 0.4 Ry lower energy than the experiment. Besides, the associated peak of the HFR plot is too broad, hence the HFR results markedly exceed the measured values at energies below the Cooper minimum. The energetic position of the Cooper minimum as well as the shape of the peak are known to be rather sensitive to many-electron correlations (Kämmerling et al. 1994) as these correlations can have different effects on the $3p \rightarrow \varepsilon s$ and $3p \rightarrow \varepsilon d$ transitions (Lörch et al. 1999). At energies ε larger than about 2.0 Ry the measured values (including error bars) and the HFR results differ only slightly. Though the HFR calculation is not capable of reproducing all measured values satisfactorily, it really constitutes a significant improvement in accuracy compared with the HS calculation (Berezhko et al. 1978a). The latter yields values which are considerably larger than the HFR results at energies between ionization threshold and the Cooper minimum that is placed at about 0.7 Ry too low. As for magnesium, an almost linear increase of the HS plot with increasing energy is inconsistent with experiment. Even above the Cooper minimum, the HS results deviate from the measured values; acceptable agreement is not obtained until

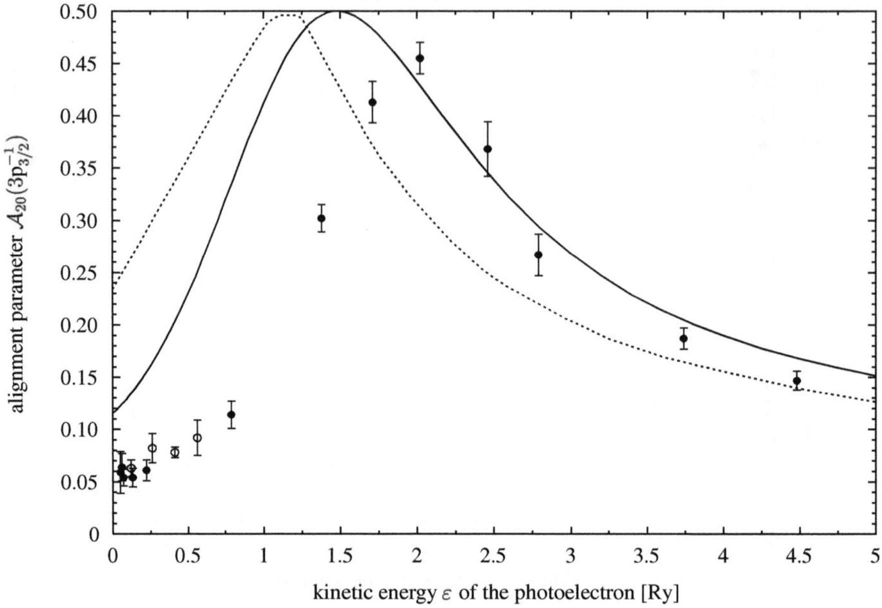

Fig. 3.16 Alignment parameter $\mathcal{A}_{20}$ of the photoionized calcium $3p_{3/2}$ inner shell hole state as a function of the kinetic energy ε of the photoelectron; numerical results: (——) HFR (Kleiman and Lohmann 2003), (······) HS (Berezhko et al. 1978a); experimental data: (●) Kämmerling et al. (1994), (○) Fanis et al. (1999); after Kleiman and Lohmann (2003)

the energy exceeds 3.5 Ry. An almost excellent reproduction of all measured values has been achieved using a threshold-adapted RRPA calculation which accounts for 3s and 4s inter-shell correlations; see Kämmerling et al. (1994). Note, that relativistic effects play an unimportant role in the 3p photoionization of calcium (Kämmerling et al. 1994; de Fanis et al. 1999; Lörch et al. 1999).

The 3d photoionization of krypton constitutes a very interesting case since both the $3d_{3/2}$ and the $3d_{5/2}$ subshell have been investigated experimentally. Besides, not only the alignment parameter $\mathcal{A}_{20}$ but also the orientation parameter $\mathcal{A}_{10}$ has been detected (Kämmerling et al. 1989, 1992a; Schmidtke et al. 2001; Snell et al. 2002). A 100 % circularly polarized photon beam, $\eta_2 = 1$, has been assumed for obtaining the orientation data. The experimental data are given in Figs. 3.17 and 3.18. The values of Schmidtke et al. (2001) merge quite well into the lower energy values recently obtained by Snell et al. (2002), where the latter represent mean values over both subshells. The HFR results reproduce the experimental values excellently, this even holds in the near-threshold region suggesting the many-electron correlations to be of minor importance. This has been found earlier by means of HF and random-phase approximation with exchange (RPAE) calculations (Kabachnik and Lee 1989). Furthermore, no evidence for post-collision interaction (PCI) has been obtained (Snell et al. 2002). A good agreement between experiment and theory should be expected at least for higher energies. Since the radial dipole integral

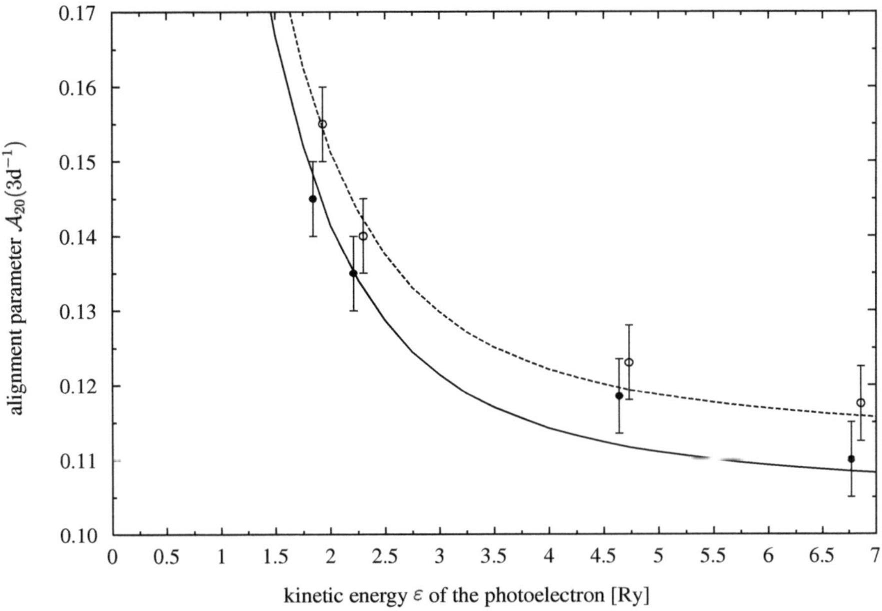

Fig. 3.17 Alignment parameter $\mathcal{A}_{20}$ of the photoionized krypton 3d inner shell hole state as a function of the kinetic energy ε of the photoelectron; numerical results: (——) $3d_{3/2}$ and (– – –) $3d_{5/2}$ HFR calculation (Kleiman and Lohmann 2003); experimental data: (●) $3d_{3/2}$ and (○) $3d_{5/2}$ (Kämmerling et al. 1989, 1992a); after Kleiman and Lohmann (2003)

$R_{\varepsilon f,3d}$ of the main transition dominates for these energies, a slight deviation from the exact numerical ratio $R_{\varepsilon f,3d}/R_{\varepsilon p,3d}$ hardly shows an effect on the orientation and alignment parameters. The HS results by Berezhko et al. (1978a) match the experimental alignment data quite well which is demonstrated by Schmidtke et al. (2001). The same holds for the RPAE results (Kabachnik and Lee 1989) with respect to the orientation data, but the RPAE results are by no means more accurate than the HFR results (see Schmidtke et al. 2001; Snell et al. 2002).

The 4d photoionization of xenon has been investigated experimentally focussing on the alignment parameter $\mathcal{A}_{20}$ as well as on the orientation parameter $\mathcal{A}_{10}$. The measured values obtained by the various experimental groups (Kämmerling et al. 1990; Kämmerling and Schmidt 1991, 1993; Schaphorst et al. 1997; Schmidtke et al. 2000b; Snell et al. 1999a,b, 2000b, 2002; Whitfield et al. 1992) are collected in Figs. 3.19 and 3.20. Most of the data have been obtained by Snell and co-workers (Snell et al. 2000b, 2002) where the alignment values have been obtained within two different experiments using time-of-flight (TOF) electron spectrometers and a hemispherical electron analyzer (HEA), respectively. Some of the alignment data significantly differ from each other. The HFR calculation reproduces the experimental alignment and orientation data very well for energies in the near-threshold region and beyond the Cooper minimum, and there is an almost excellent agreement with the orientation data. However, large deviations of the HFR results from

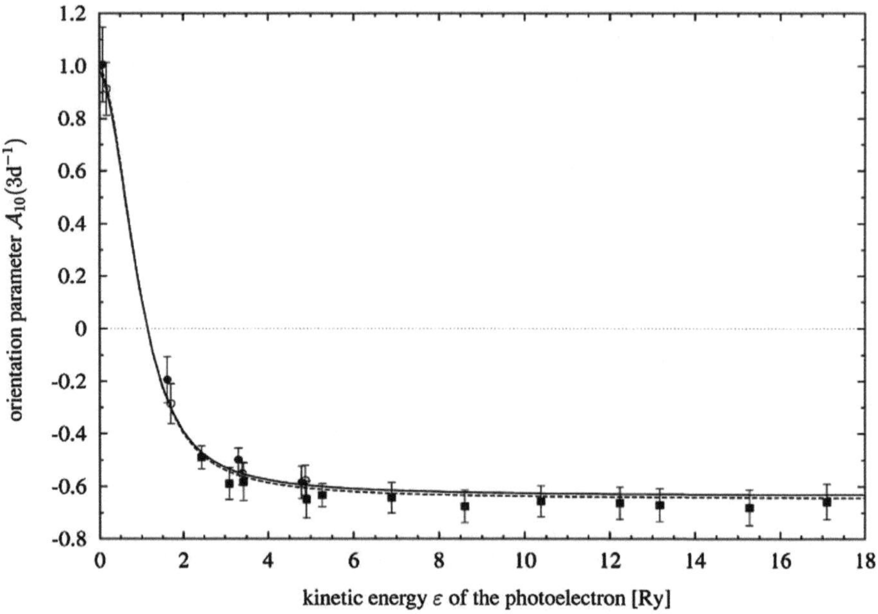

Fig. 3.18 Orientation parameter $\mathcal{A}_{10}$ of the photoionized krypton 3d inner shell hole state as a function of the kinetic energy ε of the photoelectron; numerical results: (——) $3d_{3/2}$ and (– – –) $3d_{5/2}$ HFR calculation (Kleiman and Lohmann 2003); experimental data: (●) $3d_{3/2}$ and (○) $3d_{5/2}$ (Schmidtke et al. 2001), (■) 3d mean value (Snell et al. 2002); after Kleiman and Lohmann (2003)

the measured data occur for energies from about 4.0 Ry up to 7.5 Ry. Since RPAE calculations (Berezhko et al. 1978b; Kabachnik and Lee 1989) agree much better with the measured alignment and orientation data than the HFR calculation (see Snell et al. 2002), the discrepancy arises primarily due to correlations among the 4d electrons; inter-shell correlations can be regarded as insignificant (Berezhko et al. 1978b). As has been already discussed for the case of calcium, electron correlations may affect the energetic position of the Cooper minimum and the shape of the peak. The HFR calculation places the Cooper minimum at 7.3 Ry, whereas the RPAE calculations yield a Cooper minimum at 7.6 Ry (Berezhko et al. 1978b) and at 8.1 Ry (Kabachnik and Lee 1989), respectively. Following the experimental data (Snell et al. 2000b, 2002) the Cooper minimum is located around $7.5 - 8.2$ Ry. On the other hand, an earlier alignment experiment performed by Southworth et al. (1983) does not exhibit a Cooper minimum.[8] Apart from energies close to the ionization threshold, i.e. $\varepsilon \leq 1.0$ Ry, the HS calculation (Berezhko et al. 1978a) is not capable of describing the experimental situation satisfactorily, as it yields the Cooper minimum at 5.8 Ry (see Fig. 3.19).

Eventually, the comparison of the numerical results with various experimental alignment and orientation data indicates that the HFR approach is capable of roughly reproducing the experimental data. In general, the more advanced HFR approach

[8] Since these data have large uncertainties, they are not plotted in Fig. 3.18 (see Snell et al. 2000b).

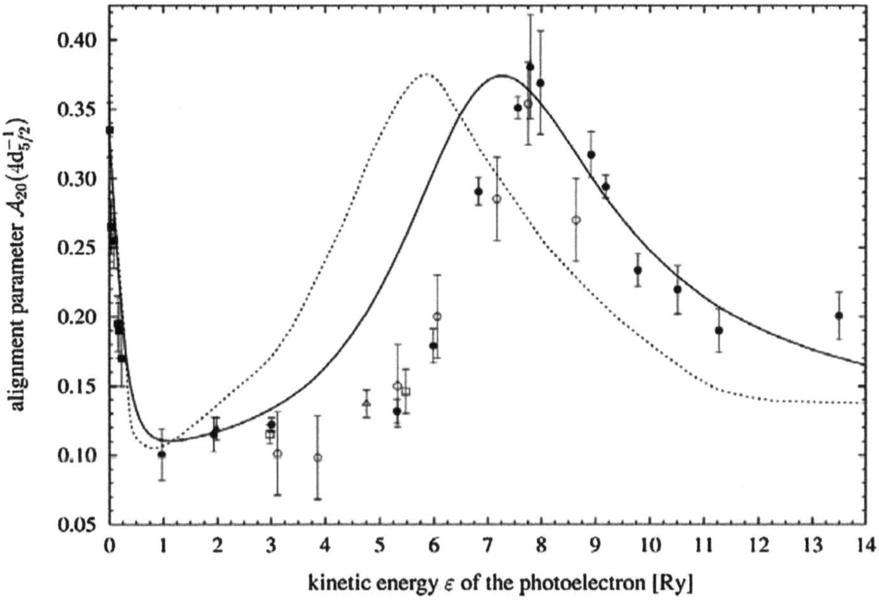

Fig. 3.19 Alignment parameter $\mathcal{A}_{20}$ of xenon after photoionization of the inner $4d_{5/2}$ sub-shell as a function of the kinetic energy ε of the photoelectron; numerical results: (——) HFR (Kleiman and Lohmann 2003), (– – –) HS (Berezhko et al. 1978a); experimental data: (●) TOF exp. (Snell et al. 2000b), (○) HEA exp. (Snell et al. 2000b), (■) Whitfield et al. (1992), (□) Kämmerling et al. (1990), (▲) Kämmerling and Schmidt (1991, 1993), (△) Schaphorst et al. (1997); after Kleiman and Lohmann (2003)

yields better results than the HS approach by Berezhko et al. (1978a), particularly if a Cooper minimum occurs. However, one should keep in mind that almost all experimental data presently available refer to photoionization processes involving outer shells, or the experimental data have been taken in the near-threshold region where many-electron correlations generally cannot be ignored. Thus, further experiments which investigate intermediate and inner shells are necessary to reliably evaluate the accuracy of the HFR approach.

3.5.5 *Photoionization of Open Shell Atoms*

The theory has been further extended by Kleiman (2006) to open shell atoms. Applying a simple but useful model, they have been able to derive a relation between the open- and closed-shell case which allows for predictions of open-shell alignment and orientation data.

Let us consider photoionization of an open shell system via transition from an initial state $|(J_0 j_v)J_i\rangle$ to a final state $|([j_f j_v]Jj)J_f\rangle$. Here, J_0 and j_v denote the total angular momenta of an initial inner electronic shell, which must not necessarily be

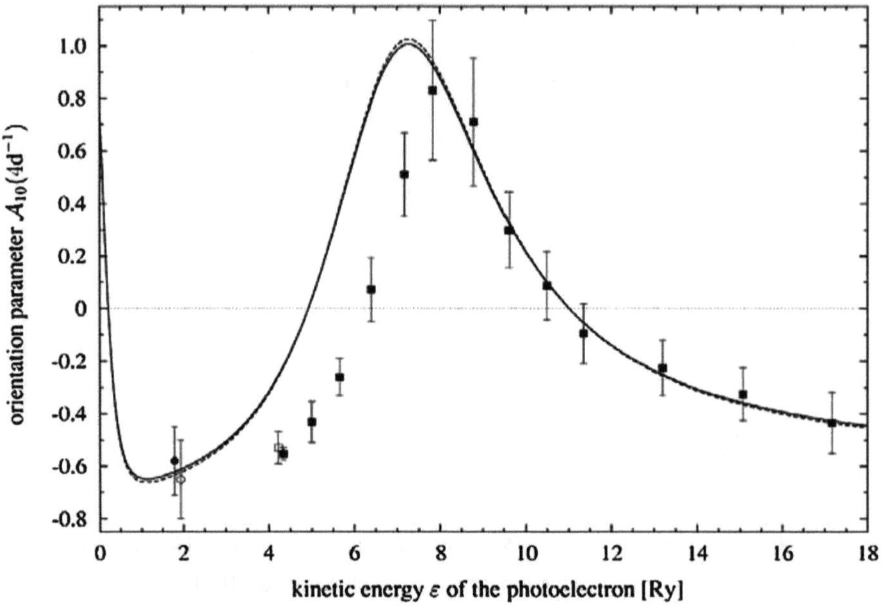

Fig. 3.20 Orientation parameter $\mathcal{A}_{10}$ of xenon after photoionization of the inner 4d shell as a function of the kinetic energy ε of the photoelectron; numerical results: (——) $4d_{3/2}$ and (– – –) $4d_{5/2}$ HFR calculation (Kleiman and Lohmann 2003); experimental data: (●) $4d_{3/2}$ and (○) $4d_{5/2}$ (Snell et al. 1999a, b), (■) 4d mean value (Snell et al. 2002), (□) $4d_{5/2}$ Schmidtke et al. (2000b); after Kleiman and Lohmann (2003)

closed, and of an remaining open outer shell, respectively. Both are coupled to the total angular momentum J_i of the initial state. The electronic inner shell configuration changes its total angular momentum to j_f via photoemission while the outer open shell remains undisturbed. Thus, j_f and j_v are coupled to the total angular momentum J of the final ionic state which we assume as fixed. The total angular momentum of the photoelectron is denoted as j which eventually couples with J to the total angular momentum J_f of the final state.

Assuming the outer valence shell not disturbed by the inner shell ionization we may recouple the final state ket-vector and express the reduced open shell dipole matrix element in terms of the recoupled reduced transition matrix elements of photoionization as

$$\big\langle (([j_f j_v]J j)J_f \| \mathbf{d} \| (J_0 j_v)J_i \big\rangle = \sum_{J'} (-1)^{j+j_f+J_f+j_v} \begin{Bmatrix} j & j_f & J' \\ j_v & J_f & J \end{Bmatrix} \times \sqrt{(2J'+1)(2J+1)} \big\langle (([j_f j]J' j_v)J_f \| \mathbf{d} \| (J_0 j_v)J_i \big\rangle \,. \tag{3.96}$$

Applying angular momentum recoupling theory (e.g. Brink and Satchler 1962) to the final state allows for making the reduced open shell dipole matrix element independent of the outer shell total angular momentum j_v and of the total angular momenta

J_i and J_f of the initial and final states, respectively.

$$\begin{aligned}\big\langle([j_f j_v]Jj)J_f\|\mathbf{d}\|(J_0 j_v)J_i\big\rangle &= \sum_{J'}(-1)^{j+j_f+J_0+1}\big\langle(j_f j)J'\|\mathbf{d}\|J_0\big\rangle \\ &\times\sqrt{(2J_i+1)(2J_f+1)(2J'+1)(2J+1)} \\ &\times\begin{Bmatrix} j & j_f & J' \\ j_v & J_f & J\end{Bmatrix}\begin{Bmatrix} J_0 & 1 & J' \\ J_f & j_v & J_i\end{Bmatrix}. \end{aligned} \tag{3.97}$$

Equation (3.97) is based on the fact that, as a single electron operator, **d** acts in the Hilbert subspace of the electron undergoing the photoemission, only. Suppose that the matrix elements on the right hand side of (3.97) are known, we are then able to calculate the open shell matrix element on the left hand side. In principle, the approach by Lohmann and Kleiman (2006 is able to relate known matrix elements obtained for an arbitrary configuration, which also includes other open shell configurations, to the required matrix elements for the open shell configuration of interest.

This leads to analytic equations for orientation and alignment of open shell atoms having the dipole matrix elements independent of the outer open shell.

Using the open shell notation of quantum numbers in (3.79), i.e. replacing

$$\big\langle(Jj)J_1\|\mathbf{d}\|J_0\big\rangle \equiv \big\langle([j_f j_v]Jj)J_f\|\mathbf{d}\|(J_0 j_v)J_i\big\rangle, \tag{3.98}$$

inserting (3.97) twice into (3.79) and applying angular momentum algebra (Varshalovich et al. 1988), the summation over the final state quantum numbers J_f and J_f' of the combined electronic–photoionic system can be carried out which yields the open shell photoionization parameters

$$\begin{aligned} B_K^{J,j_v} &= (2J+1)\sum_{\ell j J' J''}\big\langle(j_f j)J'\|\mathbf{d}\|J_0\big\rangle\big\langle(j_f j)J''\|\mathbf{d}\|J_0\big\rangle^* \\ &\times\sqrt{(2J'+1)(2J''+1)}\sum_{Y}(-1)^{K+2j-J-Y}(2Y+1) \\ &\times\begin{Bmatrix} 1 & J & Y \\ J & 1 & K\end{Bmatrix}\begin{Bmatrix} J' & J_0 & 1 \\ j_f & j_v & J \\ j & J_i & Y\end{Bmatrix}\begin{Bmatrix} J'' & J_0 & 1 \\ j_f & j_v & J \\ j & J_i & Y\end{Bmatrix}. \end{aligned} \tag{3.99}$$

Equation (3.99) is a general result and is valid for arbitrary filled shells. It therefore includes e.g. photoionization from an open inner shell in the vicinity of one or more other open shells. Note, that the summation Y is independent on the matrix elements and is solely over the geometrical factors (nj-symbols).[9]

[9] The sum over Y can be contracted in terms of a $12j$-symbol. However, the representation of $12j$-symbols is not unique which makes it tedious to extract further physical information.

The main focus is, however, on photoemission taking place from an initially closed inner (sub-)shell of an open shell atom, i.e. $J_0 = 0$. This yields the selection rules $J' = J'' = 1$ and $J_i = j_v$. Thus, the sum over J' in (3.97) vanishes and we remain with

$$\langle ([j_f j_v]Jj)J_f \|\mathbf{d}\| (0 j_v) j_v \rangle = (-1)^{j+j_f+j_v+J_f} \begin{Bmatrix} j & j_f & 1 \\ j_v & J_f & J \end{Bmatrix} \times \sqrt{(2J_f+1)(2J+1)} \langle (j_f j)1 \|\mathbf{d}\| 0 \rangle. \quad (3.100)$$

Inserting (3.100) twice into (3.79) and applying the relevant selection rules the summation over J_f and J_f' can be carried out (Varshalovich et al. 1988) and we get the open shell photoionization parameters [10]

$$B_K^{J,j_v} = \frac{2J+1}{2j_v+1} \sum_{\ell j} (-1)^{J+j_v+j} |\langle (j_f j)1 \|\mathbf{d}\| 0 \rangle|^2 \times \begin{Bmatrix} J & J & K \\ j_f & j_f & j_v \end{Bmatrix} \begin{Bmatrix} 1 & 1 & K \\ j_f & j_f & j \end{Bmatrix}. \quad (3.101)$$

Inspecting (3.101), we find the open shell coefficients B_K^{J,j_v} directly proportional to the parameters B_K^{cl} closed shell case which can be obtained by combining (3.79) and (3.98), inserting $j_v = 0$. Realizing that the summation of (3.101) does not depend on the angular momentum j_v and comparing both expressions yields

$$B_K^{J,j_v} = (-1)^{J+j_v+j_f-K} \frac{2J+1}{2j_v+1} \begin{Bmatrix} J & J & K \\ j_f & j_f & j_v \end{Bmatrix} B_K^{cl}. \quad (3.102)$$

Using (3.80) and (3.81) we end up with general expressions relating the orientation and alignmentparameters of an open shell atom directly to the orientation and alignment of the next closed (sub-)shell configuration of the same or other elements of the periodic system.

$$\mathcal{A}_{K0}^{J,j_v} = (-1)^{J+j_v+j_f-K} \sqrt{(2J+1)(2j_f+1)} \begin{Bmatrix} J & J & K \\ j_f & j_f & j_v \end{Bmatrix} \mathcal{A}_{K0}^{cl}, \quad (3.103)$$

where $K = 1, 2$, for orientation $\mathcal{A}_{10}$ and alignment $\mathcal{A}_{20}$, respectively. Note, that (3.103) allows for a change of sign for the open shell orientation and/or alignment compared to the closed shell parameters.

It is worth noting that the approach by Lohmann and Kleiman (2006) is generally valid, irrespective of the computer package used for the evaluation of the dipole transition matrix elements. For illustrating their approach, Lohmann and Kleiman (2006) performed model calculations for some selected cases of photoionization of

---

[10] Applying all relevant selection rules, the same result can be obtained from (3.99).

open shell atoms utilizing different numerical approaches. Applying the semi-relativistic Hartree–Fock program by Cowan (1981) for generating the required configuration dipole matrix elements, using the program PHOTO (Kleiman 2001) for obtaining $\mathcal{A}^{cl}_{K0}$, and applying (3.103), they obtain the relevant open shell orientation and alignment parameters. As for including polarization or charge cloud deformation effects caused by the additional open shell structure the calculations have been performed in the field of the related open shell atom, i.e. using the appropriate open shell configuration potential; this is denoted as calc. (a). This calculation is correct for photoionization from closed inner shells of open shell atoms. In more complicated cases, (3.99) must be applied. Considering the full open shell approach (a) as best, but realizing that a variety of numerical data are already available for closed shell atoms or cations (e.g. Kleiman et al. 1999b; Kleiman 2002), it might often be easier to calculate the open shell data using the ionic potential of the closed (sub-)shell ion; in the following called calc. (b). Even if both calculations cannot be performed, or the relevant matrix elements are not available, it might be sufficient for a first approximation to use the potential and dipole matrix elements of the next neighbouring closed shell atom; named calc. (c). In either case the model may be applied which offers another advantage of the derived equations.

3.5.5.1 Open Shell Alignment

Numerical alignment results for potassium, gold and manganese are shown in Figs. 3.21, 3.22, and 3.23, respectively.

Considering photoionization of the L_3 shell of the K $(3p^6 4s)\,^2S_{1/2}$ open shell ground state configuration, yields an additional splitting of the fine structure. Fig. 3.21 displays the alignment $\mathcal{A}^{J,1/2}_{20}$ for the two final state fine structure components

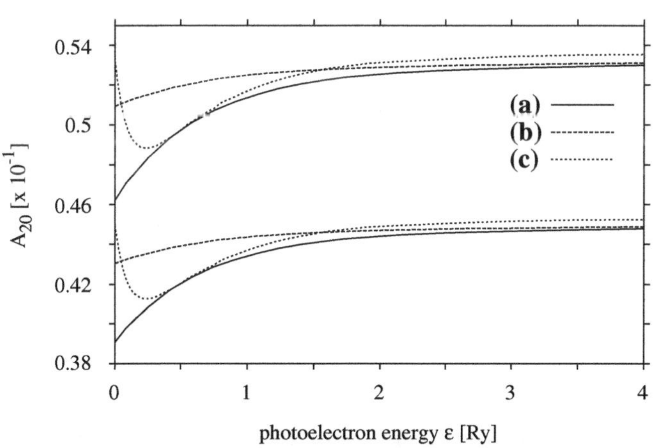

Fig. 3.21 Potassium alignment $\mathcal{A}^{J,1/2}_{20}$ for the photoionized K $2p^{-1}_{3/2}4s_{1/2}$; $J = 1$ (*bottom*) and $J = 2$ (*top*) fine structure states using the K (*calc. a*), K^+ (*calc. b*), and Ar (*calc. c*) configuration potentials; after Lohmann and Kleiman (2006)

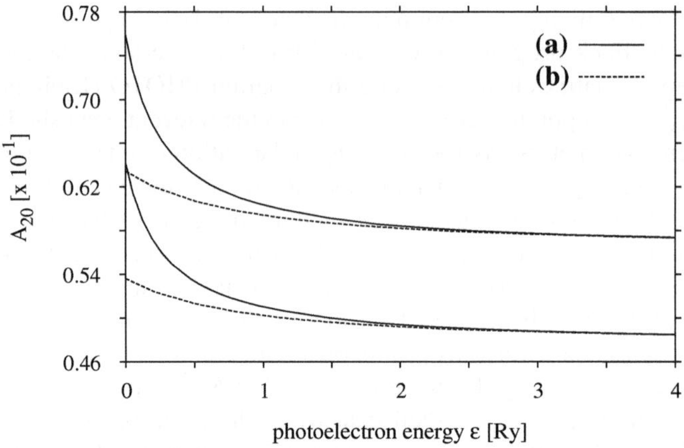

Fig. 3.22 Alignment $\mathcal{A}_{20}^{J,1/2}$ of gold for the photoionized Au $2p_{3/2}^{-1}6s_{1/2}$; $J = 1$ (*bottom*) and $J = 2$ (*top*) fine structure states using the Au (*calc. a*) and Au^{+} (*calc. b*) configuration potentials (after Lohmann and Kleiman 2006)

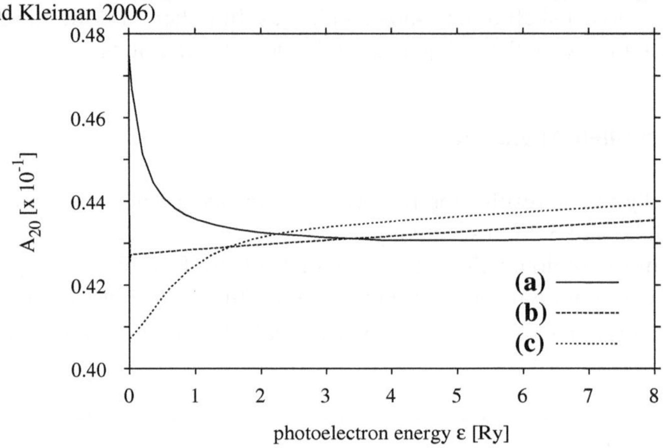

Fig. 3.23 Alignment $\mathcal{A}_{20}^{3,3/2}$ of manganese for the L_3 photoionization of the Mn $2p_{3/2}^{-1}4s^{2}3d_{3/2}^{5}$; $J = 3$ fine structure state using the Mn (*a*), Mn^{5+} (*b*), and Ca (*c*) configuration potentials; after Lohmann and Kleiman (2006)

$J = 1, 2$. Lohmann and Kleiman (2006) calculated numerical data up to 100 Ry, not shown in the figures, in order to focus on the more pronounced effects close to threshold. The orientation can be derived from the alignment utilizing the interdependence relation (3.93) in conjunction with (3.103); see the discussion in Sect. 3.5.3. In order to demonstrate the effect of the different potentials three calculations have been performed, obtaining the potassium alignment either by using the configuration potential of the neutral K atom (calc. a), the potential of the K^{+} ion stripped of its outermost electron with ionic configuration ($3p^6$) (calc. b), or the potential of the next lower closed shell neighbour, i.e. Ar which is iso-electronic to K^{+} (calc. c).

Table 3.16 Ionization thresholds for atoms/ions calculated for the numerical approaches (a-c); after Lohmann and Kleiman (2006)

	Calc. a		Calc. b		Calc. c	
Shell	Element	Energy [eV]	Element	Energy [eV]	Element	Energy [eV]
2p	K	302.02	K^+	309.29	Ar	249.69
2p	Au	12503.57	Au^+	12512.81	–	–
3d	In	455.38	In^+	463.90	Cd	414.99
4f	Tl	126.61	Tl^+	134.87	Hg	107.53
2p	Sc	412.84	Sc^+	429.29	Ca	359.04
2p	Mn	658.33	Mn^{5+}	782.78	Ca	359.04

The numerical approaches (a-c) result in different ionization thresholds, showing slight deviations for the potassium case between calculations (a), 302.02 eV, and (b), 309.29 eV, while it is more apart for case (c), 249.69 eV. However, as for the effects on alignment and orientation the interest is on their behaviour in the continuum. Threshold energies for selected open shell calculations are given in Table 3.16. The ionization edges by Lohmann and Kleiman (2006) are close to the numerical values which can be derived from the data bank of Deslattes et al. (2003), averaging over the L_2 and L_3 edges. Even their data show deviations from the experiments (Deslattes et al. 2003).

As can be seen from Fig. 3.21, the asymptotic behaviour for increasing energies of all three calculations remains approximately the same. This can be explained by the fact that the asymptotic tail is mainly influenced by the Coulombic part of the potential whereas the angular momentum terms of the three numerical approaches are more pronounced close to threshold. This is confirmed by the calculations indicating a totally different behaviour of the alignment data close to threshold, resulting e.g. in threshold values for the alignment of ~0.039 (a), ~0.043 (b), and ~0.045 (c), for the $J = 1$ fine structure state, which are due to shape resonances below and above threshold and to the different screening of the core depending on the different potentials of calculations (a-c). Calculation (c) shows a minimum for the alignment close to threshold, caused by a shape resonance above threshold (Berezhko et al. 1978a; Schmidt 1992), whereas the data of calculation (a) and (b) are slowly decreasing for decreasing photoelectron energy, having the shape resonance below threshold. The shape resonance is shifted to smaller energies along iso-electronic sequences like Ar and K^+ (Kleiman and Becker 2005). The $J = 2$ fine structure state shows a similar behaviour.

Comparing the numerical results by Lohmann and Kleiman (2006) to previous closed-shell alignment data of Ar (Kleiman and Lohmann 2003), we find the magnitude of the open-shell data generally smaller, irrespective of the calculational model (a-c). This can be explained by interpreting the $6j$-symbol in (3.103) as an additional perturbation coefficient, causing depolarization, using a picture of time-dependent density matrix theory (Blum 1996). The behaviour of the alignment in the asymptotic tail and its discussed threshold values can be seen as similar to earlier results of several studies in photoionization of closed shell atoms which established that even in the independent particle approximation, e.g. Hartree–Fock, replacing the potential of the neutral atom by the ionic one, with one electron missing, can result in a drastic change of the dipole matrix elements in the threshold vicinity, whereas the results are rather close to each other at higher photon energies (Amusia 1990; Dill et al. 1975).

In contrast to potassium as a comparatively light element, let us consider photoionization of the L_3 shell of the Au($5d^{10}6s$) $^2S_{1/2}$ open shell ground state configuration. As in the previous case, this yields an additional fine structure splitting into $J = 1, 2$. The alignment $\mathcal{A}_{20}^{J,1/2}$ is shown in Fig. 3.22. The asymptotic behaviour for both, calculation (a) and (b), show almost no deviations for photoelectron energies above 2.5 Ry for the two different fine structure states. The alignment at threshold is strongly increasing for energies below 1 Ry caused by a weaker repulsive barrier in the centrifugal potential (Berezhko et al. 1978a; Cooper 1964; Manson and Cooper 1968), resulting in threshold values for the alignment of ~0.076 (a) and ~0.063 (b) for the $J = 1$, and ~0.064 (a) and ~0.054 (b) for the $J = 2$ fine structure states, respectively.

The L_3 shell alignment of Au has been measured by Yamaoka et al. (2003). Deep inner shell ionization of Au requires relatively high ionization energies and, as has been pointed out by Lohmann and Kleiman (2006), their approach should be considered as independent of the energy regime. However, as the Cowan code is semi-relativistic, it inherently applies the LS selection rules[11] which are more stringent than in the more general jj coupling case. Therefore, deviations from the experimental data due to the strong jj coupling character which is prominent for inner shell ionization of heavy elements like the L_3 ionization of Au may occur. This can be seen by comparing the results of Fig. 3.22 to the experimental data. Yamaoka et al. (2003) obtained the alignment from the analysis of the angular distribution of the L_ℓ and L_{α_2} fluorescence radiation emitted after ionization by linearly polarized light. Their data yield different alignment values giving $\mathcal{A}_{20}^{L_\ell} = 0.09$ and $\mathcal{A}_{20}^{L_{\alpha_2}} = 0.14$, respectively.[12] Calculating the alignment, statistically averaging over the final state fine structure for the experimentally used excess energy of 79.6205 Ry, we obtain $\mathcal{A}_{20}^{\mathrm{av}} = 0.053$ independent of the used approach (a) or (b) in this energy regime. This result is close to the theoretical value of $\mathcal{A}_{20} = 0.06$ given by Yamaoka et al. (2003) calculated within a non-relativistic RPAE approach. They also investigated the validity of the dipole approximation in the high energy regime, including quadrupole terms in their closed shell approximation resulting in an effective alignment $\mathcal{A}_{20}^{\mathrm{eff}} = 0.061$ demonstrating that non-dipole contributions are negligible. An earlier Hartree–Slater approximation (Scofield 1976) has yielded an alignment of 0.075. The calculations (a-b), as well as the discussed numerical values, underestimate the experimental data which must be expected from the above discussion. These discrepancies reveal the necessity of further numerical and experimental investigations.

---

[11] See 3.87 and the discussion in Sect. 3.5.3.

[12] The results by Yamaoka et al. (2003) imply a quantization axis along the electric field of the linearly polarized incoming radiation and have been divided by a factor of –2.0 for comparison with the data by Lohmann and Kleiman (2006).

Concerning numerical improvements, we concur with Yamaoka et al. (2003) that relativistic effects are the main reason for the occurring deviations. However, as for the calculations by Lohmann and Kleiman (2006), these deviations are caused by the limitations of the applied semi-relativistic Cowan Hartree–Fock package and not by the model approach itself. Calculating the relevant dipole matrix elements within a more advanced relativistic scheme, e.g. MCDF, and applying the derived open shell approach of Lohmann and Kleiman (2006) would be the method of choice in order to obtain more precise numerical alignment data for heavy open shell atoms.

In order to illustrate the prominent effects related to the specific choice of the potential for open shell atoms with a half-filled valence shell, we consider the alignment of the L_3 photoionization of Mn having a $(3p^6 4s^2 3d^5)\,^6S_{5/2}$ ground state configuration, into one of its various final fine structure states; see Fig. 3.23. The behaviour close to threshold shows dramatic deviations. For the Mn potential (a), the alignment is strongly increasing below 1 Ry (centrifugal barrier), while using the Ca potential (c) yields the opposite behaviour (shape resonance). For the potential of the stripped Mn^{5+} ion (b), the alignment is almost a constant linear function of the photoelectron energy only slightly decreasing for lower energies. These pronounced effects can be understood since Ca is comparatively far from Mn in the periodic table. Therefore, large differences in their potentials can be expected. On the other hand, the stripped Mn^{5+} ion sees a stronger attractive core potential reducing the diameter of the electronic charge cloud, whereas the Mn potential is more influenced by the angular momentum of the half-filled valence shell.

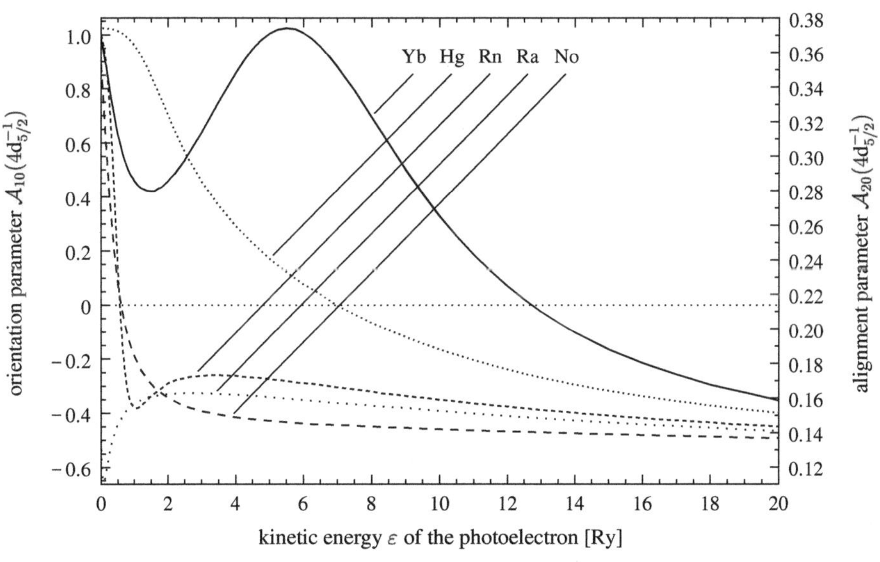

Fig. 3.24 Alignment $\mathcal{A}_{20}$ and orientation $\mathcal{A}_{10}$ data of the inner shell $4d_{5/2}$ hole state photoionization of ytterbium and mercury as a function of the kinetic energy ε of the photoelectron in comparison to the alignment and orientation of the radioactive elements radon, radium, and nobelium; after Kleiman and Lohmann (2003)

3.5.6 Photoionization of the Radionucleides

Besides the, so-called, well-investigated elements like the rare gases, the alkalis, or the earth alkalis, information from the electron shell of the radioactive elements is still rare. This is understandable considering the fact of their high toxicity. On the other hand, such data are most interesting, since these elements are extremely heavy and therefore allow for an intensive check of relativistic predictions. For instance, jj coupling should be prominent in heavy elements. As has been discussed in Sect. 3.5.3, Kleiman and Lohmann (2003) performed a detailed scan through the periodic system and provided a wealth of numerical alignment and orientation data, including the radionucleides. Considering photoionization of the $4d_{5/2}$ shell, Fig. 3.24 shows strong deviations between the non-radioactive elements mercury and ytterbium in contrast to the radioactive elements radon, radium, and nobelium. Expressive structures are obvious, both, in the alignment and orientation data. The radionucleides show a negative orientation, and small alignment, for most of the energetic scale, indicating the dominance of the main transition, which is even more pronounced due to a shape resonance for Rn and Ra at 1.1 Ry and at threshold, respectively. A shape resonance also occurs for Yb at 1.4 Ry. In the near-threshold region, below ~1 Ry, alignment and orientation tend to their upper limits for Rn and No, as well as for the non-radioactive elements Hg and Yb, which is caused by the repulsive barrier of the related effective potential. On the other hand, the photoionization of radium is not affected by the centrifugal barrier resulting in a steep decrease of alignment and orientation to their lower bounds. The non-radioactive elements like mercury or ytterbium, both, change their sign of orientation far beyond the near-threshold region at photoelectron energies of ~7 and ~13 Ry, respectively. Moreover, ytterbium exhibits an intermediate maximum in the orientation and alignment data around 5 Ry, due to a Cooper minimum not prominent in the Hg data.

Impressive alignment data are obtained for the photoionization of the $5f_{7/2}$ shell of nobelium, shown in Fig. 3.25. The upper limits of Table 3.15 are reached twice, for both, the orientation $\mathcal{A}_{10} \sim 0.98$, and the alignment $\mathcal{A}_{20} \sim 0.325$, due to the strong centrifugal barrier prominent below 1 Ry (near-threshold region), and, due to the pronounced Cooper minimum around 21 Ry.

A systematic study of orientation and alignment of the heavy elements reveals to be cautious about adopting the wide-spread opinion in the literature, that at high kinetic energies of the photoelectron — far beyond the Cooper minimum, if any — the main transition $nL \rightarrow \varepsilon L+1$ dominates the photoionization process. A remarkable exception to this dominance is the outer shell photoionization of heavy atoms. Fig. 3.26 illustrates the orientation and alignment after photoionization of the $6p_{3/2}$ state of the heavy radioactive elements radon, radium, and nobelium where, apart from the near-threshold region which is governed by a shape resonance below threshold, the main transition appears to be much weaker than the other, even at high kinetic energies like 20 – 30 Ry. For Ra and Rn the Cooper minimum yields a maximum alignment of $\mathcal{A}_{20} \sim 0.5$, as well as a maximum orientation $\mathcal{A}_{10} \sim 1.1$ approximately at 10 Ry, while for No the Cooper minimum is shifted towards ~15 Ry.

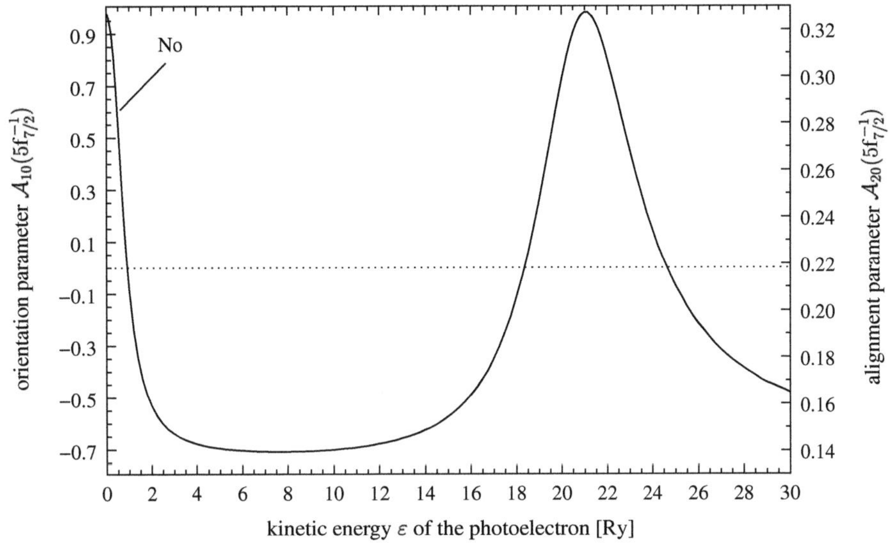

Fig. 3.25 Alignment $\mathcal{A}_{20}$ and orientation $\mathcal{A}_{10}$ of the inner shell $5f_{7/2}$ hole state photoionization of nobelium as a function of the kinetic energy ε of the photoelectron. The maximum of the alignment around 21 Ry should be noted; after Kleiman and Lohmann (2003)

3.5.7 State Multipoles of Electron Impact Excitation

Let us consider the following two-step process,

$$\begin{aligned} e + A(J_0) \longrightarrow\ & A^*(J) + e' \\ & \hookrightarrow A^{+^*}(J_f) + e_{Auger}. \end{aligned} \tag{3.104}$$

In the first step, after a primary electron impact excitation the exciting electron is not detected while another electron is excited from a deep inner shell into a Rydberg state. After a certain lifetime, this Rydberg state decays via resonant Auger decay and the emitted Auger electron is eventually detected. The validity of this approach has been proved in a variety of experiments; e.g. see the review by Mehlhorn (1990) and refs. therein. Applying the two-step model the set of dynamic parameters, describing the excitation/scattering and the subsequent emission process, can be factorized into parameters of orientation and alignment of the total angular momentum J of the intermediate excited state A^*, and angular distribution and spin polarization parameters of the Auger electrons, respectively. The alignment and orientation parameters contain the information about the electron impact excitation while the latter yield information about the Auger decay dynamics.

In contrast to the case of photoexcitation (see Sect. 3.5.2) the total set of parameters $\mathcal{A}_{KQ}$, describing the excitation–emission process, is not limited by the dipole approximation. Therefore, the maximum rank of irreducible statistical tensors, i.e. state multipoles, or in other words, measurable quantities is given by the general

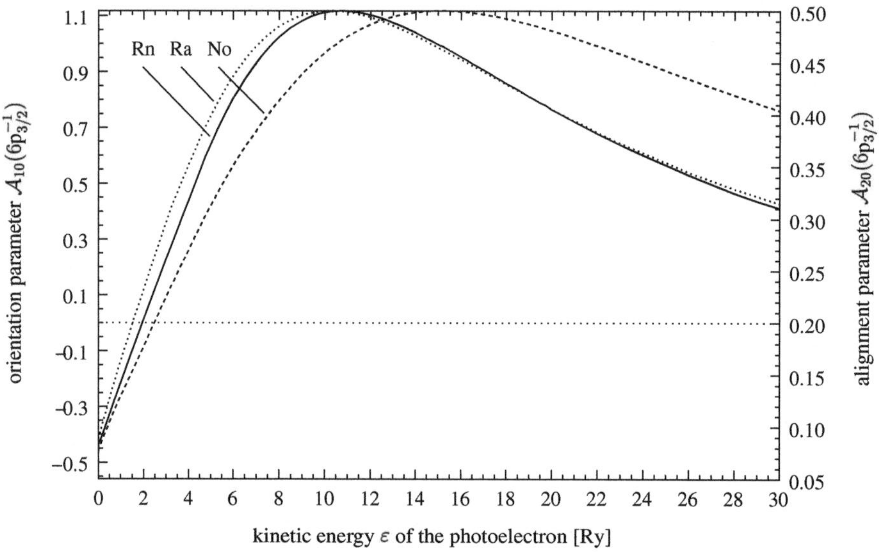

Fig. 3.26 Alignment $\mathcal{A}_{20}$ and orientation $\mathcal{A}_{10}$ of the radionucleides radon, radium, and nobelium after photoionization of the inner shell $6p_{3/2}$ state as a function of the kinetic energy ε of the photoelectron; after Kleiman and Lohmann (2003)

restriction $K \leq 2J$, only (Blum 1996), while the magnetic components obey $|Q| \leq K$. For our specific case of electron impact excitation, or more generally for excitation processes with spin 1/2 particles, it has been shown that general selection rules further restrict the values of the magnetic components to $|Q| \leq 1$ (Lohmann 1984, 1998, 2008), provided the scattered electron is not detected and the quantization axis is chosen along the incident beam axis. However, even with this restriction, the general equations of angular distribution and spin polarization after electron impact excitation still remain rather complicated. Explicit expressions have been given by Lohmann (1998, 2008). Similar expressions for related cases of interest have been given by Balashov et al. (2000).

Electron impact excitation can populate intermediate states with $J > 1$ which, within the limits of the dipole approximation, cannot be accessed via photoexcitation. For the considered case of deep inner shell excitation, we are focusing on $J = 1$ intermediate excited states in order to present the formalism. With respect to closed shell atoms, $J_0 = 0 \rightarrow J = 1$ excitations are the general case for resonant Auger processes. Therefore, quantities of a maximum rank of $K = 2$ can occur, only. In order to simplify the discussion we assume an excitation process with either longitudinally or unpolarized electrons, only.

This simplification yields the advantage that the expressions for angular distribution and spin polarization of the emitted Auger electrons remain the same as has been derived for the case of photoexcitation with circularly polarized light (Lohmann

1999b, 2008).[13] In particular, we obtain the angular distribution as

$$I(\theta) = \frac{I_0}{4\pi}\Big(1 + \alpha_2\, \mathcal{A}_{20}\, P_2(\cos\theta)\Big), \tag{3.105}$$

where I_0 denotes the total intensity, and $P_2(\cos\theta)$ is the second Legendre polynomial.

The cartesian components of the spin polarization vector, with respect to the helicity system of the emitted Auger electrons, i.e. z-axis $\parallel$ $\mathbf{k}_{Auger}$ (see Fig. 3.8), can be written as:

$$P_x(\theta) = \frac{\xi_1\, \mathcal{A}_{10}\, \sin\theta}{1 + \alpha_2\, \mathcal{A}_{20} P_2(\cos\theta)}, \tag{3.106}$$

$$P_y(\theta) = \frac{-\frac{3}{2}\,\xi_2\, \mathcal{A}_{20}\, \sin(2\theta)}{1 + \alpha_2\, \mathcal{A}_{20} P_2(\cos\theta)}, \tag{3.107}$$

and

$$P_z(\theta) = \frac{\delta_1\, \mathcal{A}_{10}\, \cos\theta}{1 + \alpha_2\, \mathcal{A}_{20} P_2(\cos\theta)}. \tag{3.108}$$

The dynamics of the electron impact excitation into the intermediate excited A* state can be generally described by a set of *state multipoles* or *statistical tensors* $\mathcal{A}_{KQ}$ (see the previous sections). Particularly, $\mathcal{A}_{10}$ and $\mathcal{A}_{20}$ are known as orientation and alignment parameters. The orientation parameter $\mathcal{A}_{10}$ can be non-zero, only, if a longitudinally polarized electron beam is used for the primary excitation process whereas the alignment parameter $\mathcal{A}_{20}$ can be different from zero even for an unpolarized electron beam.

The Auger decay dynamics is described by the angular distribution parameter α_2 and spin polarization parameters δ_1 and ξ_1, referring to the transferred spin polarization (TSP), and the parameter ξ_2 related to the dynamic spin polarization (DSP). From their structure (3.105)–(3.108) are similar to relations obtained for the emission of photoelectrons.

3.5.7.1 Alignment and Orientation Parameters

For the case of electron impact excitation, both, alignment and orientation parameters are functions of the electron impact transition matrix elements and therefore become dependent of the electron impact energy. Applying a relativistic distorted wave approximation (RDWA), Lohmann et al. (2002) calculated the primary excitation cross section as a function of the electron impact energy in order to identify energy ranges where a comparatively high cross section coincides with large values for the alignment and orientation parameters, respectively. Results for the cross section of the electronically excited $(2p_{3/2}^{-1}4s_{1/2})_{J=1}$ state of argon are plotted in Fig. 3.27a as

[13] As a matter of fact, (3.105)–(3.108) are special cases of (3.61)–(3.64) related to an arbitrarily oriented coordinate frame; see Sect. 3.4 for further details.

a function of the electron impact energy. As can be seen, the cross section is comparatively large close to threshold (248.63 eV) from where it rapidly decreases to its minimum at an electron impact energy of ~300 eV. Then, it continuously increases to a maximum even higher than its threshold value at an energy of ~800 eV. For larger energies we find the cross section slowly decreasing.

Orientation and alignment parameters $\mathcal{A}_{10}$ and $\mathcal{A}_{20}$ are shown in Fig. 3.27b,c (Lohmann et al. 2002). Considering the alignment parameter, for the case of photoexcitation its value remains constant at $\mathcal{A}_{20} = 1/\sqrt{2}$ whereas the orientation parameter takes a value of $\mathcal{A}_{10} = \sqrt{3/2}$. As can be seen from Fig. 3.27 this is no longer the case for the case of electron impact excitation. Moreover, both parameters vary over a wide range of electron impact energy.

Though the orientation parameter varies as a function of electron impact energy and does even change sign at ~300 eV, its magnitude does hardly exceed values of 10 %. Close to threshold negative values can be found whereas at ~400 eV the orientation takes its maximum of ~11 % after changing sign. For larger energies the orientation smoothly decreases to small values. For the calculation a 100 % longitudinally polarized electron beam has been assumed.

In contrast to the photoexcitation case, the alignment has always negative magnitude but shows a dramatic behaviour. Close to threshold an alignment of ~−0.5 has been found which slightly decreases to ~−0.4 at 300 eV electron impact energy. Then, it increases to a maximum value of ~−0.69 at ~550 eV, from where it smoothly decreases to smaller numbers.

3.5.7.2 Dynamic Spin Polarization

The numerical results for the angular distribution and spin polarization parameters for the resonant $\mathrm{Ar}^*(4s_{1/2})L_3M_{2,3}M_{2,3}$ Auger transition have been extensively discussed (Lohmann et al. 2002; Kleinpoppen et al. 2005; Lohmann 2008), while the resonant Auger spectrum has been investigated by Aksela and Mursu (1996) and Farhat et al. (1997). Here, we only mention the large angular distribution parameter $\alpha_2 = -1.12$ for the $^4\mathrm{P}_{1/2}$ and the comparatively large dynamic spin polarization parameter $\xi_2 = -0.228$ for the $^2\mathrm{D}_{5/2}$ final states, respectively.

For the considered type of experiment the dynamic spin polarization parameter can be accessed via observation of the p_y-component of the spin polarization vector, only. Inserting α_2 and ξ_2 values for the $^2\mathrm{D}_{5/2}$ state into (3.107) one needs to identify a range of electron impact energy which yields a large alignment $\mathcal{A}_{20}$ in coincidence with a comparatively large cross section. The cross section gets its maximum around 800 eV where we still have an alignment of ~−0.60. At an energy of 500–600 eV the cross section is still large but we are closer to the maximum of the alignment value of −0.69. Inserting both values of $\mathcal{A}_{20}$ into (3.107) the p_y-component of the spin polarization vector has been plotted as a function of the Auger emission angle. As can be seen from Fig. 3.28, a maximum degree of spin polarization of ~19 % has been obtained for the first, and ~22 % for the latter alignment of the $(2\mathrm{p}_{3/2}^{-1}4s_{1/2})_{J=1}$ excited argon state, respectively.

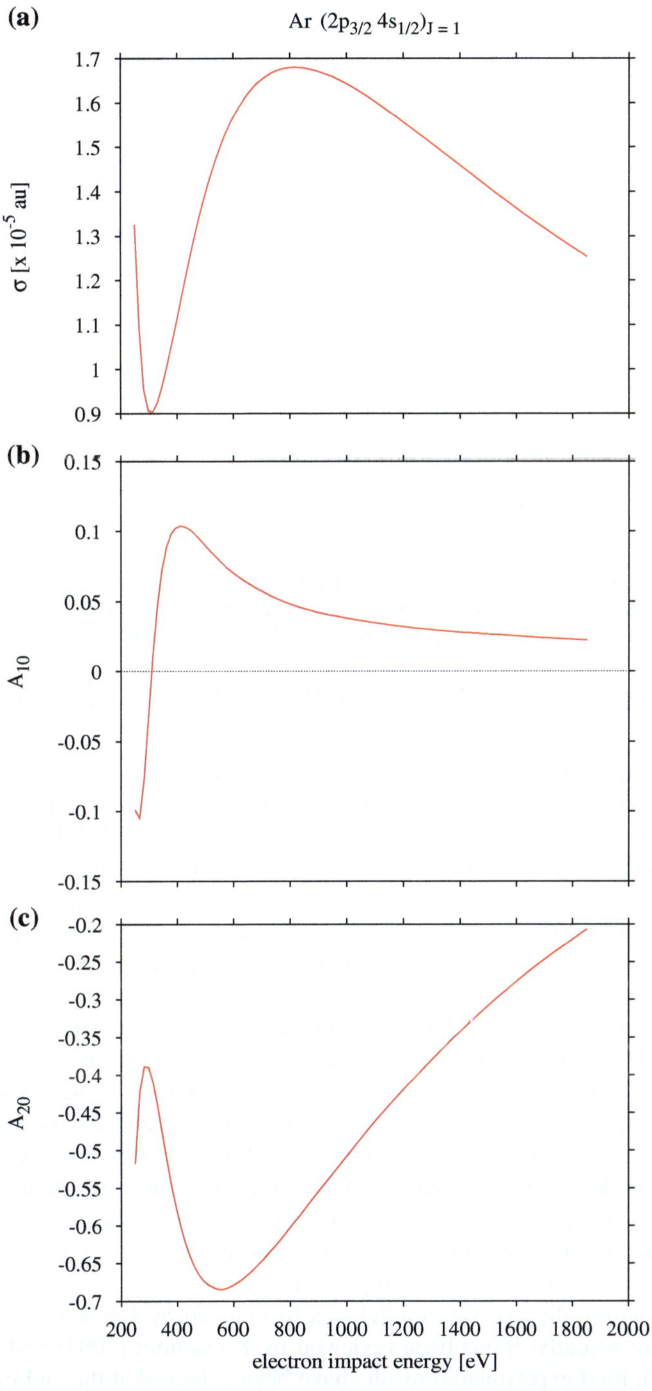

Fig. 3.27 **a** The cross section σ, **b** orientation parameters $\mathcal{A}_{10}$, and **c** alignment parameters $\mathcal{A}_{20}$ for the electronically excited $\mathrm{Ar}^*(2p_{3/2}^{-1}4s_{1/2})_{J=1}$ state; after Lohmann et al. (2002)

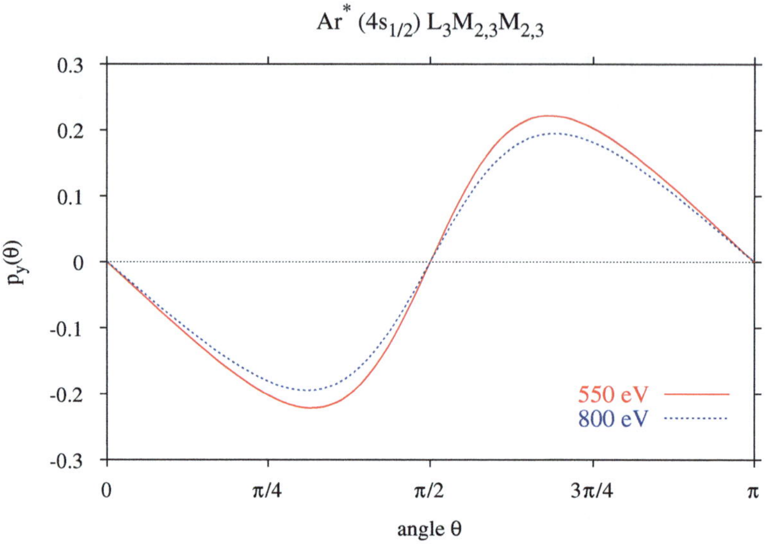

Fig. 3.28 The P_y-component of the spin polarization vector for the resonant $Ar^*(4s_{1/2})L_3M_{2,3}M_{2,3}$ $^2D_{5/2}$ final state Auger transition as a function of the Auger emission angle θ. (——): electron impact energy 550 eV, (· · · · · ·): electron impact energy 800 eV; after Lohmann et al. (2002)

3.6 Tasking Complete Experiments in Atomic Auger Decay

Auger decay (3.1) is a typical example of a half collision process. Similar to the full collision discussed in Chap. 2, one can set the task of a complete experiment for the half collision. Such kind of investigation is meaningful provided the disintegration of the decaying system can be treated independently from its preceding excitation, and thus, the process can be described within the two-step model. This two-step approximation is often appropriate, especially, if the data are integrated over the Auger line in the spectrum. On the far wings of the Auger line, the cross section of the two-step process drops down and the contribution from the direct process leading to the same final state (direct double ionization for the normal Auger decay) cannot be neglected. Besides, post-collision interaction (PCI) between the knocked-out electron and the final state of the system is neglected in the two-step model.

The ultimate goal of a complete experiment on the Auger decay is to obtain from the experiment ambiguously all the complex amplitudes of the Auger decay. Though studies of the Auger process have a long history, reviewed by Mehlhorn (2000), complete Auger decay experiments as a task for determining the decay amplitudes is much younger than the concepts of complete scattering and complete photoionization experiments, respectively. For the Auger decay of atoms, a concept of a complete experiment within the two-step model was formulated by Kabachnik and Sazhina (1990). More recently, it has been reviewed by Kabachnik (2004) and Kabachnik et al. (2007). First experimental results have been achieved at the end of the 1990s for both normal (Grum-Grzhimailo et al. 1999) and resonant (Hergenhahn et al. 1999; Ueda et al. 1998, 1999; West et al. 1998) Auger processes.

The complete experiment normally requires angle-resolved, spin-resolved and coincidence measurements with low counting rate. To enhance the counting rate, the resonant Auger process is often utilized. Complete experiments for the resonant Auger decay, like (3.38) or (3.60), with only one outgoing electron, are a particular case of complete experiments for photoionization. This will be dealt with in Sect. 4.8.2 where a complete experiment for the resonant Auger decay of the $Ar^*(2p_{3/2}^{-1}4s)_J = 1$ state is discussed, while examples of the resonant Auger decay for the $Xe^*(4d_{5/2}^{-1}6p)_{J=1}$ state are considered in Sect. 4.3.3. Here, we concentrate on the normal Auger decay, where the hole is created after ionization by particle or photon impact in an inner atomic shell, with the subsequent Auger decay leading to a doubly charged ion.

In the full collision the total angular momentum of the combined system of the atom and the scattering particle is not fixed, because the particle is usually characterized by a large number of partial waves. This makes the partial wave representation of the scattering amplitudes not practical for the purpose of complete scattering experiments and the representation of projections of angular momenta is used. The objects of complete scattering experiment therefore are angle dependent scattering amplitudes of transitions between different magnetic substates in the combined atom and scattering particle system, or equivalent sets of parameters. Such an approach has been a basis for the analysis in Chap. 2. In contrast, for the Auger decay, the total angular momentum J of the system is fixed. Then, the partial-wave representation of the Auger electron is convenient and complex decay amplitudes,

$$V_{J \to J_f, \ell j} \equiv \langle (J_f, \ell j) J | H - E | J \rangle \equiv V_{\ell j} \exp(i\delta_{\ell j}), \tag{3.109}$$

are becoming the objects of interest for complete experiments on the Auger decay (we omit for brevity other quantum numbers specifying the states). In this equation, the transition operator $H - E$, where H and E are the total atomic Hamiltonian and the total atomic energy, in practical calculations often reduces to the Coulomb interaction V between the atomic electrons. Assuming states with sharp values of J and J_f, we introduced brief notations for the absolute values $V_{\ell j}$, and phases $\delta_{\ell j}$ of the decay amplitudes. The number of independent amplitudes (3.109) governing the Auger decay is restricted and determined by the triangular conditions

$$\mathbf{J}_f + \mathbf{j} + \mathbf{J} = \mathbf{0}, \tag{3.110a}$$

$$\boldsymbol{\ell} + \mathbf{j} + \frac{\mathbf{1}}{\mathbf{2}} = \mathbf{0}, \tag{3.110b}$$

and by conservation of parity,

$$(-1)^{\ell} \pi_f = \pi, \tag{3.111}$$

where π and π_f are the parities of the initial (J) and final (J_f) states, respectively.

Since the absolute Auger decay width can be measured independently, the main efforts in the complete experiment are directed to measurements of dimensionless dynamical parameters of the Auger decay (intrinsic parameters). With enough number of measured intrinsic parameters, subsequent extraction of the absolute ratios, $V_{\ell j}/V_{\ell' j'}$, and the phase differences, $\delta_{\ell j} - \delta_{\ell' j'}$, of the Auger decay amplitudes (3.109) is possible. Measurable quantities include relative intensities of Auger lines; asymmetry parameters of the angular distribution of Auger electrons and components of their spin polarization (for example, the parameters α_2, ξ_1, ξ_2, δ_1 of Sect. 3.4.1); polarization and anisotropy of the angular distribution of secondary products of decay of the residual ionic state after the Auger decay, which carry information on polarization of the residual ion; angular correlations between the Auger electron and the secondary products. Expressions for the observable quantities contain bilinear combinations of the decay amplitudes (3.109). The general formulas as well as expressions for states with particular values of J, π and J_f, π_f can be found in the literature, e.g. Balashov et al. (2000); Berezhko and Kabachnik (1977); Blum et al. (1986); Kabachnik and Sazhina (1984); Lohmann (1990, 1998), and many others. Particularly, we mention the book by Lohmann (2008).

In the general case of the Auger decay of the initial state with the total angular momentum J to the final state with the total angular momentum J_f the number of complex partial amplitudes (3.109) describing uniquely the Auger decay is $2J + 1$ for $J_f > J$ and $2J_f + 1$ for $J_f < J$. For the case $J_f > J$, therefore, $4J + 1$ real parameters need to be determined from a set of measurements (the overall phase of the amplitudes has no physical meaning). If only the Auger electrons from a particular transition are considered and if one measures their angular distribution and spin polarization, then the total number of the dynamical parameters which may be derived from the measurements is also $4J + 1$. One of these parameters can be taken as the absolute Auger rate, while others $4J$ are dimensionless parameters. For example, in the decay of an Auger state with $J = 1$, see Sect. 3.4.1, the four dynamical parameters are the anisotropy parameter α_2 and the three spin-polarization parameters ξ_1, ξ_2, δ_1. However, recently it was found that the dynamical parameters describing angular distribution and spin polarization of the Auger electrons are not independent. There are pure algebraic relations between them based on the parity and angular momentum conservation in the decay, which reduces the number of measurable independent parameters. The first of the relations was found by Schmidtke et al. (2000a, b, 2001), which followed by further relations (Kabachnik and Grum-Grzhimailo 2001; Kabachnik and Sazhina 2002) and, finally, Kabachnik (2005) derived most general relation between the intrinsic parameters of the Auger decay. For illustration, the intrinsic parameters α_2, ξ_1, ξ_2, δ_1 describing angular distribution and spin polarization of the resonant Auger electrons, which have been discussed in Sect. 3.4.1 for the decay of $J = 1$ states, are interrelated by

$$3(\xi_1)^2 + (3\xi_2)^2 = \frac{1}{\sqrt{2}}\left(1 - \sqrt{2}\alpha_2\right)\left(\sqrt{2} + \alpha_2 + \sqrt{3}\delta_1\right). \quad (3.112)$$

As a result, investigations which solely rely on the detection of the Auger electron can constitute the complete experiment in the simplest case of two decay channels, only, for example for J or J_f equal to 1/2. In all other cases, additional information about the polarization state of the residual ion is necessary. One way to gain this information is a measurement of the angular correlations in the cascades, which essentially depend on the polarization of the intermediate states. Another possibility is to measure the angular distributions of either the secondary Auger electrons or the angular distribution, or polarization, of the secondary fluorescence light. Unlike the parameters of the angular distribution and spin polarization of the Auger electrons, the angle-integrated quantities are not related to the phase differences between the decay amplitudes into different channels and therefore allow for an extraction of the relative partial widths $\Gamma_{\ell j}/\Gamma_{\ell' j'}$, where

$$\Gamma_{\ell j} = 2\pi \left| V_{\ell j} \right|^2 . \qquad (3.113)$$

The realization of a complete experiment for Auger decay is simplified, if the number of unknown amplitudes is reduced by more or less plausible model assumptions, for example, by assuming the extreme LSJ coupling approximation for the levels involved, or supposing that the relativistic phase shift splitting in the continuum, $\delta_{\ell j} - \delta_{\ell j'}$ is negligible.

Despite the large variety of measurable quantities, a few factors make the complete experiment for the Auger decay very difficult. In practice, parts of the necessary experiments are hard to perform, and/or parts of the quantities to be measured show a very small magnitude. For instance, measurements of spin polarization of Auger electrons and different kinds of coincidence studies are difficult because of the low counting rate. The spin component of the Auger electron perpendicular to the reaction plane, which is related to the dynamic spin polarization, is usually small (see Sect. 3.4.4. The spin polarization of the Auger state in the normal Auger process is usually also small, but needs to be determined in order to find the intrinsic Auger decay parameters. The polarization of the residual ions is often disturbed by external depolarization effects due to radiative cascades and intraatomic hyperfine interactions, which should be taken into account in the analysis.

Displayed in Fig. 3.29 are results for the normal Auger decay of the 2s hole state $Na^+(2s2p^6 4p\,^3P)$, generated by fast electron impact from sodium atom excited by optical laser pumping (Grum-Grzhimailo et al. 1999, 2001). Here, the Auger decay reduces to the two-channel case with emission of εs and εd electrons, only, after turning to the LSJ coupling approximation, which can be assumed as valid for the sodium case with high accuracy. The absolute ratio of the amplitudes, V_s/V_d, has been deduced from the relative line intensities in the Auger fine structure multiplet, while the angular anisotropy parameter α_2 gave the second relation between the decay amplitudes. The calculations converge to the experiment when improving the theoretical description of the ionic core. Note, that rather sophisticated approximations have been used for the calculation of the Auger decay amplitudes. In particular, close-coupling in the continuum and relaxation of the discrete wavefunctions have

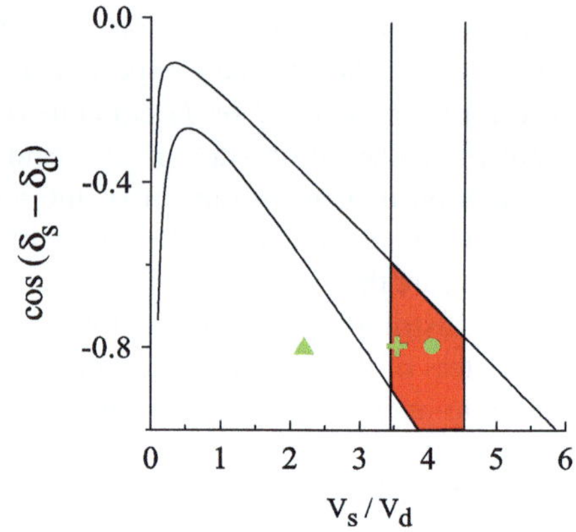

Fig. 3.29 Results of a complete experiment for the Auger decay $Na^{+}(2s2p^{6}4p\,^{3}P) \rightarrow Na^{2+}(2s^{2}2p^{5}\,^{2}P) + e_{Aug}(\epsilon s, \epsilon d)$ (see text). Calculations by Zatsarinny (1995): Hartree-Fock ion core (▲); correlated ionic core (+); correlated ionic core with polarization potential (●). After Kleinpoppen et al. (2005)

been accounted for. The various approximations of wavefunctions show almost no effect on the relative phase of the Auger decay amplitudes. Thus, it seems to be, that the determination of the relative phases less stringently tests the theoretical models than the absolute ratio of the amplitudes. A possible reason for this weak dependence on the wavefunction approximation is that the relative phase is a property integrated from the inner part of the potential of the atom to infinity whereas the moduli depend directly on the quality of wavefunctions being active in the Auger decay.

The above example shows that complete experiments for Auger decay processes are feasible. Such experiments can provide the most detailed information about the mechanisms of the Auger decay. Theoretical descriptions of the complete experiments require highly sophisticated models of the Auger decay.

3.7 Molecular Auger Processes: Angle Resolved Auger Emission from CO Molecules

Investigating the Auger decay of molecules, the two-step model (Mehlhorn 1990),[14] can be well applied to both, atomic or molecular targets, treating the generation of a primary inner shell hole independent of the subsequent Auger emission. Theoretical descriptions for the angular distribution of Auger electrons emitted from molecules have been derived (Bonhoff et al. 1998, 1996, 1997; Bonhoff 1998; Bonhoff et al. 1999; Busalla and Blum 1997; Lehmann et al. 1997; Lehmann and Blum 1997), while Kabachnik et al. (2007) reviewed the field. The key parameters for describing the ionization or excitation of molecules and the subsequent Auger emission are the *anisotropy coefficients*, depending on the Auger decay amplitudes and phases, and the *order parameters*, describing the anisotropy of the excited/ionized molecular ensemble. The order parameters depend on the symmetry of the molecules and on the dynamics of the corresponding excitation process.

[14] We also refer to the discussion in Sect. 3.5.

Considering photoexcited resonant Auger decay of linear molecules in initially unpolarized states, the order parameters become independent of the excitation dynamics and can be obtained analytically. Thus, within the two-step model, the resonant Auger emission of such molecules depends on the anisotropy parameters of angular distribution, only. Therefore, resonant Auger transitions can be seen as an ideal test-case for probing the angle resolved molecular Auger dynamics.

Let us consider the resonant Auger transition of CO molecules in the gas phase after excitation of the $(2\sigma^{-1}2\pi^{1})\,{}^{1}\Pi$ state by linearly polarized photons, where we particularly focus on the calculation and discussion of the Auger angular distribution. This case of diatomic molecules has been numerically investigated by Bonhoff et al. (1999) while measurements have been performed by Hemmers et al. (1993) which allow for a comparison between theory and experiment of the relevant parameters.

Particular emphasis is on the symmetry conditions of the excitation process and its implications for the excited electronic Π states. The relevant anisotropy parameters are defined and their relation to shape and spatial orientation of the electronic orbitals is discussed. The physical importance of the coherence terms is discussed in detail.

The case of Auger processes from *fixed-in-space* molecules is also considered which has been first discussed by Zähringer et al. (1992). Both cases, Auger processes of rotating molecules and of molecules with fixed axis direction, can be treated within the same general framework (Bonhoff et al. 1997). This allows to apply the numerical results to the case of fixed-in-space molecules, too. The obtained data by Bonhoff et al. (1999) allow for some insight into the *stereodynamics* of the Auger decay, that is, the dependence of the Auger emission on the direction of the molecular axis and on shape and spatial orientation of the initial electronic Π states.

3.7.1 Theoretical Framework

3.7.1.1 Experimental Conditions and Theoretical Assumptions

The resonant Auger process can be treated as a two-step process

$$\gamma + \mathrm{CO} \longrightarrow \mathrm{CO}^{*}(2\sigma^{-1}2\pi^{1})\,{}^{1}\Pi \tag{3.114a}$$

$$\hookrightarrow \mathrm{CO}^{+} + e_{Auger} \tag{3.114b}$$

In the first step (primary process) the molecules are excited by a completely linearly polarized photon beam γ into the $(2\sigma^{-1}2\pi^{1})\,{}^{1}\Pi$ state; the inner shell C(1s) electron is excited to the $2\pi^{*}$ orbital C(1s) $\rightarrow 2\pi^{*}$. In the subsequent resonant Auger decay the Auger electron is emitted during the transition of the molecule into a singly ionized CO^{+} molecule in various possible final states.

The CO molecules are initially in the ${}^{1}\Sigma^{+}$ ground state and freely rotating with isotropic axis distribution. After the primary excitation (3.114a) the axis distribution

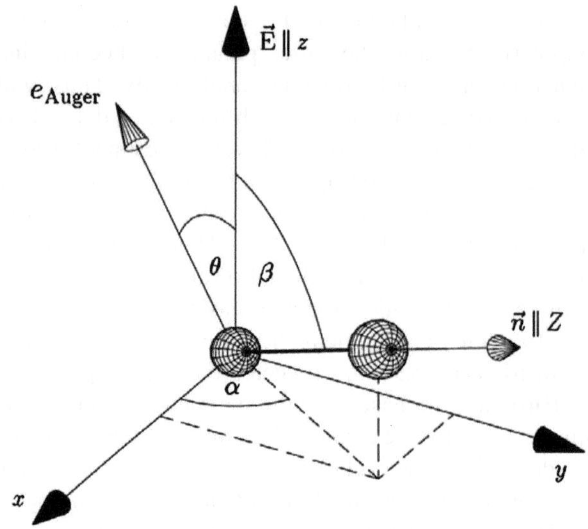

Fig. 3.30 Angles for molecular axes and Auger electrons related to the space-fixed frame xyz; after Bonhoff et al. (1999)

will, in general, be anisotropic (Dill et al. 1980). In addition, shape and spatial orientation of the excited orbitals will influence the subsequent Auger decay (3.114b) and must be taken into account (Bonhoff et al. 1996, 1998). The basic approximation is that the molecular rotation between excitation (3.114a) and decay (3.114b) can be neglected.

3.7.1.2 The Excitation Process

Let us first consider the process (3.114a). It is assumed that the incident light is linearly polarized. The direction of the electric field vector is taken as the z axis of the laboratory coordinate system, x and y can be chosen arbitrarily but orthogonal to each other. The molecular axis $\mathbf{n}$ is chosen as the Z axis of a molecule fixed system. Its direction is specified by the polar angle β and the azimuthal angle α with respect to the laboratory system (see Fig. 3.30). The third Euler angle γ fixes the orientation of the molecular X–Y plane. It is customary to put $\gamma = 0$. This implies that the molecular X axis lies in the z–Z plane. For example, for $\beta = \pi/2$ the X axis would be antiparallel to the laboratory z axis.

Considering the experimental conditions, and taking the dipole approximation into account, the excited molecular ensemble possesses the following symmetry conditions (e.g., see Bonhoff et al. 1996, and refs. therein):

1. The axis distribution is axially symmetric with respect to the laboratory z axis.
2. The ensemble remains invariant under a reflection in any plane through z, in particular any z–Z plane.

The states of molecules with axis $\mathbf{n}$ pointing in the direction (β, α) will be described by state vectors $|\Omega, \mathbf{n}\rangle$. Here, Ω is the component of the total electronic angular

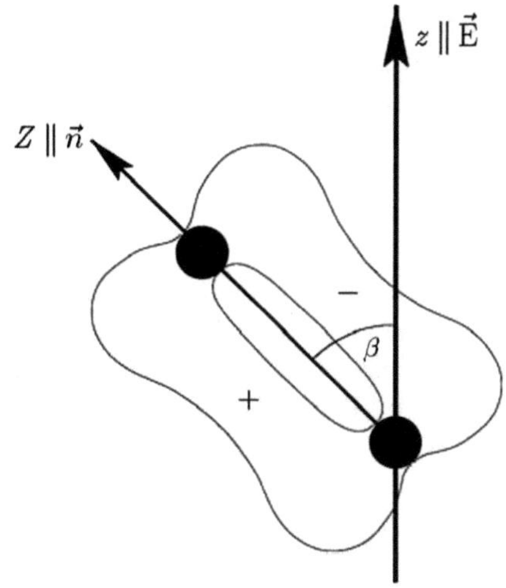

Fig. 3.31 Projection of the excited Π_X state on the z–Z plane, spanned by the molecular axis **n** and the electric field vector **E**. + and − mark the relative sign of the wavefunction; after Bonhoff et al. (1999)

momentum with regard to **n**. For $^1\Pi$ states this means $\Omega = \pm 1$. The corresponding charge distributions are axially symmetric around **n**. Any other electronic $^1\Pi$ state can be written as a linear superposition of these two basis states. In order to depict the directional properties of $^1\Pi$ states it is often convenient to use the following two states as basis

$$\left|\Pi_X\right\rangle, \mathbf{n} = -\frac{1}{\sqrt{2}}\Big(\left|1,\mathbf{n}\right\rangle - \left|-1,\mathbf{n}\right\rangle\Big), \tag{3.115a}$$

$$\left|\Pi_Y,\mathbf{n}\right\rangle = \frac{\mathrm{i}}{\sqrt{2}}\Big(\left|1,\mathbf{n}\right\rangle + \left|-1,\mathbf{n}\right\rangle\Big). \tag{3.115b}$$

The symmetry plane of the Π_X state lies in the z–Z plane, the Π_Y state is perpendicular to it.

Condition 2. has an important consequence for the allowed orbitals. It has been shown that the Π_Y state cannot be excited under the given experimental conditions (Bonhoff et al. 1998; Blum 1996, Sect. 7.7.5), and any molecule is in the Π_X state (note that this result does not hold for ionization processes). The corresponding charge distribution is therefore not axially symmetric around the molecular axis (see Fig. 3.31), and this anisotropy will be reflected in the angular distribution of the Auger electrons.

3.7.1.3 Angular Distribution of Emitted Auger Electrons

Let us now discuss the Auger process (3.114b). The Auger angular distribution $I(\theta)$ is given by the expression (Dill et al. 1980; Bonhoff et al. 1998)

$$I(\theta) = \frac{I_0}{4\pi}\Big(1 + \beta P_2(\cos\theta)\Big)\,, \tag{3.116}$$

where θ is the angle between z and the emission direction of the electron (see Fig. 3.30). The asymmetry parameter β can be expressed in terms of the anisotropy parameters $A_K(\Omega', \Omega)$

$$\beta = -\frac{1}{5}\frac{A_2(1,1) + \sqrt{6}A_2(-1,1)}{A_0(1,1)} . \tag{3.117}$$

The parameters A_K solely characterize the dynamics of the Auger emission. Thus, they are not influenced by the primary excitation (3.114a). Explicit expressions for anisotropy parameters of angular distribution for molecular Auger emission have been derived by Bonhoff et al. (1996, 1998). In particular, for the considered diatomic case (CO), we write

$$\begin{aligned} A_K(\Omega', \Omega) = {} & \frac{(2K+1)}{4\pi k_{Auger}} \sum_{\substack{\Omega_f m_s \\ \ell' \ell m' m}} \mathrm{i}^{\ell+\ell'} (-1)^{\ell'+m} \sqrt{(2\ell+1)(2\ell'+1)} \\ & \times \begin{pmatrix} \ell' & \ell & K \\ m' & -m & \Omega - \Omega' \end{pmatrix} \begin{pmatrix} \ell' & \ell & K \\ 0 & 0 & 0 \end{pmatrix} \\ & \times \langle \Omega_f \ell' m' m_s | \mathbf{T} | \Omega' \rangle \langle \Omega_f \ell m m_s | \mathbf{T} | \Omega \rangle^* . \end{aligned} \tag{3.118}$$

Specializing to $^1\Pi$ states we have $\Omega', \Omega = \pm 1$. ℓ and ℓ' denote the partial waves of the Auger electrons, m and m' are the corresponding components with respect to **n**. The symmetry restriction

$$A_K(\Omega', \Omega) = A_K(-\Omega', -\Omega) \tag{3.119}$$

follows from reflection invariance in the molecular X–Z plane. Equations (3.117) and (3.118) are the basis for the numerical calculation. The rank K in (3.118) is limited by the number $2\ell_{max}$, only, where ℓ_{max} is the largest angular momentum of the partial waves of the emitted Auger electron. Generally, ℓ_{max} is not restricted by any triangle rules and can be taken finite only due to the decreasing magnitude of the decay amplitudes with increasing angular momentum.

It is important to note that two anisotropy parameters, $A_2(1, 1)$ and $A_2(-1, 1)$, contribute to the angular distribution parameter β. This is due to the fact that the molecules are in pure Π_X states which are represented by the linear superposition (3.115a). In all bilinear expressions containing these state vectors, coherence terms with $\Omega' = 1$ and $\Omega = -1$ (and vice versa) will necessarily occur. This coherence between the two basis Ω states is responsible for the term $A_2(-1, 1)$. This anisotropy parameter would vanish, if the molecules were in states with sharp Ω or in incoherent superposition states of the two basis Ω states.

These coherence terms have sometimes been overlooked in the literature, and have been omitted in general equations. The geometric importance of these terms will be discussed in Sect. 3.7.3 and their meaning will be illustrated by some examples.

3.7.2 Numerical Results

The basic numerical problem is the calculation of the transition matrix elements $\langle \Omega_f \ell m m_s | \mathbf{T} | \Omega \rangle$ in 3.118. We will briefly describe the procedure.

The electronic ground state of CO consists of 14 electrons in the configuration of electronic orbitals: $(1\sigma^2 2\sigma^2 3\sigma^2 4\sigma^2 1\pi^4 5\sigma^2)\,^1\Sigma^+$. The program developed by Schimmelpfennig and co-workers (Schimmelpfennig et al. 1995; Schimmelpfennig and Peyerimhoff 1996; Bonhoff et al. 1997) has been used for calculating the anisotropy parameters of several resonant Auger transitions in CO.

The orbitals have been optimized in a self-consistent-field (SCF) calculation for the decaying $(2\sigma^{-1} 2\pi^1)\,^1\Pi$ state to describe all involved states with configuration interaction (CI) functions.

The multi-centre basis sets are similar to those which have been used in the calculations of angle-integrated Auger spectra of CO (Schimmelpfennig and Peyerimhoff 1996). For the description of the carbon atom an uncontracted (10s, 6p, 1d) basis of primitive Gaussians given by Huzinaga (1965) has been employed, where four s, five p, nine d, and three f functions have been added, optimized to model Bessel functions. For the oxygen atom also, a (10s, 6p, 1d) Huzinaga basis set has been used at the equilibrium geometry $R = 2.132$ au (Herzberg 1965), where three s and two d functions have been added.

We investigate Auger transitions of a C(1s) vacancy. Therefore the molecule fixed coordinate system XYZ is chosen in such a way that the origin of the system coincides with the carbon atom and that the Z axis points in the direction of the oxygen atom (see Fig. 3.32).

As performed by Schimmelpfennig and Peyerimhoff (1996) the CI space has been restricted to configurations which distribute electrons in the five σ, the first π, plus the virtual 6σ and the 2π orbitals.

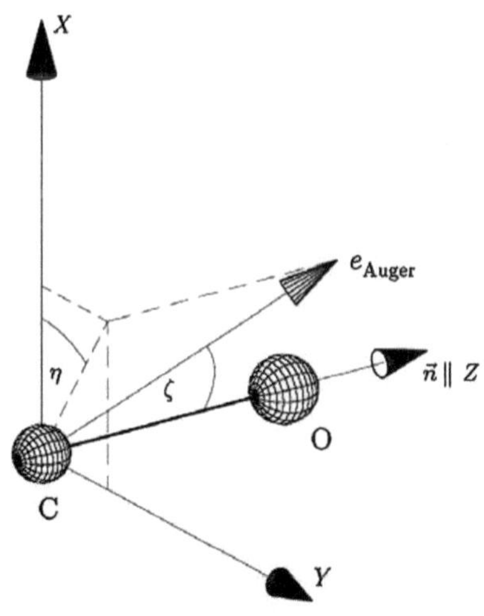

Fig. 3.32 Angles of the Auger electron emission related to the molecular frame (after Bonhoff et al. 1999)

To perform the partial wave expansion of the Auger electron in a suitable way, a one-centre basis expansion has been introduced, where a (14s, 11p, 10d, 3f) basis has been employed for the carbon atom and (5s, 3p, 5d, 8f, 10g, 8h, 5i) Gaussian functions have been added up to $\ell = 6$. Further technical details of the numerical method may be found in the literature (Schimmelpfennig et al. 1995; Schimmelpfennig and Peyerimhoff 1996; Bonhoff et al. 1997).

The various final states are listed in Table 3.17 together with the numerically obtained energies E_{Auger} and the total intensities I_0 of the Auger electrons.

For the calculation of the asymmetry parameter β only $A_2(1, 1)$ and $A_2(-1, 1)$ are required according to (3.117). In Table 3.18 the anisotropy parameters are presented up to $K = 8$. The additional coefficients will be needed for the discussion of the Auger decay of *fixed-in-space* molecules in Sect. 3.7.3.

Using the results for the coefficients A_K we calculated the asymmetry parameter β by means of (3.117). The results are given in Table 3.17 and compared with experimental results by Hemmers et al. (1993). They measured an angle-resolved Auger spectrum of the $(2\sigma^{-1}2\pi^1)\,^1\Pi$ initial state after excitation by linearly polarized light and determined the parameter β for the energetically separated participator transitions (a)–(c). Numerical and theoretical values are in good agreement except for the transition to the $^2\Pi$ final state.

The direct comparison of angle resolved Auger spectra yields further results. Hemmers et al. (1993) determined the energy resolved Auger spectrum for two different polar angles: $\theta = 0°$ and $\theta = 54.7°$. It follows from (3.116) for the Auger angular distribution

$$I(0°) = \frac{I_0}{4\pi}(1+\beta)\,, \tag{3.120a}$$

$$I(54.7°) = \frac{I_0}{4\pi}\,. \tag{3.120b}$$

Using (3.120) and the numerical results of Table 3.17, the Auger spectra have been calculated for these two polar angles which can directly be compared with the exper-

Table 3.17 Calculated final states for the $(2\sigma^{-1}2\pi^1)\,^1\Pi$ initial vacancy of CO

Initial state $(2\sigma^{-1}2\pi^1)\,^1\Pi$						
	Final state	E_{Auger}	I_0	β	β_{exp}	FWHM
a	$(5\sigma^{-1})^2\Sigma^+$	271.8	1.88	0.820	0.79	0.35
b	$(1\pi^{-1})^2\Pi$	267.9	3.99	0.997	0.67	0.35
c	$(4\sigma^{-1})^2\Sigma^+$	265.1	0.90	1.170	1.10	0.30
d	$(1\pi^{-1}5\sigma^{-1}2\pi^1)^2\Sigma^-$	263.0	0.59	−1.000		0.70
e	$(1\pi^{-1}5\sigma^{-1}2\pi^1)^2\Delta$	262.9	1.41	−0.051		0.35
f	$(5\sigma^{-2}2\pi^1)^2\Pi$	262.9	2.63	−0.535		0.70
g	$(1\pi^{-1}5\sigma^{-1}2\pi^1)^2\Sigma^+$	260.9	1.75	1.000		1.20

Auger energy E_{Auger} in eV and total intensity I_0 in 10^{-4} au. Asymmetry parameter β for the Auger transitions, compared with the experimental data obtained by Hemmers et al. (1993), line width (FWHM in eV) used in the fit for the angle resolved Auger spectra of CO; see Fig. 3.33 (after Bonhoff et al. 1999)

Table 3.18 Anisotropy parameters $A_K(1, 1)$ and $A_K(1, -1)$ (in 10^{-5} au) for the Auger transitions of the $(2\sigma^{-1}2\pi^1)\,^1\Pi$ initial state of CO; after Bonhoff et al. (1999)

Initial state $(2\sigma^{-1}2\pi^1)\,^1\Pi$

	a: $(5\sigma^{-1})\,^2\Sigma^+$		b: $(1\pi^{-1})\,^2\Pi$		c: $(4\sigma^{-1})\,^2\Sigma^+$	
K	$A_K(1,1)$	$A_K(1,-1)$	$A_K(1,1)$	$A_K(1,-1)$	$A_K(1,1)$	$A_K(1,-1)$
0	1.493 064	–	3.172 947	–	0.713 700	–
1	0.757 647	–	−0.440 562	–	−0.502 893	–
2	0.672 008	−2.773 359	−3.636 175	−4.973 663	−0.303 543	−1.580 755
3	−0.685 771	−1.343 900	0.000 286	−0.437 634	0.479 229	0.905 191
4	−1.928 709	−3.273 790	2.454 639	4.293 749	−0.338 910	−0.603 455
5	0.217 702	0.095 669	0.051 055	−0.783 550	−0.041 984	−0.011 009
6	0.002 266	−0.148 275	−0.321 254	−0.803 853	−0.016 001	−0.056 906
7	−0.276 341	−0.369 037	0.332 763	0.045 681	0.075 915	0.093 795
8	−0.222 844	−0.290 066	0.251 303	0.011 848	−0.046 831	−0.063 399

	d: $(1\pi^{-1}5\sigma^{-1}2\pi^1)\,^2\Sigma^-$		e: $(1\pi^{-1}5\sigma^{-1}2\pi^1)\,^2\Delta$		f: $(5\sigma^{-2}2\pi^1)\,^2\Pi$	
K	$A_K(1,1)$	$A_K(1,-1)$	$A_K(1,1)$	$A_K(1,-1)$	$A_K(1,1)$	$A_K(1,-1)$
0	0.472 510	–	1.123 147	–	2.096 299	–
1	−0.255 755	–	−0.637 162	–	−2.026 476	–
2	0.138 398	0.908 007	0.277 636	0.003 669	4.439 762	0.474 941
3	0.262 687	−0.470 739	0.649 162	0.002 616	−2.150 517	0.266 477
4	−0.597 542	0.957 477	−1.344 792	−0.001 331	4.372 643	−0.788 663
5	0.003 366	0.002 939	0.001 677	−0.000 912	0.212 403	0.087 913
6	0.005 697	0.004 309	0.008 658	0.000 288	0.877 839	0.146 337
7	−0.009 722	0.013 043	−0.011 294	−0.000 116	0.395 541	0.028 208
8	−0.018 414	0.023 655	−0.063 850	−0.000 064	0.932 627	−0.064 801

	g: $(1\pi^{-1}5\sigma^{-1}2\pi^1)\,^2\Sigma^+$	
K	$A_K(1,1)$	$A_K(1,-1)$
0	1.394 768	–
1	−0.874 831	–
2	0.001 787	−2.846 329
3	0.898 450	1.610 152
4	−1.279 726	−2.134 258
5	−0.037 929	−0.044 102
6	0.002 168	−0.072 522
7	0.021 380	0.024 136
8	−0.108 703	−0.142 737

imental data by Hemmers et al. (1993); see Fig. 3.33. Here, Gaussian functions have been used with a full width half maximum (FWHM) between $0.3 - 1.2$ eV (see Table 3.17) to get a continuous spectrum fitting the experimental curves. The results show that the numerical calculations are able to predict the main features of the Auger decay very satisfactorily.

These results are based on the investigation of the axis distribution and the shape and spatial orientation of the electronic lobes. Omitting the second point, we would get an expression for the parameter β where the term $\sqrt{6}A_2(-1, 1)$ vanishes. Though, the numerical results clearly show that this term gives an important contribution to

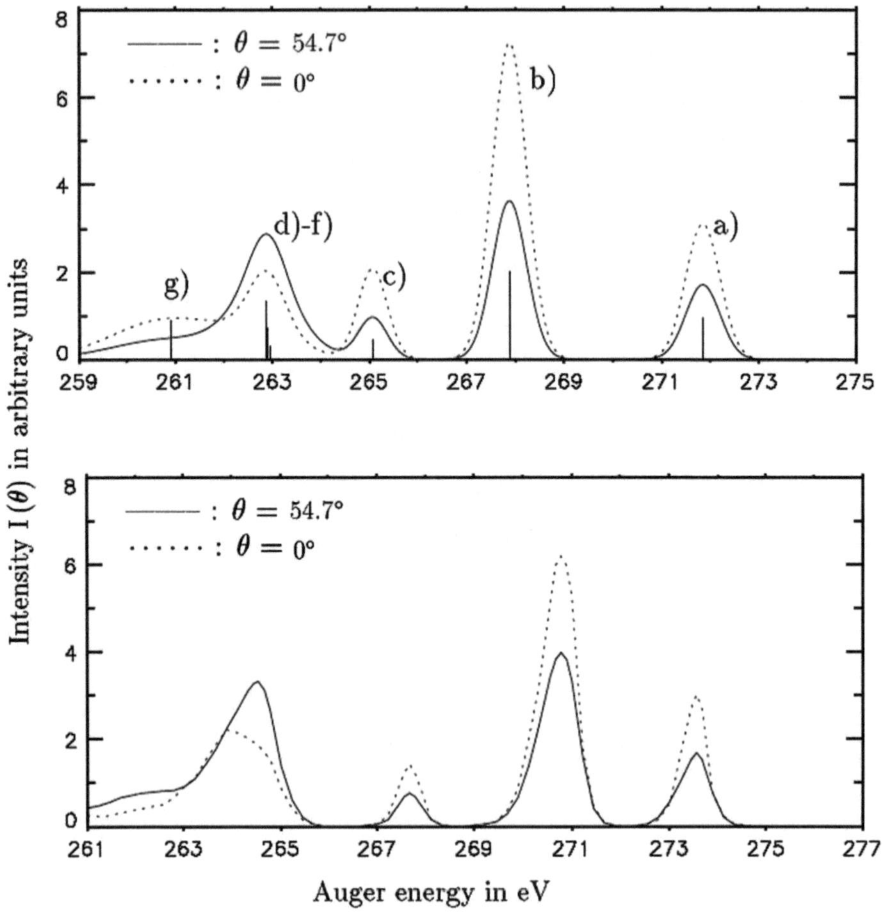

Fig. 3.33 Auger spectrum for the transitions of the $(2\sigma^{-1}2\pi^{1})\,^{1}\Pi$ vacancy of CO created by absorption of linearly polarized photons. (*Top*) calculated spectrum (Bonhoff et al. 1999), the labels refer to Table 3.17. (*Bottom*) experimental spectrum by Hemmers et al. (1993)

the β parameter and, hence, to the angular distribution of the Auger electrons. If, for example, for the $(5\sigma^{-1})\,^{2}\Sigma^{+}$ final state this coherent term is not taken into account, the asymmetry parameter would get the value $\beta = -0.090$, which would not agree with the experimental data.

3.7.3 Auger Emission from Fixed-in-Space Molecules

It is interesting to consider the Auger emission (3.114b) from a sub-ensemble of the excited molecules, namely those molecules whose axes **n** are pointing in one and the same direction (β, α). In principle, the Auger decay from such a sub-ensemble can be observed in fragmentation processes where one of the molecular fragments and the Auger electron are detected in coincidence (Zheng et al. 1995). The results for these *fixed-in-space* molecules will be most helpful in order for clarifying and illustrating the meaning of the anisotropy coefficients $A_K(1, 1)$ and $A_K(-1, 1)$.

Using the general results of Bonhoff et al. (1997) one obtains for the angular distribution of Auger electrons

$$I_{\mathbf{n},|\Pi_X\rangle}(\zeta,\eta) = \sum_K A_K(1,1)P_K(\cos\zeta) - \cos 2\eta \sum_K A_K(-1,1)\, d^{(K)}_{20}(\zeta)\,. \quad (3.121)$$

The indices $\mathbf{n}$ and Π_X indicate that (3.121) describes the decay of molecules in $^1\Pi_X$ states and with axes $\mathbf{n}$ pointing in one direction. ζ and η are the polar and azimuthal angles of the emission direction with respect to the molecular system (see Fig. 3.32). As described in Sect. 3.7.1, the molecular X axis lies in the z–Z plane (the space-fixed z axis is not shown in Fig. 3.32). The $d^{(K)}_{20}(\zeta)$ are elements of the reduced rotation matrices (Zare 1988).

It should be noted that measurements of $I_{\mathbf{n},|\Pi_X\rangle}$ for several ζ and η values allows for determining the parameters $A_K(1, 1)$ and the coherence parameters $A_K(-1, 1)$, separately. The anisotropy coefficients A_K are given by (3.118).

Equation (3.121) differs from (3.116); coefficients of higher rank than $K = 2$ contribute and, both, even and odd K values are present. As has been mentioned, no general upper limit of K can be given. Numerically, Bonhoff et al. (1999) have found that coefficients become in general small for higher K values. The results given in Table 3.18 show that coefficients up to $K = 4$ are dominant. Similar results have been obtained for HF (Bonhoff et al. 1997, 1998).

The dominance of terms with $K \leq 4$ underpins the reliability of a one-center atomic model for calculating the molecular Auger decay matrix elements, at least for diatomic molecules (Fink et al. 2000). This has been confirmed for the resonant Auger decay of the photoexcited $CO^*(2\sigma^{-1}2\pi^1)\,^1\Pi$ states by Fink et al. (2009), who obtained numerical results close to the discussed data.

More precisely, (3.121) contains the complete information on the *stereodynamics* of the Auger decay, that is the dependence of the Auger emission on the direction of the molecular axis and on shape and spatial orientation of the electronic orbitals. In particular, the second term in (3.121) reflects the anisotropy of the initial Π_X orbital. This term stems from the coherent excitation of the two states with opposite sign of Ω_1. For singlet states this leads to an η dependence of the form $\cos(\Omega_1 - \Omega_1')\eta$ and to a ζ dependence of the form $d^{(K)}_{\Omega_1-\Omega_1' 0}(\zeta)$. This shows clearly the influence of the coherence terms on the η and ζ dependence.

For molecules in states with sharp Ω_1 (or an incoherent superposition of the two Ω_1 states) the second term in (3.121) therefore vanishes. The Auger angular distribution would then be independent on η, reflecting the axial symmetry of the initial state with respect to $\mathbf{n}$.

In order to illustrate these points the angular distribution (3.121) has been calculated for CO molecules initially in $(2\sigma^{-1}2\pi^1)\,^1\Pi$ states and decaying to various

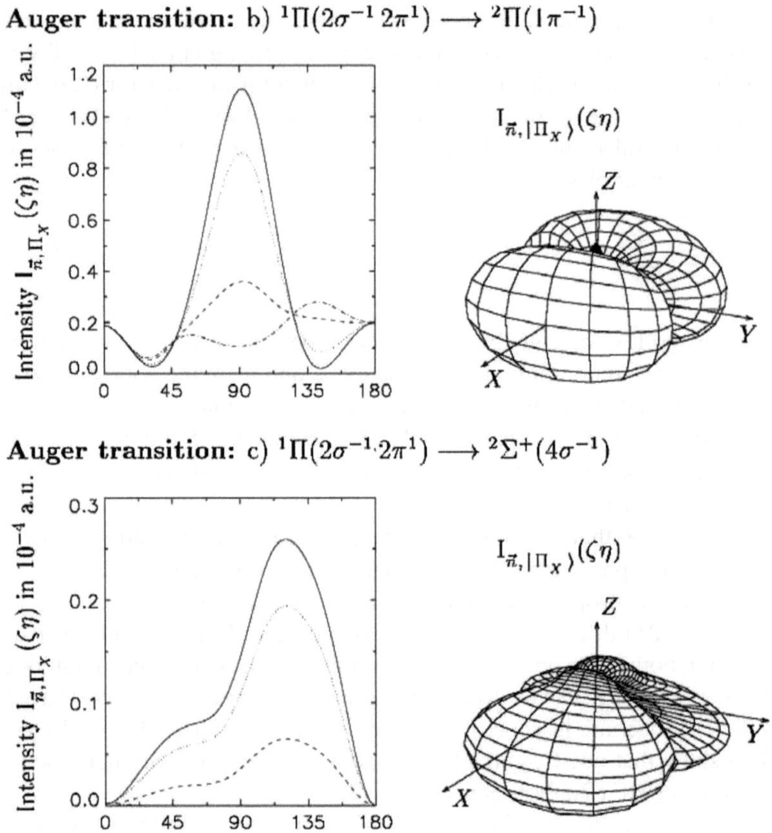

Fig. 3.34 Angular distribution of the emitted Auger electrons from axis-fixed CO molecules; the molecules are initially excited in a pure $|\Pi_X\rangle$ state. (*Left column*): $I_{\mathbf{n},|\Pi_X\rangle}$ as a function of ζ for various fixed values of η. (*Right column*): $I_{\mathbf{n},|\Pi_X\rangle}(\zeta,\eta)$ as a spherical plot. (——): $\eta = 0°$, (······) : $\eta = 30°$, (− − −) : $\eta = 60°$, (− · −) : $\eta = 90°$. Labels **b** and **c** refer to Table 3.17; after Bonhoff et al. (1999)

final states, using the anisotropy parameters of Table 3.18. The results are shown in Figs. 3.34 and 3.35. Here, the Auger intensity $I_{\mathbf{n},|\Pi_X\rangle}$ is shown as a function of ζ for various η values.

The results show in general a strong η dependence. The angular distribution of the Auger electrons is not cylindrically symmetric around the molecular axis. This is illustrated by the polar plots on the right-hand side of Figs. 3.34 and 3.35. As before, XYZ denote the molecular coordinate system. The electric field vector of the exciting light lies in the X–Z plane with angle β to Z (shown in Fig. 3.31).

Figures 3.34 and 3.35 reveal the *stereodynamics* of the Auger decay. In the case of the $(1\pi^{-1})\,^2\Pi$ final state (Fig. 3.34b) the emission is dominant for $\zeta = 90°$ and $\eta = 0°$, that is perpendicular to the molecular axis and parallel to the initial Π_X state. Since the excitation probability, process (3.114a), is proportional to $\sin^2\beta$ it follows that the Auger electrons are preferentially emitted parallel to the electric

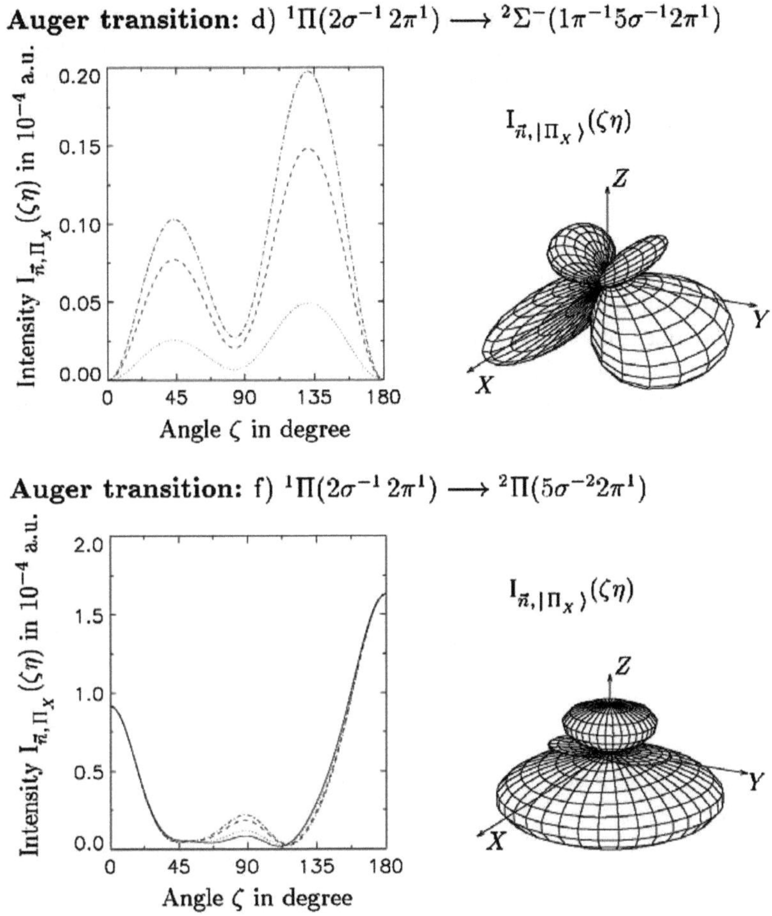

Fig. 3.35 Angular distribution of the emitted Auger electrons from axis-fixed CO molecules (continued). Labels **d** and **f** refer to Table 3.17; after Bonhoff et al. (1999). See Fig. 3.34 for explanations

field direction, corresponding to a large positive value for the asymmetry parameter β (Table 3.17). The same arguments hold for the $(4\sigma^{-1})\,{}^2\Sigma^+$ final state in Fig. 3.34c.

For the transition to the $(5\sigma^{-2}1\pi^1)\,{}^2\Pi$ final state (Fig. 3.35f) emission parallel or antiparallel to the molecular axis is favoured; the η dependence is very small. This leads to an emission preferentially perpendicular to the electric field vector **E** (because the axes of most excited molecules are perpendicular to **E**); the asymmetry parameter β has a negative value.

An interesting case is the transition to the $(1\pi^{-1}5\sigma^{-1}2\pi^{1})\,^2\Sigma^-$ final state (Fig. 3.35d). Because of symmetry conditions, the emission into the direction of the molecular X–Z plane is forbidden (compare Figs. 3.34 and 3.35, for details see Bonhoff 1998). Because of the defined shape and spatial orientation of the molecular orbitals after photoabsorption, s no electron can be emitted parallel to the electric field vector. So the asymmetry parameter must get the extreme value $\beta = -1$ for analytical reasons.

More examples of angular distributions for the decay of the considered CO Auger states may be found in Fink et al. (2009). Measurements by Rolles et al. (2008) confirm our general understanding of such type of Auger transitions. Though, spin resolved electron–photon coincidence experiments as have been proposed by Bederson and Fano (e.g. Bederson 1969a, b; Fano 1957) are still at the brink of research concerning the full, i.e. complete, analysis of all available physical information.

Chapter 4
Complete Experiments in Atomic Photoionization

4.1 General Theoretical Background

4.1.1 Formalized Definition of Complete Experiments

Quantum reactions are completely characterized by the absolute magnitudes and phase differences of the complex amplitudes of transitions between the initial state and each of the degenerate final states (channels). A complete set of experimental data is one from which all the amplitudes can be extracted within a certain theoretical model, as was first discussed at the end of the 1960s and the beginning of the 1970s by Bederson (1969a), Kleinpoppen (1971), Eminyan et al. (1973). From this information any observable can be predicted and the most comprehensive test of theory can be provided up to the degree allowed by quantum mechanics. Discussions and implementations of the complete photoionization experiments started later in the papers by Duncanson et al. (1976); Cherepkov (1979); Heinzmann (1980a, b); Kessler (1981). Many experimental and theoretical investigations of atomic photoionization have been motivated by the quest for such an experiment. Before going into further details on complete experiments in photoionization it is instructive to put the definition of complete experiment on more formalistic grounds.

Let a state of a system be characterized before the reaction by the density operator ρ_0. After the reaction the system is transformed into a state described by the density operator

$$\rho = \mathbf{S}\,\rho_0\,\mathbf{S}^+, \tag{4.1}$$

where $\mathbf{S}$ is the scattering operator. Suppose that the detector(s) of the final state(s) is/are *tuned* to a state described by the density operator ε, which means that the registration occurs only, if the detector is affected by the system in the state ε. With appropriate normalization of the density operators, according to the quantum mechanical postulates, the intensity of the detected signal is

H. Kleinpoppen et al., *Perfect/Complete Scattering Experiments*,
Springer Series on Atomic, Optical, and Plasma Physics 75,
DOI: 10.1007/978-3-642-40514-3_4,

$$I = \mathrm{tr}\; \rho\varepsilon. \tag{4.2}$$

Choosing convenient basis sets, $|\xi_i\rangle$ and $|\chi_f\rangle$, for the initial and final states of the system, respectively, we obtain the intensity as

$$\begin{aligned} I &= \sum_{ff'} \langle \chi_f | \mathbf{T}\rho_0\mathbf{T}^+ | \chi_{f'}\rangle \langle \chi_{f'} |\varepsilon| \chi_f\rangle \\ &= \sum_{ff'ii'} \langle \xi_i |\rho_0| \xi_{i'}\rangle \langle \chi_{f'} |\varepsilon| \chi_f\rangle\, T_{i\to f}\, T^*_{i'\to f'}, \end{aligned} \tag{4.3}$$

where $\mathbf{T} = \mathbf{S} - \mathbf{1}$ is the transition operator, $\mathbf{1}$ is the unity operator. We have introduced the transition amplitudes $T_{i\to f} \equiv \langle \chi_f |\mathbf{T}| \xi_i\rangle$. The coefficients in the bilinear combination (4.3) of the amplitudes $T_{i\to f}$ are to be determined within a certain theoretical framework. Tuning the detector to a different state ε', another bilinear combination of the same amplitudes must be measured

$$I' = \sum_{ff'ii'} \langle \xi_i |\rho_0| \xi_{i'}\rangle \langle \chi_{f'} |\varepsilon'| \chi_f\rangle\, T_{i\to f}\, T^*_{i'\to f'}, \tag{4.4}$$

and so on. To perform the complete experiment means to choose the detector states $\varepsilon, \varepsilon', \ldots$ and/or to change the initial state ρ_0 in order to find all the independent amplitudes $T_{i\to f}$, relevant to the process under study, from the set of equations provided by the measurements of $I, I', \ldots$. The total number of the amplitudes, and therefore recipes for the complete experiment, are model dependent.

4.1.2 Multipole Expansion of Photoionization Amplitude

We consider single photoionization of atom A in the *weak* electromagnetic field

$$\underbrace{h\nu + A}_{|a\rangle} \longrightarrow \underbrace{A^+ + e^-_{ph}}_{|b\rangle}. \tag{4.5}$$

We imply that the field can be treated classically and the process is well described within the first-order perturbation theory. Then, the amplitude of the photoionization can be written as

$$T_{a\to b} = C\left\langle b\left|\sum_n \left((\mathbf{p_n A}) + \frac{1}{2}(\sigma_n[\nabla \times \mathbf{A}])\right)\right| a\right\rangle, \tag{4.6}$$

where $|a\rangle$ and $|b\rangle$ are the initial and the final states of atom as indicated in (4.5), the summation is over atomic electrons. The factor C can be neglected in the further discussion. Note, that in the case of photoionization of molecules equation (4.6) remains valid within the assumption of 'fixed-nuclei' approximation, when the ionization proceeds much faster than the nuclear motion. In (4.6), **A** is the vector potential of the plane electromagnetic wave with the polarization vector **e** and the wave vector **k**; **p** and $\boldsymbol{\sigma}$ are the operators of the linear momentum and spin of the electron, respectively. Let us expand the vector potential in multipoles

$$\mathbf{A} = \mathbf{e}\exp(\mathrm{i}\mathbf{kr}) = \sum_{\substack{\pi=0,1\\ L=1,\ldots}} \sum_{\lambda=-L}^{L} A_\lambda^{\pi L}, \tag{4.7}$$

where the number π takes values of zero or unity for electric and magnetic multipoles, respectively, L is the photon multipolarity and its projection λ takes the values $\lambda = \pm 1$ (helicity) provided the quantization axis is chosen along the photon beam, but generally can take the values $\lambda = 0, \pm 1, \ldots, \pm L$ in any other coordinate system. The convenience of the multipole expansion (4.7) is determined, besides a small expansion parameter (see below), by the fact that its each term transforms under rotation as an irreducible tensor and possesses definite parity. The parity of the multipole operators is determined by $P_{\pi L} = (-1)^{L+\pi}$. In accordance with (4.7), the amplitude (4.6) is expanded in terms of the multipole amplitudes

$$T_{a\to b} = \sum_{\pi L \lambda} T_\lambda^{\pi L} = \underset{E1}{\begin{bmatrix} \pi = 0 \\ L = 1 \end{bmatrix}} + \underset{M1}{\begin{bmatrix} \pi = 1 \\ L = 1 \end{bmatrix}} + \underset{E2}{\begin{bmatrix} \pi = 0 \\ L = 2 \end{bmatrix}} + \cdots . \tag{4.8}$$

The first term in (4.8) corresponds to the electric dipole $E1$ approximation with the transition operator (in the length form of the long-wave length approximation)

$$T_\lambda^{E1} = D_\lambda = \sum_n (r_n)_\lambda = \sqrt{\frac{4\pi}{3}} \sum_n r_n\, Y_{1\lambda}(\theta_n, \phi_n). \tag{4.9}$$

The 'first-order' corrections to the dipole approximation are given by the second, magnetic dipole $M1$ term with the transition operator

$$T_\lambda^{M1} = -\mathrm{i}\, M_\lambda = -\mathrm{i}\frac{\alpha}{2} \sum_n \big[(\ell_n)_\lambda + 2(s_n)_\lambda\big], \tag{4.10}$$

and by the third, electric quadrupole $E2$ term with the transition operator

$$T_\lambda^{E2} = \frac{\mathrm{i}\alpha\omega}{2\sqrt{3}} Q_\lambda = \frac{\mathrm{i}\alpha\omega}{2\sqrt{3}} \sqrt{\frac{4\pi}{5}} \sum_n r_n^2\, Y_{2\lambda}(\theta_n, \phi_n). \tag{4.11}$$

In (4.9)–(4.11), D_λ, M_λ, and Q_λ are the components of the atomic electric dipole, magnetic dipole, and electric quadrupole operators, respectively; $\mathbf{r}_n = \{r_n, \theta_n, \phi_n\}$ are the positions of atomic electrons; $\boldsymbol{\ell}_n$ and $\mathbf{s}_n$ are their orbital momenta and spin; α is the fine structure constant ($\alpha \approxeq 1/137.036$), and ω is the photon frequency.

A measurable quantity, I, is generally expressed in terms of the bilinear combinations of the amplitudes as in (4.3) and (4.4). One can write down symbolically

$$\begin{aligned} I = & \sum_{\substack{\pi L \lambda \\ \pi' L' \lambda'}} X(\pi L \lambda, \pi' L' \lambda')\, T^{\pi L \lambda}\, (T^{\pi' L' \lambda'})^* \\ \approx & \left|T^{E1}\right|^2 \\ & + \left(T^{E1} T^{E2*} + \text{c.c.}\right) + \left(T^{E1} T^{M1*} + \text{c.c.}\right) \\ & + \left|T^{E2}\right|^2 + \left|T^{M1}\right|^2 + \left(T^{E1} T^{E3*} + \text{c.c.}\right) + \left(T^{E1} T^{M2*} + \text{c.c.}\right) \\ & + \cdots, \end{aligned} \tag{4.12}$$

where $X(\pi L \lambda, \pi' L' \lambda')$ are the angular coupling coefficients and "c.c." denotes the complex conjugate terms. After integration over the photoelectron emission angles, the interference terms between the amplitudes with different πL vanish provided the system A in (4.5) possesses a definite parity. An appropriate but not unique tool to derive expressions of the type (4.12) with inclusion of arbitrary polarized initial atomic target, detecting products of the photoionization and their polarization, including coincidence experiments, together with a full multipole expansion of the arbitrary polarized incident radiation in electric and magnetic moments, is the density matrix and statistical tensor formalism, as reviewed, for example, by Devons and Goldfarb (1957), Fergusson (1965), Blum (1996), Balashov et al. (2000). In this approach the observable quantities are expressed in terms of angle-independent coefficients, describing the dynamics of photoionization through the partial wave representation of the photoionization amplitude, and known geometrical factors describing the angular dependence of the cross sections.

Theoretical basis for understanding of the vast majority of the photoionization processes is the electric dipole ($E1$) (or simply, dipole) approximation. This is possible because, for the bulk of the processes studied in atomic physics, the photoionization cross section is large in the range, where the wavelength of the radiation is much larger than the effective size of the interacting system (see Fig. 4.1). Equivalent statements are that in the dipole approximation one can neglect the changing in the phase of the electromagnetic wave on the length of the interacting object, or neglect the retardation effects. We first discuss the main photoionization phenomena, which are well understood within the dipole approximation, and turn to the non-dipole effects in Sect. 4.9.

4.1.3 What is the Complete Experiment on Atomic Photoionization?

In the dipole approximation the dynamics of atomic photoionization is determined by only a few angle-independent amplitudes due the fact that, in contrast to the

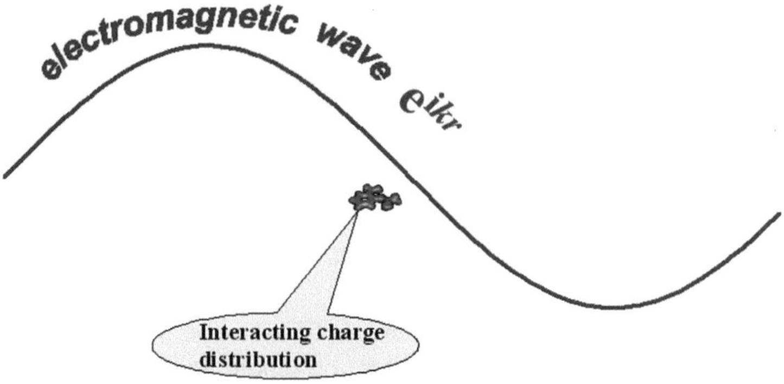

Fig. 4.1 An illustration to the dipole approximation

particle impact, the photon in the dipole approximation carries only one unit of the angular momentum. Therefore the partial wave representation of the photoionization amplitudes is practical, similar to the complete experiment for Auger decay (Sect. 3.6) and the purpose of the complete experiment in photoionization is to find these partial amplitudes.

Here we consider photoionization to particular ionic states. A powerful experimental method of such studies is photoelectron spectroscopy: detecting of the photoelectron energy relates the photoionization process to particular final state of the residual ion. Thus, the relative intensities of lines in the photoelectron spectra directly show the relative probabilities of creating different ionic states in course of the photoionization. The photoelectron spectroscopy is a very well developed field. The advent of permanently improving synchrotron radiation facilities brought the photoelectron spectrometry on the high level of sophistication. Further refinement of the method is related to investigations of additional characteristics of the photoionization products, as well as with controllable preparation of particular initial states of the atomic target and polarization state of the incoming radiation beam.

Consider the atomic photoionization

$$h\nu + A(\alpha_i J_i) \longrightarrow A^+(\alpha_f J_f) + e^-_{ph}(\varepsilon\ell j), \tag{4.13}$$

where both the initial atomic and the final ionic states are characterized by sharp total electronic angular momenta, J_i and J_f, respectively. All other quantum numbers that are necessary for specifying the states are denoted by α_i and α_f. The natural basis set for the initial atomic state is $|\xi_i\rangle = |\alpha_i J_i M_i\rangle$, where M_i is the projection of the angular momentum J_i. For the final state of process (4.13) the basis set $|\chi_f\rangle = |(\alpha_f J_f, \varepsilon\ell j)\, JM\rangle$ is convenient in order to use directly the conservation of the total angular momentum and parity. Here, ε, ℓ and j are the energy, orbital and total angular momenta of the photoelectron; J is the total angular momentum of the final state in (4.13), $\mathbf{J} = \mathbf{J}_f + \mathbf{j}$, and M is its projection. Due to the Wigner-Eckart theorem,

$$\langle(\alpha_f J_f, \varepsilon\ell j)\, JM\, |D_\lambda|\, \alpha_i J_i M_i\rangle = \frac{(J_i M_i, 1\lambda|JM)}{\sqrt{2J+1}}\, D_\varepsilon(J_f\ell j J), \tag{4.14}$$

the dependence of the amplitudes on the projections M_i, M and λ is explicitly given by the Clebsch-Gordan coefficient $(J_i M_i, 1\lambda|JM)$. Measurement of the partial wave amplitudes

$$D_\varepsilon(J_f\ell j, J) \equiv \langle(\alpha_f J_f, \varepsilon\ell j)\, J \| D \| \alpha_i J_i\rangle \equiv d^{\varepsilon}_{J_f\ell jJ} \exp \mathrm{i}\, \delta^{\varepsilon}_{J_f\ell jJ}, \tag{4.15}$$

which contain the dynamics of the atomic photoionization, is the purpose of the complete photoionization experiment within the dipole approximation. In (4.15), $d^{\varepsilon}_{J_f\ell jJ}$ is the absolute value of the amplitude and $\delta^{\varepsilon}_{J_f\ell jJ}$ is its phase. We will omit the energy index ε in the notations of the amplitudes and phases for brevity. Expressions of the type (4.3) and (4.4) for observable quantities in terms of the photoionization amplitudes (4.15) can be obtained by standard methods with the use of the statistical tensor formalism or alternative methods. Many of such expressions are presented, for example, in Balashov et al. (2000) or Kupliauskiene et al. (2001).

Photoionization with only one active electron is normally considered a favorable object for a complete analysis due to a large probability of this process and relative simplicity of its theoretical description. In contrast, in photoionization with excitation, the absorption of a single photon leads to one electron being ionized and the second electron being excited. Since the electromagnetic interaction operator responsible for the photoionization is a single-particle operator, photoionization with excitation is forbidden in the independent particle model and the occurrence of photoionization with excitation is a consequence of electron-electron correlations. Studies by means of the complete experiment can provide very detailed information on such correlations and on the mechanisms of the process.

The photoionization with excitation shows up in photoelectron spectra as discrete satellites, described by different types of configuration interaction (Schmidt 1997). For example, the shake-up satellites can be considered as a result of final ionic state configuration interactions. Furthermore, at certain photon energies the satellites can be strongly enhanced by an autoionizing/Auger state decaying to the excited final ionic states; this resonance process (resonant Auger decay) is described in the framework of the final state configuration interactions between the discrete atomic state and the adjacent continuum (Fano 1961; Fano and Cooper 1965; Mies 1968). The complete experiments require measurements of angle-differential and even spin-resolved cross sections which imply low counting rates. For this reason the complete experiment for the photoionization with excitation has been realized to date only for the resonantly enhanced satellites. Since particular mechanisms of the photoionization are not implied in the amplitudes (4.15), the general approaches to complete experiments for the resonant and direct photoionization are the same.

4.1.4 Counting the Number of Independent Amplitudes

The number of the independent amplitudes (4.15) for the process (4.13) is generally determined by conservation of total angular momentum due to the triangular conditions

$$\mathbf{J}_f + \mathbf{j} + \mathbf{J} = \mathbf{0}, \tag{4.16a}$$

$$\mathbf{J}_i + \mathbf{J} + \mathbf{1} = \mathbf{0}, \tag{4.16b}$$

$$\boldsymbol{\ell} + \mathbf{j} + \frac{\mathbf{1}}{\mathbf{2}} = \mathbf{0}, \tag{4.16c}$$

and by conservation of parity,

$$(-1)^{\ell+1}\,\pi_f = \pi_i, \tag{4.17}$$

where π_i and π_f are the parities of the initial atomic and final ionic states, respectively. Furthermore this number depends crucially on the model, which one uses to describe the process.

For illustration, let us consider photoionization from excited calcium atom into the excited state of the calcium ion:

$$\begin{aligned} h\nu + \mathrm{Ca}^*(3\mathrm{p}^6 4\mathrm{s}4\mathrm{p},\ J_i^{\pi_i} = 1^-) \\ \hookrightarrow \mathrm{Ca}^{+*}(3\mathrm{p}^6 3\mathrm{d},\ J_f^{\pi_f} = 5/2^+) + e^-_{ph}(\varepsilon\ell j). \end{aligned} \tag{4.18}$$

This reaction was studied by Wedowski et al. (1995, 1997), where the initial calcium excited state was prepared by the laser optical pumping from the ground Ca $3\mathrm{p}^6 4\mathrm{s}^2$ state. Nine partial amplitudes can be distinguished within the dipole approximation, with contributing channels as listed in Table 4.1.

Let us now neglect the spin-orbit interaction, i.e. we turn to the non-relativistic approximation, which implies that

1. spin and orbital angular momenta are considered (and conserved) separately, and
2. radial wavefunctions do not depend on the value of total angular momenta J, j, J_i, and J_f.

This model allows to write the reaction (4.18) in the way

$$h\nu + \mathrm{Ca}^*(3\mathrm{p}^6 4\mathrm{s}4\mathrm{p}\ {}^1\mathrm{P}^{\mathrm{o}}) \longrightarrow \mathrm{Ca}^{+*}(3\mathrm{p}^6 3\mathrm{d}\ {}^2\mathrm{D}^{\mathrm{e}}) + e^-_{ph}(\varepsilon\ell), \tag{4.19}$$

Table 4.1 Channels of ionization in reaction (4.18)

$\mathbf{J} = \mathbf{J}_f + \mathbf{j}$	Part. waves $\varepsilon\ell j$	Number of amplitudes
$J = 0$	$\varepsilon \mathrm{d}_{5/2}$	1
$J = 1$	$\varepsilon \mathrm{d}_{3/2}$, $\varepsilon \mathrm{d}_{5/2}$, $\varepsilon g_{7/2}$	3
$J = 2$	$\varepsilon \mathrm{s}_{1/2}$, $\varepsilon \mathrm{d}_{3/2}$, $\varepsilon \mathrm{d}_{5/2}$, $\varepsilon g_{7/2}$, $\varepsilon g_{9/2}$	5

Table 4.2 Channels of ionization in reaction (4.18) in a simplified LS coupling approximation (4.19)

$\mathbf{L} = \mathbf{L}_f + \ell$	Part. waves $\varepsilon\ell$	Number of amplitudes
$L = 0$	εd	1
$L = 1$	εd	1
$L = 2$	εs, εd, εg	3

with five relevant partial photoionization amplitudes of transitions into the final states given in Table 4.2, where L is the total orbital angular momentum of the system *residual ion + photoelectron*. The total channel spin conserves and is zero. A further approximation, i.e. neglecting dependence of the radial wavefunctions of the final state on the total orbital angular momentum L, results in only three partial amplitudes with the photoelectron in the εs, εd, and εg channels.

Since the reaction (4.18) or (4.19) is a two-electron transition, it can proceed only through the electron correlations. In the cases of single-electron transitions, the similar extreme reduction is known as the Cooper–Zare model (Cooper and Zare 1968, 1969). It implies the LS coupling scheme and the independence of the electron wavefunctions on the global quantum numbers of spin and orbital angular momentum. The dipole amplitudes (4.15) are then reduced to the single-particle amplitudes:

$$d_{J_f \ell j J} \; \exp \mathrm{i}\delta_{J_f \ell j J} \longrightarrow R_\ell \exp \mathrm{i}\delta_\ell \tag{4.20}$$

with the three-parameter description of the photoionization: two real single-electron ionization amplitudes (radial dipole integrals) R_ℓ and $R_{\ell'}$, and the asymptotic relative phase of the partial waves $\Delta_{\ell\ell'} = \delta_\ell - \delta_{\ell'}$, with $\ell = \ell_0 - 1$, $\ell' = \ell_0 + 1$, where ℓ_0 is the orbital angular momentum of the active electron in the initial state. For $\ell_0 = 0$ only a single channel remains in this model, and therefore only a single parameter remains, which is equivalent to the integral photoionization cross section.

As an important example, for ionization of the closed-shell atoms ($J_i = 0$) the number of dipole amplitudes (4.15) is two for final ionic states with $J_f = 1/2$ and three for those with $J_f > 1/2$. This leads to three and five independent real parameters, respectively, characterizing the absolute values of the amplitudes and their relative phases. One of the parameters can be excluded if we consider relative amplitudes instead of their absolute values; the latter are determined by the absolute integral cross section, which can be found separately. In the Cooper–Zare model there are three independent parameters for the closed-shell atom, regardless to the value of J_f.

The independent dynamical parameters are extracted from observable quantities, as described in the next sections. These quantities can depend either on real or on imaginary part of the interference terms between the amplitudes (4.15); i.e. on the cosine or sine of the phase differences between the amplitudes. Furthermore the relations between the observable quantities and the amplitudes are non-linear. Strictly speaking this makes the number of independent observable quantities needed for unambiguous extraction of the photoionization amplitudes more than the number of the independent parameters, characterizing the process. We will nevertheless speak about the *complete experiment* even in the case of this restricted unambiguities. The latter often can be eliminated by simple theoretical estimates. Sometimes the term *almost complete experiment* is used in the literature in such a situation.

4.1.5 Theoretical Methods for Calculating Photoionization Amplitudes

Methods to get the photoionization amplitudes (4.15) describing the dynamics of the process constitute a core of the atomic photoionization theory. Although satisfactory results are often obtained for the photoionization parameters within the one-electron picture of the process, in many cases it is complicated by the presence of electron correlations. Such phenomena as photoionization with excitation, photoionization in the region of autoionizing/Auger resonances, electron correlations in the initial atomic and final ionic states, interchannel coupling between final continuum states cannot be treated in the independent-particle model and more advanced theoretical methods should be used to predict the set of the amplitudes and/or the results of complete photoionization experiment.

One class of these methods, such as multiconfigurational Hartree–Fock (MCHF) or Dirac–Fock (MCDF) approaches, and close-coupling approach, concentrates on accurate evaluation of the initial- and final-state atomic wave functions to use them in calculation of the amplitudes (4.15). For a many-electron atom this task is complex and only an approximate wave functions can be obtained, which quality is often questioned until comparison of the calculated quantities with experiment is done. Especially difficult is to treat the final continuum atomic state, where its multi-channel origin and appropriate channel asymptotic have to be taken into account. Such a wave function is generally constructed by the methods of multichannel scattering theory. Mixing of configurations in continuum, characteristic for these methods, leads to formally infinite set of coupled equations (*close-coupling* equations), which can be solved with different level of sophistication. Including of the closed channels (discrete states) into the configuration expansion of the continuum wave function allows to treat autoionizing/Auger resonances in the photoionization amplitudes.

Other classes of powerful methods, like many-body perturbation theory (MBPT), random phase approximation with exchange (RPAE) with its relativistic extension (RRPA), and other generalizations, are based on perturbation theory with respect to the Coulomb interaction between the atomic electrons. Instead of improving atomic wave functions as the primary task in calculating the amplitudes (4.15), these methods suggest a set of coupled equations for the photoionization amplitudes, which can be solved by iterations with respect to the Coulomb interelectron interaction starting from a zero-order photoionization amplitude in the independent-particle (Hartree–Fock or Dirac–Fock) approximation. The diagram technique allows to visualize the included classes of electron correlations. Certain types of correlations can be summed up to infinite order.

Generally, modern theoretical instrumentary makes it possible to calculate photoionization amplitudes for atoms throughout the periodic table with different ranges of accuracy. Detailed theoretical description of particular transitions most often needs individual calculations. We refer to many excellent books and reviews, which contain description of modern theoretical models used in the photoionization studies, for example, (Starace 1982; Amusia 1990, 1996; Kutzner 1996; Aymar et al. 1996; Froese-Fischer et al. 1997; Connerade 2005). Program packages for the photoionization calculations have been developed based on the multiconfiguration Hartree–Fock (Cowan 1981; Froese-Fischer et al. 1997; Froese-Fischer 2000) and Dirac–Fock (Parpia et al. 1996; Fritzsche 2001) approximations, on the random phase approximation with exchange (Amusia et al. 1997), on different modifications of an R-matrix (close-coupling) approach (Berrington et al. 1995; Zatsarinny 2006) and others. Some photoionization parameters calculated in simple approximations or for selected targets are tabulated, for example, by Band et al. (1979), Huang et al. (1981), Yeh and Lindau (1985), Derevianko et al. (1999), Trzhaskovskaya et al. (2001, 2002).

4.2 Photoelectron Spectrometry

4.2.1 Angular Distribution of Photoelectrons

Angular distribution of photoelectrons produced from an isotropic atom by linearly polarized radiation in the dipole approximation has been given by Cooper and Zare (1968)

$$\frac{d\sigma}{d\Omega} = \frac{\sigma}{4\pi}\Big(1 + \beta P_2(\cos\theta)\Big), \tag{4.21}$$

where σ is the angle-integrated cross section, $P_2(x)$ is the second Legendre polynomial, θ is the angle between the photoelectron direction and the photon polarization and β is the asymmetry (or anisotropy) parameter. For fixed final ionic state with sharp value of the total angular momentum J_f, the cross section is given in terms of the absolute values of the partial wave amplitudes (4.15) by

$$\sigma = \frac{4\pi^2 a_0^2\,\alpha\,\omega}{2(2J_i+1)} \sum_{\ell j J} \left|d_{J_f \ell j J}\right|^2, \tag{4.22}$$

where $a_0 = 5.2918 \times 10^{-9}$ cm is the Bohr radius. For unpolarized or circularly polarized photons, (4.21) remains with θ being the angle between photon and photoelectron directions of propagation and with the substitution $\beta \to -\beta/2$. More generally, the arbitrary and partially polarized light with the Stokes parameters P_1, P_2, and P_3, see expressions (3.69)–(3.71) and the discussion in Sect. 3.5.1,[1] leads to

$$\frac{d\sigma}{d\Omega} = \frac{\sigma}{4\pi}\left\{1 - \frac{\beta}{2}\left[P_2(\cos\theta) - P_1\frac{3}{2}\cos(2\phi)\sin^2\theta\right]\right\}, \tag{4.23}$$

where θ is counted from the photon beam direction and ϕ is counted from the direction of the main axis of the polarization ellipse of the ionizing radiation (Schmidt 1973; Samson and Starace 1975). Therefore, the asymmetry parameter β is the entire dynamical parameter, additional to the cross section σ, which can be obtained from the photoelectron angular distributions in the case of unpolarized target. The dimensionless β parameter depends on the ratio of the complex partial wave ionization amplitudes, i.e. it generally depends on the relative phases of the amplitudes (4.15). General expressions for the asymmetry parameter in terms of photoionization amplitudes (4.15) can be found in the literature (Balashov et al. 2000). They contain bilinear combinations with the terms $Re\big[D(J_f \ell j, J)\, D(J_f \ell' j', J')^*\big]$ and therefore include interference of the amplitudes (4.15). A characteristic example is the Cooper–Zare model for ionization of electron with initial orbital momentum ℓ_0:

$$\begin{aligned}\beta = &\left((2\ell_0+1)\left[\ell_0\, R^2_{\ell_0-1} + (\ell_0+1)R^2_{\ell_0+1}\right]\right)^{-1} \\ &\times (\ell_0+1)\Big[\ell_0\, R^2_{\ell_0-1} + (\ell_0+2)R^2_{\ell_0+1} \\ &- 6\,\ell_0\, R_{\ell_0+1}R_{\ell_0-1}\cos(\delta_{\ell_0+1} - \delta_{\ell_0-1})\Big]\,.\end{aligned} \tag{4.24}$$

The β parameter is a known constant, when the photoionization proceeds via a single channel. The simplest example is photoionization from the s-subshell in non-relativistic single-electron approximation with ejection of an εp photoelectron. In this case $\beta = 2$, as follows from (4.24). When turning to the relativistic model, the εp channel splits into the $\varepsilon p_{1/2}$ and $\varepsilon p_{3/2}$ channels with the two corresponding amplitudes. The asymmetry parameter is then expressed as

$$\beta = \frac{\sigma_{3/2} + 2\sqrt{2\sigma_{1/2}\sigma_{3/2}}\cos(\delta_{3/2} - \delta_{1/2})}{\sigma_{1/2} + \sigma_{3/2}}, \tag{4.25}$$

where σ_j is the partial photoionization cross section into the εp_j channel, proportional to the absolute square of the corresponding amplitude, and δ_j is the phase of the amplitude. The asymmetry parameter (4.25) is sensitive to the relative phase of the partial amplitudes and turns into $\beta = 2$, when the two relativistic amplitudes are equal, and hence $\sigma_{3/2} = 2\sigma_{1/2}$, $\delta_{3/2} = \delta_{1/2}$. The relativistic splitting of the continuum is important in the regions of the Cooper minima, where the two amplitudes

---

[1] The Stokes parameters P_i ($i = 1, 2, 3$) are defined as in quantum electrodynamics: In the helicity representation, the photon 2×2 density matrix is of the form $\langle\lambda\,|\rho^\gamma|\,\lambda'\rangle = \frac{1}{2}\left[\delta_{\lambda,\lambda'}(1+\lambda P_3) + (1-\delta_{\lambda,\lambda'})(-P_1 + i\lambda P_2)\right]$, where the helicity $\lambda = +1$ ($\lambda = -1$) corresponds to right (left) circularly polarized photons.

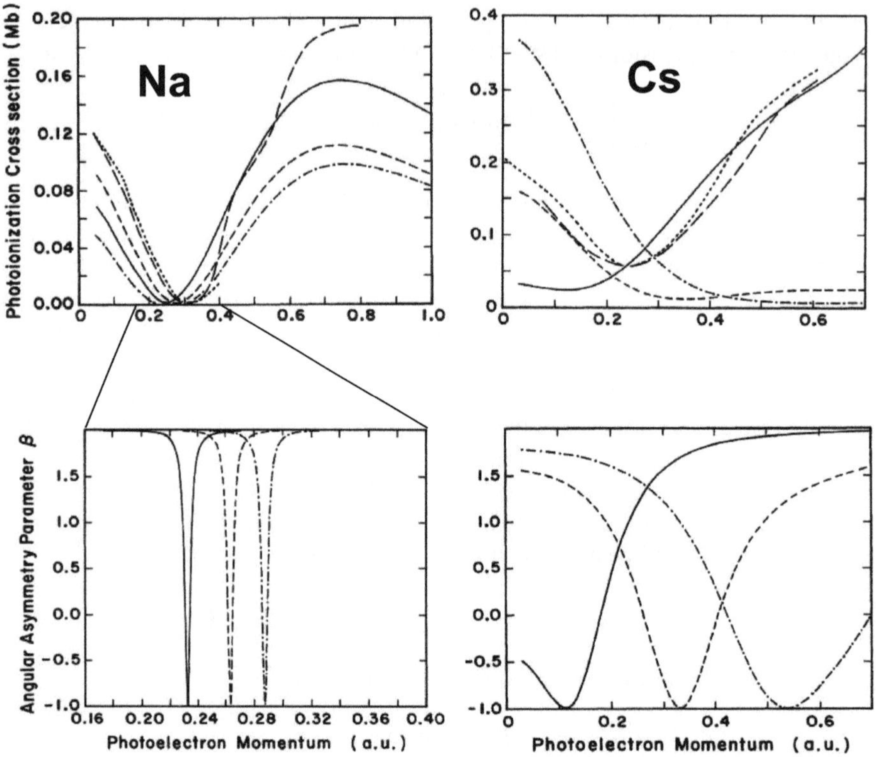

Fig. 4.2 Photoionization of the 3s electron in Na (*left column*) and of the 6s electron in Cs (*right column*). (*Upper panels*): integral cross section σ, (*lower panels*): asymmetry parameter β. For the β parameter: (——) relativistic random phase approximation (RRPA); (– · –) relativistic Dirac-Fock approximation (DF), length form; (– – –) DF, velocity form. Additional curves for the cross sections in the upper panels are experiments by Hudson (1964) and Hudson and Carter (1967) for Na and by Marr (1963) for Cs. Adapted from Fink and Johnson (1986)

cross zero at slightly different energies. Around the Cooper minimum the ratio of the two amplitudes changes drastically, causing sharp changes in the β parameter, as demonstrated by Fig. 4.2. A small spin-orbit interaction leads to a pronounced effect, although in the region of small cross section.

It is possible to introduce another set of the channel quantum numbers, when the asymmetry parameter is expressed as incoherent sum of the channel asymmetry parameters weighted with the channel cross sections. This representation in terms of *angular momentum transfer* amplitudes was suggested by Fano and Dill (1972), Dill and Fano (1972). The relevant quantity is the angular momentum transfer

$$\mathbf{j}_t = \mathbf{j}_\gamma - \ell = \mathbf{J}_i - \mathbf{J}_f, \tag{4.26}$$

where $\mathbf{j}_\gamma$ is the angular momentum of the incident radiation ($j_\gamma = 1$ for the photon in the dipole apparoximation). Then, the β parameter can be presented as an incoherent sum

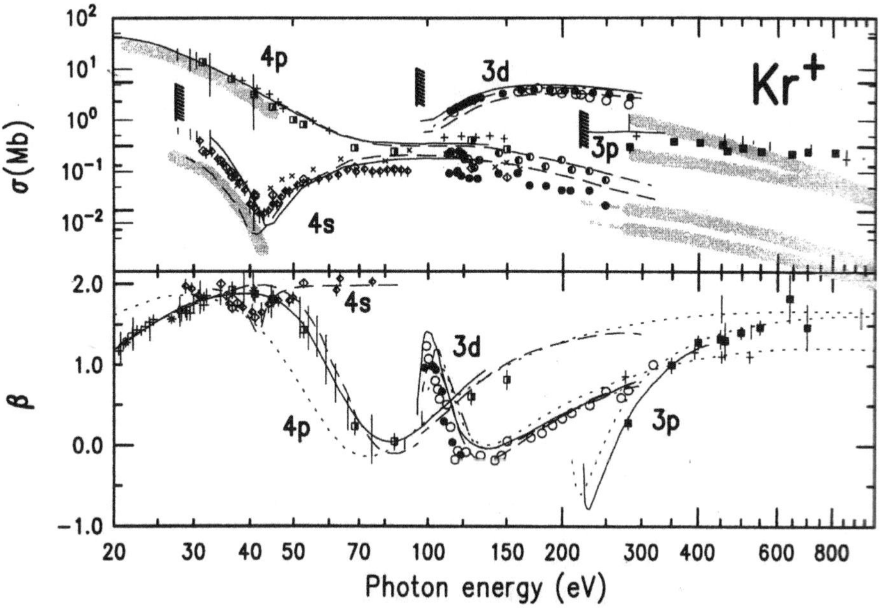

Fig. 4.3 Photoionization cross section (*upper panel*) and angular distribution asymmetry parameter β (*lower panel*) for the 4p, 4s, and 3d subshells of krypton. Theory: (······): local potential (Yeh 1993), and (—·—): Hartree–Fock (Kennedy and Manson 1972) calculations. (— — —): multiconfiguration Dirac–Fock calculation (Tullki et al. 1992); (——): relativistic random phase approximation (Huang et al. 1981; Shanti 1988). Experiment: (*grey bands*): Berkowitz (1978); for 3p subshell, (■): Lindle et al. (1986), (*vertical bars*) and (+): Krause (1969) and Krause and Carlson (1966); for 3d subshell, (○): Lindle et al. (1986), (●) for cross section: Tullki et al. (1992), and for β: Carlson et al. (1982), (+): Krause (1969); for 4s subshell, (◇): Berrah et al. (1993), (●): Tullki et al. (1992), (×): Aksela et al. (1987), (*vertical bars*): Samson and Gardner (1974), (◁▶) for cross section: Ehresmann et al. (1994), and for β: Derenbach and Schmidt (1984); for 4p subshell, (□■): Berrah et al. (1993), (◐): Tullki et al. (1992), (∗): Dehmer et al. (1975); (+): cross section (Aksela et al. 1987), and β (Southworth et al. 1986); (*vertical bars*): cross section by Samson and Gardner (1974), and β by Miller et al. (1977). After Becker and Shirley (1996)

$$\beta = \left(\sum_{j_t} \sigma(j_t) \right)^{-1} \sum_{j_t} \sigma(j_t)\, \beta(j_t), \tag{4.27}$$

where $\sigma(j_t)$ and $\beta(j_t)$ are the integrated cross section and the anisotropy parameter for each j_t, respectively. We refer to the literature (Fano and Dill 1972; Dill 1973; Manson and Dill 1978; Manson and Starace 1982) for more details on the formalism of the angular momentum transfer.

Photoelectron angular distribution is a subject of many reviews (for example Manson and Starace 1982; Becker and Shirley 1996; Reid 2004). Figure 4.3 shows a compilation of data for photoionization of krypton. Some general features are seen for the asymmetry parameters β: deviation from 2 for ionization of the 4s subshell in the Cooper minimum, sharp variation and dropping down to zero for ionization from the 4p subshell when the partial photoionization cross section into the εd channel crosses zero (see (4.24)). Other data compilations on the angular distribution of photoelectrons can be found in Becker and Shirley (1996).

Experiments on the photoelectron angular distributions for ions became feasible with the advent of intense VUV femtosecond pulses from free electron laser (Braune et al. 2007; Moshammer et al. 2007; Kurka et al. 2009). The high photon flux within the short pulse allows by first photon to ionize neutral atom and by second photon from the same pulse to ionize the residual ion. The ions stay together within the time of the pulse despite of their Coulomb repulsion, therefore keeping high enough density of the gaseous ionic target. The method has been applied to the outer np^6 subshells of noble gas atoms in parallel with the corresponding theoretical developments (Kheifets 2007, 2009; Fritzsche et al. 2008, 2009; Kurka et al. 2009; Grum-Grzhimailo et al. 2009; Gryzlova et al. 2010). In the stepwise model of such a sequential photoionization, alignment of the intermediate ionic state has to be taken into account.

The asymmetry parameter β is generally more sensitive to the dynamics of the photoionization than the integral cross section. On the other hand, β is a dimensionless quantity, in which possible systematic inaccuracies in the calculated amplitudes can cancel in their ratio. From the viewpoint of the complete photoionization experiment, the value of β is not enough to extract the photoionization amplitudes even in the case of two ionization channels. Hence, measurements of other independent parameters of photoionization are needed to achieve this goal.

4.2.2 Spin Polarization of Photoelectrons

Studies of spin polarization of photoelectrons is a universal method of obtaining additional dynamical photoionization parameters. Historically, this was the first method to complete the photoionization experiment in the VUV (Heinzmann 1980a, b). Spin polarization measurements are difficult, because spin-sensitive electron analyzers decrease the counting rate by 2–3 orders of magnitude. Therefore, such experiments often suffer from poor energy resolution and low statistics, which results in unresolved photoelectron lines and large experimental inaccuracies. Furthermore, a full power of the method is achieved with the circularly polarized synchrotron radiation beams, which are not easy accessible.

Theory of the spin polarization phenomenon in atomic photoionization has been developed in a variety of approaches, for example in the papers of Jacobs (1972), Cherepkov (1972, 1973), Lee (1974), Kabachnik and Sazhina (1976), Huang (1980), whereas the field was reviewed by Cherepkov (1983), Kessler (1985), Heinzmann and Cherepkov (1996). Photoelectrons emitted from unpolarized atoms can be polarized when the incoming radiation is circularly polarized, linearly polarized and even unpolarized. The spin polarization of photoelectron emission from unpolarized atoms occurs due to spin-orbit interaction in atomic continuum or discrete spectrum. For example, in the well-known Fano effect (Fano 1969a, b; Kessler and Lorenz 1970; Heinzmann et al. 1970), illustrated in Fig. 4.4, the spin polarization

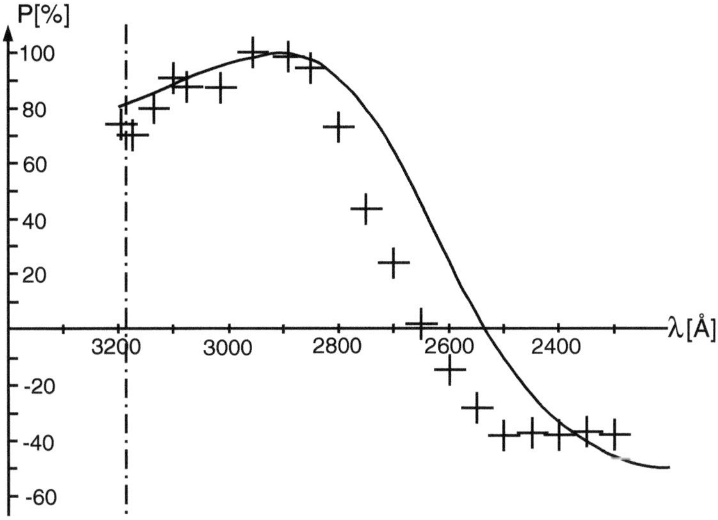

Fig. 4.4 Experimental and theoretical results for integral spin polarization of photoelectrons in ionization of the Cs atom by circularly polarized light. After Heinzmann et al. (1970)

in the angle-integrated photoelectron flux is produced by ionization of unpolarized alkali metal atoms with circularly polarized light in the region of Cooper minimum, (see Fig. 4.2) where the two relativistic channels $\varepsilon p_{1/2}$ and $\varepsilon p_{3/2}$ are strongly distinguished. The effects of spin-orbit interaction in the discrete spectrum produce, for example, polarized photoelectrons in the angle-resolved emission from energy-selected fine-structure atomic state for arbitrary polarization of the incoming radiation (Cherepkov 1972, 1973).

Consider linearly polarized incoming photon beam and define the reaction plane XZ spanned by the direction of the photon polarization and the momentum of the photoelectron (Fig. 4.5). Due to symmetry reasons, only the spin polarization component of the photoelectrons normal to the reaction plane can be non-vanishing (the so-called *dynamical* spin polarization), which is conventionally expressed as

$$P_y = \frac{\xi \sin 2\theta}{1 + \beta P_2(\cos\theta)}, \tag{4.28}$$

where β is the angular asymmetry parameter from (4.21). For circularly polarized or unpolarized radiation the substitution $\beta \to -\beta/2$ should be made in the denominator of (4.28) and θ should be counted from the photon beam direction. Expressions for the parameter ξ in terms of photoionization amplitudes include only imaginary parts of the interference terms between the amplitudes (4.15) in the numerator, with the photoionization cross section in the denominator. In the Cooper–Zare model the interference terms transform to the products $R_\ell R_{\ell'} \sin(\delta_\ell - \delta_{\ell'})$. Therefore the dynamical spin polarization can be noticeable when channels with different orbital angular momentum ℓ of photoelectrons are opened. The dynamical spin polarization

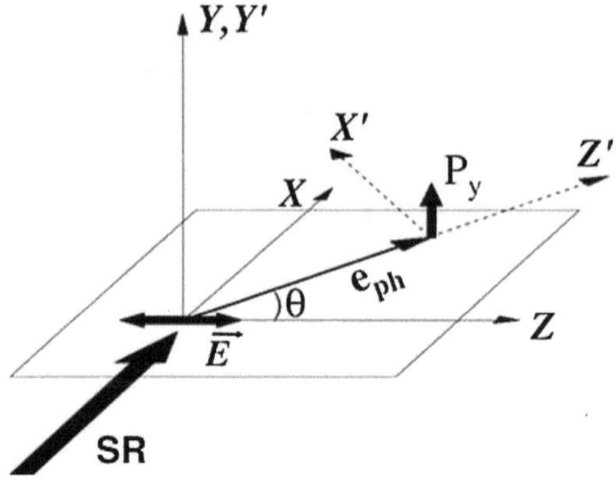

Fig. 4.5 Geometry of photoionization by linearly polarized synchrotron radiation (SR). The spin component $P_y = P_{y'}$ of the photoelectron is shown

is a pure interference effect and it reduces considerably when one channel is dominating because then the interference terms become small in comparison with the photoionization cross section.

For photoionization from a closed-shell atom to an ionic state with the total angular momentum $J_f = 1/2$, i.e. when only two amplitudes contribute to the photoionization process, the values of β and P_y, providing two real parameters, are sufficient to extract the absolute ratio and the relative phase of the two amplitudes up to an unambiguity between two solutions of the set of non-linear equations. The latter may be removed only either by additional measurements of independent characteristics or by theoretical predictions. Equivalently, by means of measurements of only the asymmetry parameter β and the dynamical spin polarization parameter ξ, the complete photoionization experiment in more general cases is possible, provided the simple three-parameter Cooper–Zare model is valid.

The method of combining data for β and ξ was originally used for the reactions (Schönhense and Heinzmann 1984)

$$h\nu + \mathrm{Hg}\,(5\mathrm{d}^{10}6\mathrm{s}^2\,{}^1\mathrm{S}_0) \hookrightarrow \mathrm{Hg}^+(5\mathrm{d}^9 6\mathrm{s}^2\,{}^2\mathrm{D}_{3/2}) + e^-_{ph}(\varepsilon\mathrm{p}_{1/2}, \varepsilon\mathrm{p}_{3/2}, \varepsilon\mathrm{f}_{5/2})\,, \quad (4.29a)$$

$$h\nu + \mathrm{Hg}\,(5\mathrm{d}^{10}6\mathrm{s}^2\,{}^1\mathrm{S}_0) \hookrightarrow \mathrm{Hg}^+(5\mathrm{d}^9 6\mathrm{s}^2\,{}^2\mathrm{D}_{5/2}) + e^-_{ph}(\varepsilon\mathrm{p}_{3/2}, \varepsilon\mathrm{f}_{5/2}, \varepsilon\mathrm{f}_{7/2})\,. \quad (4.29b)$$

The data on the relevant parameters for this reaction have been obtained within the time interval of 10 years by different groups: the total photoionization cross section (Cairns et al. 1970; Dehmer and Berkowitz 1974), the branching ratios for ionization into the ionic states with $J_f = 5/2$ and $3/2$ (Shannon and Codling 1978), the asymmetry parameters β_{J_f} (Schönhense 1981) and spin polarization parameters ξ_{J_f} (Schönhense et al. 1982). In the framework of the Cooper–Zare model, combination of these data yields partial wave cross sections for the 5d $\rightarrow$ εf and 5d $\rightarrow$ εp transitions and the phase difference $\delta_\mathrm{f} - \delta_\mathrm{p}$ (Fig. 4.6). The complete experiment reveals the dynamical features explicitly in terms of these basic quantities, whereas the measurable parameters are rather complex functions of these basic quantities. Figure 4.6 (top panel) demonstrates a maximum in the matrix element of the 5d $\to \varepsilon$f channel, shifted from the ionization threshold due to the angular momentum barrier in the effective potential, which prevents the εf continuum orbital from penetrating into the barrier region at small energies (see e.g. Starace 1982), and a striking effect of many-electron correlations in the 5d $\to \varepsilon$p channel, specifically, the influences of the strong 5d $\to \varepsilon$f transition upon the weak 5d $\to \varepsilon$p transition. The latter influences show themselves in a drastic change of the calculated 5d $\to \varepsilon$p matrix element when turning to the RPAE (middle panel of Fig. 4.6) and a good agreement of the RPAE calculations with the results of the complete experiment. The phase difference (the lower panel of Fig. 4.6) is also sensitive to the interchannel coupling.

The in-plane spin components of the photoelectron are described by two more dynamical parameters, for example, A and α. These components are non-zero only for excitation by a photon beam with a circular component (*transferred* spin polarization). For the circular polarized radiation,

$$P_x = -\frac{3}{4}\frac{\alpha \sin 2\theta}{1 - \frac{1}{2}\beta P_2(\cos\theta)}, \tag{4.30}$$

$$P_z = \frac{A - \alpha P_2(\cos\theta)}{1 - \frac{1}{2}\beta P_2(\cos\theta)}, \tag{4.31}$$

where θ is the angle between the incident photon beam and the direction of the photoelectron emission. In contrast to the dynamical spin polarization parameter ξ, the numerator of the expressions for the parameters α and A depends on the real parts of the interference terms between the amplitudes (4.15) and partial photoionization cross sections into separate channels. Therefore the polarization transfer always occurs in photoionization by circularly polarized light with seldom exceptions. Note that an alternative parameterization of the in-plane spin components as projections on the X' and Z' axes (see Fig. 4.5) is also used in the literature; we have used it in Sect. 3.4.1. Modern experimental techniques, such as displayed in Fig. 4.7, can provide simultaneous measurements of the three spin components of the photoelectron.

For a long time it has been implied that the parameters β, ξ, A and α, or an equivalent set of dynamical parameters extracted from the angular distribution and spin polarization of photoelectrons, are independent and the methods of angular and spin resolved electron spectroscopy in principle can provide four dynamical parameters, which are enough to extract two absolute amplitude ratios and two relative phases between the amplitudes, at least up to ambiguity in solution of the corresponding set of non-linear equations. Therefore, it was believed that the (almost) complete photoionization experiment is possible by photoelectron spectrometry methods in the case of three amplitudes, i.e. beyond the Cooper–Zare model. Instead, it has been discovered by Schmidtke et al. (2000a) that the four parameters are dependent and interrelated by the relationship

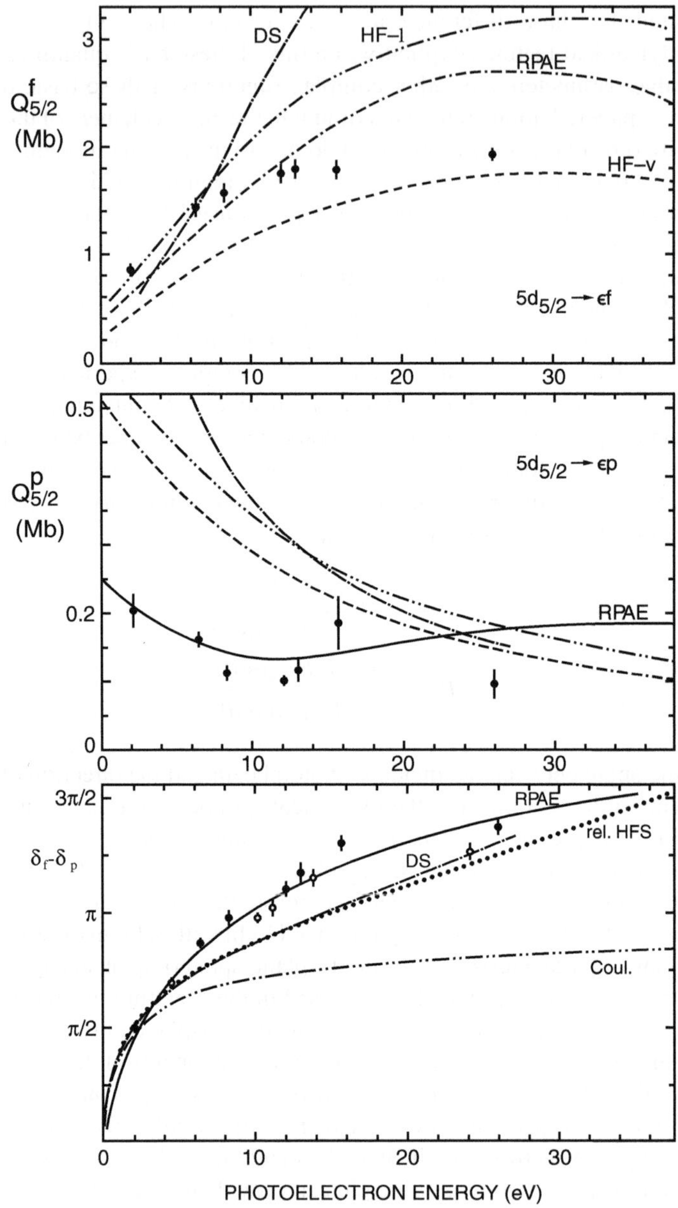

Fig. 4.6 Partial cross section for the 5d $\rightarrow$ εf and 5d $\rightarrow$ εp channels (*upper and middle panels*), respectively, with the photoion left in its final state $Hg^+\ ^2D_{5/2}$ and the phase-shift difference between the εf and the εp partial waves (*lower panel*), where full and *open symbols* correspond to $J_f = 5/2$ and $3/2$, respectively. Theory: Hartree–Fock length form (HF-ℓ), Hartree–Fock velocity form (HF-v), RPAE with many-electron correlations within one channel only and RPAE with interchannel correlations between 5d $\rightarrow$ εf and 5d $\rightarrow$ εp from Ivanov et al. (1979); Dirac-Slater (DS) model from Keller and Combet-Farnoux (1979, 1982). (*Lower panel*): relativistic Hartree–Slater from Kim et al. (1980) and the pure Coulomb phase difference are additionally shown. Adapted from Schönhense and Heinzmann (1984)

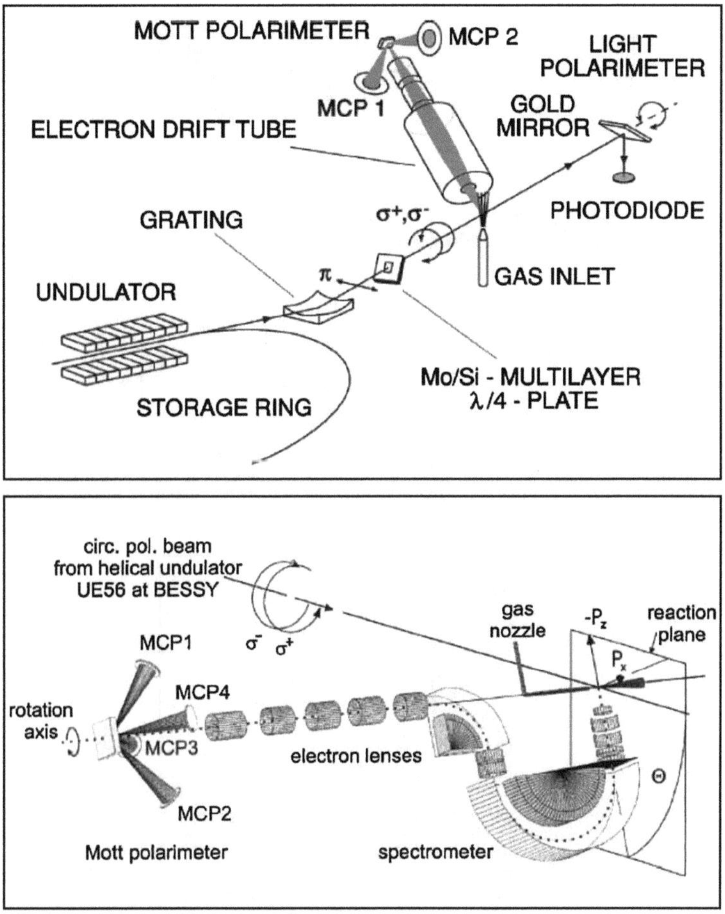

Fig. 4.7 Experimental setup for measurements of spin polarization of photoelectrons. (*Upper panel*): general scheme from Snell et al. (2001), (*lower panel*): scheme of simultaneous measurements of the photoelectron spin components after Khalil et al. (2002)

$$\left(A+\frac{1}{2}\alpha\right)^2+(2\xi)^2=\frac{1}{2}(1+\beta)\left(1-\frac{1}{2}\beta+A-\alpha\right), \tag{4.32}$$

which is valid for photoionization of either closed or one-electron atomic shells. The existence of this equation opens up the possibility of checking mutual consistency of the experimental data, but on the other hand proves that the complete photoionization experiment with extraction of three complex photoionization amplitudes is impossible by means of photoelectron angle and spin resolved spectroscopy even if all spin components of the photoelectrons are measured. Strictly speaking, for only a simple case of two amplitudes the methods of photoelectron spectrometry provide enough information for the purpose of the complete photoionization experiment.

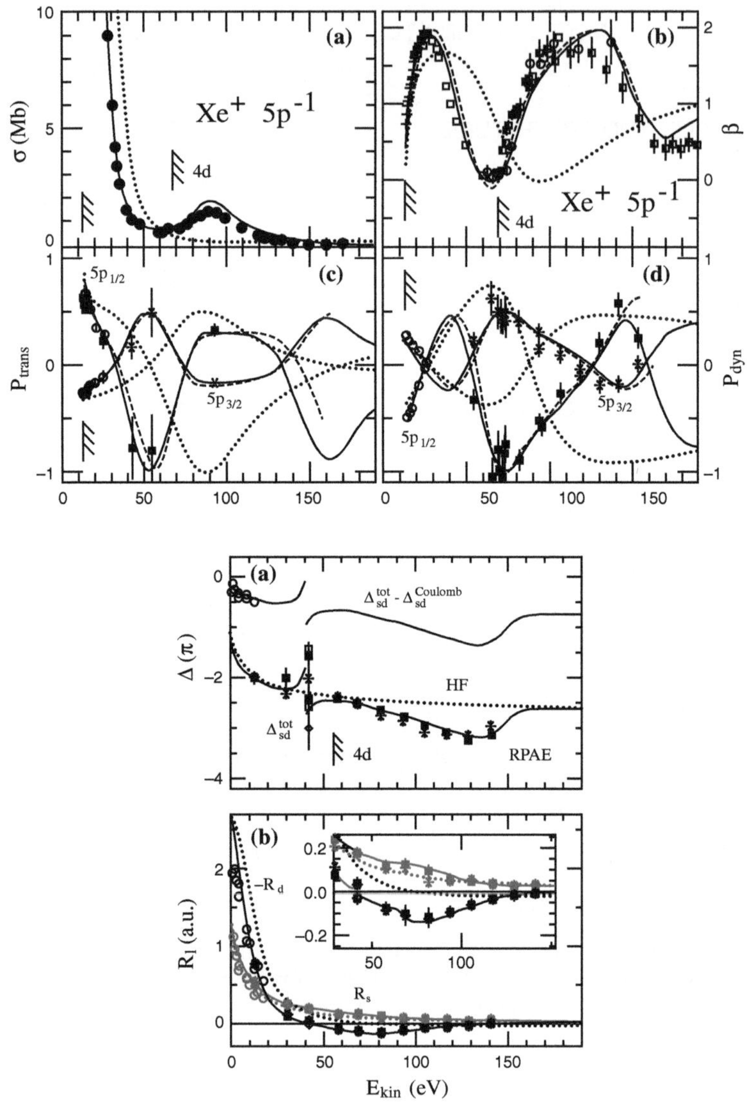

Fig. 4.8 (*Four upper panels*): parameters of Xe 5p photoionization. **a** cross section σ: (●) Becker et al. (1989); **b** anisotropy parameter β: (+) Southworth et al. (1986), (□) Krause et al. (1981), (■□) Southworth et al. (1983), (○) Torop et al. (1976); **c** transferred and **d** dynamical spin polarization: (○) Heckenkamp et al. (1986); Heinzmann (1980a, b), ($5p_{1/2}$: ■, $5p_{3/2}$: ∗) Zimmermann et al. (2001). The region between 65 eV and 70 eV is affected by autoionizing resonances. (— — —) relativistic RRPA of Huang et al. (1981). (——) non-relativistic RPAE calculation, and (······) non-relativistic HF calculation (Zimmermann et al. 2001) both using the codes of Amusia et al. (1997). (*Two lower panels*): Phase shifts **a** and matrix elements **b** of Xe 5p photoionization. (○) data from Heckenkamp et al. (1986); Schmidtke et al. (2000a); Heinzmann (1980a, b), ($5p_{1/2}$: ■, $5p_{3/2}$: ∗) Zimmermann et al. (2001). (□ and ◇) further solutions from a sign change of R_d and a phase shift by π. (——) non-relativistic RPAE calculation, and (······) non-relativistic HF calculation by Zimmermann et al. (2001) both using the codes of Amusia et al. (1997). Discontinuity in the RPAE curves is due to the chosen representation (see text). After Zimmermann et al. (2001)

A combination of different experimental techniques or a coincidence technique is required to complete the photoionization experiment for a number of the amplitudes larger than two. In practice, even for larger number of the amplitudes, some of the parameters can be taken as their limiting values due to specific features of the photoionization process in particular cases. For example, the phases between relativistically splitted amplitudes containing photoelectrons in $\varepsilon\ell j$ $(j = \ell - \frac{1}{2})$ and $\varepsilon\ell j$ $(j = \ell + \frac{1}{2})$ states far from resonance regions and Cooper minima often turn out to be negligible. This still permits the complete photoionization experiment in the case of three amplitudes by means of measurements of angular distribution and spin polarization of photoelectrons.

For a closed-shell atom, which therefore cannot be polarized, the following photoionization characteristics are measurable implying the validity of the electric dipole approximation: the cross section, the asymmetry parameter β in the angular distributions of photoelectrons, three photoelectron spin components and two (for $J_f > 1/2$) polarization parameters, characterizing polarization of the residual ion. From another point of view, these seven real quantities are described in the general case in the dipole approximation by five (for $J_f > 1/2$) real parameters: three absolute values of the partial photoionization amplitudes and their two relative phases. Therefore, the seven above characteristics or, speaking about the dimensionless parameters, the six parameters, must be related by two general identities. One of them is (4.32). The second was found by Cherepkov and Semenov (2004) and is very complicated, including more than 40 terms.

We conclude this section providing an example of a complete photoionization experiment of the Xe 5p subshell:

$$h\nu + \mathrm{Xe}(5\mathrm{p}^6\,{}^1\mathrm{S}_0) \longrightarrow \mathrm{Xe}^+(5\mathrm{p}^5\,{}^2\mathrm{P}_{1/2}) + e^-_{ph}(\varepsilon\mathrm{s}_{1/2}, \varepsilon\mathrm{d}_{3/2}), \tag{4.33a}$$

$$h\nu + \mathrm{Xe}(5\mathrm{p}^6\,{}^1\mathrm{S}_0) \longrightarrow \mathrm{Xe}^+(5\mathrm{p}^5\,{}^2\mathrm{P}_{3/2}) + e^-_{ph}(\varepsilon\mathrm{s}_{1/2}, \varepsilon\mathrm{d}_{3/2}, \varepsilon\mathrm{d}_{5/2})\,. \tag{4.33b}$$

The three-parameter Cooper–Zare model is a good approximation for the Xe 5p subshell ionization (4.33). The two dipole matrix elements, R_s and R_d, corresponding to photoelectrons in the εs and εd continua, respectively, can be derived from the experimental data along with the relative phase Δ_sd of the corresponding partial wave amplitudes using the relations

$$\sigma = 4\pi^2 a_0^2\, \alpha\, \omega \frac{2}{3}\left(R_\mathrm{s}^2 + 2R_\mathrm{d}^2\right), \tag{4.34a}$$

$$\beta_{1/2} = \beta_{3/2} = \frac{2R_\mathrm{d}^2 - 4R_\mathrm{s}R_\mathrm{d}\cos\Delta_\mathrm{sd}}{R_\mathrm{s}^2 + 2R_\mathrm{d}^2}, \tag{4.34b}$$

$$P_{y,1/2} = -2P_{y,3/2} = \frac{6R_s R_d \sin \Delta_{sd}}{2R_s^2 + 5R_d^2 - 2R_s R_d \cos \Delta_{sd}}, \tag{4.34c}$$

$$P_{trans,1/2} = -2P_{trans,3/2} = \frac{-2R_s^2 + 4R_d^2 + 6R_s R_d \cos \Delta_{sd}}{2R_s^2 + 5R_d^2 - 2R_s R_d \cos \Delta_{sd}}. \tag{4.34d}$$

Here, the index 1/2 or 3/2 indicate the residual ion in the $Xe^+(5p^{-1}\,{}^2P_{J_f})$ state with $J_f = 1/2$ or $J_f = 3/2$, respectively, and P_{trans} denotes the in-plane spin polarization component perpendicular to the direction of the photoemission (i.e. along the X' axis in Fig. 4.5).

Figure 4.8 overviews experimental and theoretical results for the processes (4.33), together with the results for the extracted parameters and their comparison with calculations in the Hartree–Fock and in the RPAE approximations. The agreement between the derived data and the RPAE calculation is excellent and confirms the validity of the chosen three-parameter model for the description of the Xe 5p photoionization. The lack of interchannel interaction in the HF approximation is the main reason of disagreement with the experimental data for the εd-channel amplitude R_d and the relative phase Δ_{sd}; a strong influence of the $4d^{10}$ subshell, especially in the region of the 4d ionization threshold, and a *giant resonance* in the 4d ionization cross section are well known phenomena (Amusia 1990).

A complete experiment for photoionization of the Xe 4d subshell will be discussed in Sect. 4.3.2.

4.3 Polarimetry of the Residual Ion

4.3.1 General Features of the Residual Ion Alignment and Orientation

Analysis of polarization of the residual ion A^+ in the process (4.13), is complementary to the spin analysis of the photoelectrons. The appropriate parameters are the alignment, $\mathcal{A}_{20}(J_f)$, and orientation, $\mathcal{A}_{10}(J_f)$, of the electronic angular momentum J_f of the residual ion. The concept of alignment and orientation of the residual ion becomes meaningful for the purpose of the complete photoionization experiment provided the two-step model is appropriate. In the two-step model, the first-step photoionization process results in the alignment and/or orientation of the residual ion, whereas these parameters determine angular distribution and polarization of the second-step residual-ion decay products. General equations for the statistical tensors (state multipoles) $\mathcal{A}_{kq}(J_f)$ of the residual ion, including the case of arbitrary polarized photon and arbitrary polarized atomic target, are presented by Balashov et al. (2000). A detailed survey of alignment and orientation studies after inner shell ionization has been already outlined in Sect. 3.5, and alignment and orientation photoionization data have been discussed in Sect. 3.5.3 in more detail. Here, we outline some features appropriate for discussion of complete photoionization experiments.

For unpolarized target atoms, the set of non-vanishing polarization tensorial components of the residual ion are determined solely by the polarization state of the incoming photon. This can be understood in the context of conservation of the tensorial structure for the set of the statistical tensors of the whole system $[h\nu + A(\alpha_i J_i)]$ and $[A^+(\alpha_f J_f) + e^-_{ph}]$: unobserved photoelectrons e^-_{ph} and unpolarized target atoms $A(\alpha_i J_i)$ do not bring new tensorial components. Therefore, for an unpolarized atom,

$$\mathcal{A}_{kq}(J_f) = C_k(J_f)\, \rho^{\gamma}_{kq}(P_1, P_2, P_3), \tag{4.35}$$

where the coefficients $C_k(J_f)$ depend on the photoionization amplitudes and $\rho^{\gamma}_{k0}(P_1, P_2, P_3)$ are the statistical tensors of the incoming photons $h\nu$, expressed in terms of the Stokes parameters $P_1,\ P_2,\ P_3$ of the radiation beam. Equation (4.35) shows that for fixed rank k the q-components of the statistical tensors $\mathcal{A}_{kq}(J_f)$ are related 'kinematically', i.e. regardless to the dynamics of the photoionization. Therefore only a single q-component is meaningful for the purpose of complete photoionization experiment. The most convenient choice is the alignment parameter, $\mathcal{A}_{20}(J_f)$, and the orientation parameter, $\mathcal{A}_{10}(J_f)$. The alignment parameter is zero for $J_f < 1$. Particular values of $\mathcal{A}_{20}(J_f)$ and $\mathcal{A}_{10}(J_f)$ depend on the coordinate system. When discussing the orientation of the residual ion, normally the coordinate system with the z axis along the radiation beam is used. In this case (for $k = 1$)

$$\rho^{\gamma}_{10} = \frac{1}{\sqrt{2}}\, P_3, \qquad \rho^{\gamma}_{1\pm1} = 0\,. \tag{4.36}$$

When discussing the alignment, either the same coordinate system is implied, with

$$\rho^{\gamma}_{20} = \frac{1}{\sqrt{6}}, \qquad \rho^{\gamma}_{2\pm1} = 0, \qquad \rho^{\gamma}_{2\pm2} = -\frac{1}{2}\,(P_1 \mp \mathrm{i}P_2), \tag{4.37}$$

or a coordinate system with the z axis along the direction of the linear polarization (or along the principal axis of the polarization) of the incoming radiation. In the latter case

$$\rho^{\gamma}_{20} = -\frac{1}{2\sqrt{6}}\,(1 + 3P_1), \qquad \rho^{\gamma}_{2\pm1} = \frac{\mathrm{i}}{2}\, P_2, \qquad \rho^{\gamma}_{2\pm2} = \frac{1}{4}\,(1 - P_1). \tag{4.38}$$

For the pure linearly polarized radiation ($P_1 = +1$), (4.38) give

$$\rho^{\gamma}_{20} = -\sqrt{\frac{2}{3}} \tag{4.39}$$

with other q-components for the ranks $k = 1$ and $k = 2$ vanishing. This makes the coordinate system with the z axis along the linear polarization of the incoming photon more convenient for treating the alignment. Note that the monopole tensor $\rho^{\gamma}_{00} = 1/\sqrt{3}$ is independent of the coordinate system.

For initial atom with $J_i = 0$, expressions for $C_k(J_f)$ in (4.35) take especially simple form. Due to the dipole selection rules, the total angular momentum of the channels possesses a sharp value $J = 1$, and the polarization parameters of the residual ion are described by (4.35) with

$$C_k(J_f) = 3 \sum_{\ell j} (-1)^{J_f+j+k+1} \begin{Bmatrix} 1 & 1 & k \\ J_f & J_f & j \end{Bmatrix} \frac{\sigma_{\ell j}}{\sigma}, \tag{4.40}$$

where $\sigma = \sum_{\ell j} \sigma_{\ell j}$. Equation (4.40) together with (4.35) is equivalent to (3.85) taking into account the fact that two different coordinate systems are used. Unlike the parameters β, ξ, α, A and others, characterizing quantities related to photoelectrons, the residual ion orientation and alignment do not depend on the relative phases of the photoionization amplitudes, $\delta_{\ell j} - \delta_{\ell' j'}$; they depend only on the relative partial ionization cross sections $\sigma_{\ell j}/\sigma$ into the channel with the orbital angular momentum ℓ and total angular momentum j of the photoelectron. Note that in a general case of $J_i \neq 0$ the coefficients $C_k(J_f)$ depend on relative phases between the ionization amplitudes with different total angular momentum J, staying diagonal with respect to the quantum numbers ℓ and j.

Experimentally, the alignment $\mathcal{A}_{20}(J_f)$ may be derived from the angular distribution or linear polarization of fluorescence or angular distribution of Auger electrons from subsequent decay of the residual ion $A^{+*}(\alpha_f J_f)$, provided after the photoionization the ion is left in the excited state. From the experimental point of view the alignment method is the easiest approach to complete photoionization experiments because it requires neither target polarization nor spin-dependent measurements or coincidence experiments but simply angle-resolved electron or fluorescence spectroscopy. If one uses circularly polarized light for primary photoionization, the result will be an oriented final ionic state. The orientation $\mathcal{A}_{10}(J_f)$ may be determined by the spin polarization of subsequently emitted Auger electrons or the circular polarization of fluorescence photons (Snell et al. 1996, 1999b; Schmidtke et al. 2000a; West et al. 1996). As already mentioned, the phase information is lost in polarization of the residual ion. To keep this information, a coincidence experiment between photo- and Auger electrons or between photoelectron and fluorescence photon has to be performed, as will be described in Sects. 4.4 and 4.5, respectively.

After the pioneering papers by Flügge et al. (1972) and Jacobs (1972), and first measurements of the residual ion alignment in photoionization by Caldwell and Zare (1977), Southworth et al. (1981, 1983), Kronast et al. (1984), supplemented by first theoretical studies of Caldwell and Zare (1977), Berezhko et al. (1978a), Klar (1979), Greene and Zare (1982), Bußert and Klar (1983) and others, many experimental and theoretical investigations were published mainly for the alignment in photoionization of closed subshell atoms. A comprehensive list of references can be found in the papers by Yamaoka et al. (2002) and Lohmann (2003). Experimental data and systematic numerical calculations of the alignment of ions in photoionization of different atomic subshells (Berezhko et al. 1978a; Kleiman and Lohmann 2003; Kleiman and Becker 2005) have shown that in general the alignment of the residual ion is small, $\mathcal{A}_{20}(J_f) \sim 0.1$. For comparison, in photoexcitation of a closed-shell atom the absolute value of the alignment is much larger, $\mathcal{A}_{20} = -\sqrt{2}$. However, the alignment in photoionization may be much larger in the near threshold region as well as near the Cooper minimum (Berezhko et al. 1978a). In both these cases, the usually dominant amplitude of the transition $n_0\ell_0 \to \varepsilon(\ell_0 + 1)$ is suppressed and a large alignment is determined by the weaker transition $n_0\ell_0 \to \varepsilon(\ell_0 - 1)$. The orientation parameter $\mathcal{A}_{10}(J_f)$ is usually large $|\mathcal{A}_{10}(J_f)| \sim 0.5$–$1.0$ in a broad range of photon energies (Kabachnik and Lee 1989; Kleiman and Lohmann 2003).

Typical experimental and theoretical results presented and discussed in Sect. 3.5 show that in many cases, the agreement between theory and experiment is rather good. Note, that a systematic overestimate of the alignment in photoionization by radiation from radionucleides was derived in a few studies from anisotropy of the L X-ray fluorescence lines in comparison with other measurements and theory; this overestimate is probably connected with self-absorption of the X-rays in the solid target. Studies by Santra et al. (2007) discussed this point in detail and include a comprehensive reference list.

Resonant behavior of the alignment and orientation parameters of the residual ion is observed in the region of autoionizing or Auger resonances, reflecting sharp variations of the relative partial photoionization cross sections. We refer to Figs. 4.45 and 4.46 in Sect. 4.8.1 for examples of the resonance behavior.

4.3.2 Secondary Electron Spectrometry

In case of photoionization from the inner shell, the final ionic state is the Auger/autoionizing state, which decays preferentially by the emission of the second (Auger) electron and formation of the doubly charged ion,

$$\begin{aligned} h\nu + A(\alpha_i J_i) \longrightarrow\ & A^{+*}(\alpha_f J_f) + e^-_{ph} \\ & \hookrightarrow A^{++}(\alpha_d J_d) + e^-_{Auger}, \end{aligned} \tag{4.41}$$

Hausmann et al. (1988) suggested and performed for the first time a complete experiment on photoionization by measuring the angular distribution of the photoelectrons e^-_{ph}, following 2p photoionization of Mg, and the angular distribution of the resulting Auger electrons e^-_{Auger} in the reactions:

$$\begin{aligned} h\nu + \mathrm{Mg}(2\mathrm{p}^6 3\mathrm{s}^2\ {}^1\mathrm{S}_0) \longrightarrow\ & \mathrm{Mg}^{+*}(2\mathrm{p}^5 3\mathrm{s}^2\ {}^2\mathrm{P}_{1/2}) + e^-_{ph}(\varepsilon \mathrm{s}_{1/2}, \varepsilon d_{3/2}) \\ & \hookrightarrow \mathrm{Mg}^{++}(2\mathrm{p}^6\ {}^1\mathrm{S}_0) + e^-_{Auger}(\varepsilon \mathrm{p}_{1/2})\ , \end{aligned} \tag{4.42a}$$

and

$$h\nu + \text{Mg}(2\text{p}^6 3\text{s}^2\,{}^1\text{S}_0) \longrightarrow \text{Mg}^{+*}(2\text{p}^5 3\text{s}^2\,{}^2\text{P}_{3/2}) + e^-_{ph}(\varepsilon \text{s}_{1/2}, \varepsilon \text{d}_{3/2}, \varepsilon \text{d}_{5/2})$$
$$\hookrightarrow \text{Mg}^{++}(2\text{p}^6\,{}^1\text{S}_0) + e^-_{Auger}(\varepsilon \text{p}_{3/2}) \,. \tag{4.42b}$$

In the processes (4.42a) and (4.42b) with linearly polarized radiation beam, the angular distributions of the $L_2M_1M_1$ and $L_3M_1M_1$ Auger electrons are observed, respectively, in addition to the angular distribution of the photoelectrons. In the Cooper–Zare model, which is supposed to be adequate for this case, the goal of the complete photoionization experiment is the set of three parameters, describing photoionization into the εs and εd channels: R_s, R_d, and Δ_{sd}. The asymmetry parameters in the angular distribution of photoelectrons are expressed by (4.34a). The angular distribution of Auger electrons with subsequent $L_{2,3}M_1M_1$ Auger decay is described by

$$I(\theta_A) = \frac{I_0}{4\pi}\Big(1 + \beta_A P_2(\cos\theta_A)\Big), \tag{4.43}$$

where I_0 is the angle-integrated intensity of the Auger line, θ_A is the angle of the Auger emission counted from the direction of the linear polarization of the incident radiation beam, and the asymmetry parameter β_A is different for the $L_2M_1M_1$ and $L_3M_1M_1$ Auger transitions:

$$\beta_A(L_2M_1M_1) = 0, \tag{4.44}$$

and

$$\beta_A(L_3M_1M_1) = \frac{R_s^2 + \frac{1}{5}R_d^2}{R_s^2 + 2R_d^2} \,. \tag{4.45}$$

Vanishing parameter $\beta(L_2M_1M_1)$ is related to the fact that a state with $J_f = 1/2$ cannot be aligned. The absolute value of the amplitudes is determined by the 2p photoionization cross section given by (4.34a). The alignment $\mathcal{A}_{20}$ of the 2p hole, which is the prerequisite for a non-isotropic angular distribution of the Auger transition, is for the present case directly related to the asymmetry parameter (4.45)

$$\mathcal{A}_{20}(2\text{p}^5 3\text{s}^2\,{}^2\text{P}_{3/2}) = -\beta_A(L_3M_1M_1) \,. \tag{4.46}$$

This straightforward connection is crucial for success of the complete experiment and is related to the zero angular momentum (J_d) of the final state $\text{Mg}^{++}(2\text{p}^6)$ in (4.41): the vanishing J_d results in only one Auger decay channel. For other values of J_d, generally the angular distribution of the Auger electrons in the second step of the process (4.41) depends on the Auger decay amplitudes, which are the additional independent dynamical parameters. This dependence cancels only for a single Auger decay channel.

Figure 4.9 illustrates the results of the complete experiment within the framework of the Cooper–Zare model for the Mg 2p subshell photoionization together

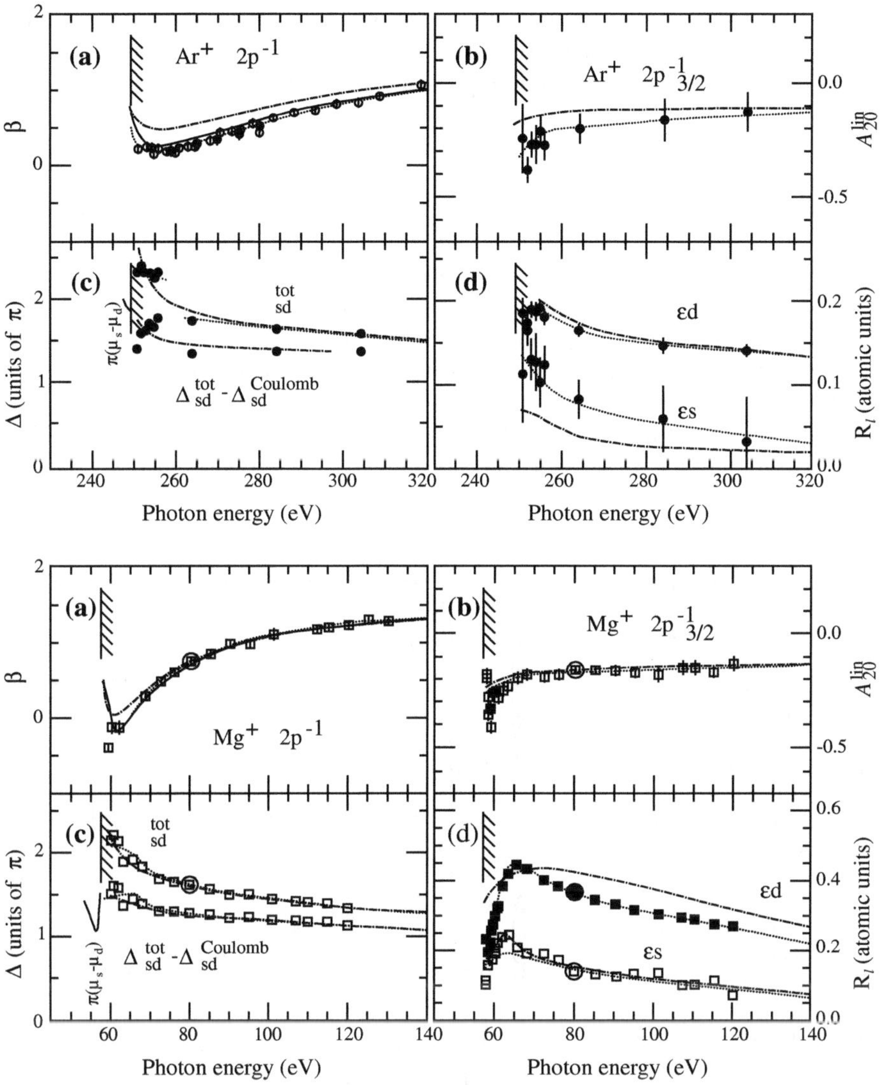

Fig. 4.9 Angular distribution and photoion alignment parameters with the derived dipole matrix elements and relative phase shifts of the outgoing εs and εd photoelectron waves following 2p photoionization of Ar (*upper panels*) and Mg (*lower panels*). Argon: Experimental data from Becker (1990); (— · —): theoretical data obtained by the Hartree–Fock–Slater method (Kennedy and Manson 1972; Berezhko et al. 1978a); (——): RRPA calculations (Lindle et al. 1988); (· · · · · ·): semi-empirical values derived from the interpolation of the experimental data. Magnesium: Experimental data from (○) Hausmann et al. (1988) and (□) Schmidt (1997); (— · —): Hartree–Fock results of Flügge et al. (1972); (——): β results obtained by the RRPA method (Desmukh and Manson 1983); (· · · · · ·): Semi-empirical values derived from a critical evaluation of the experimental data in conjunction with theoretical cross sections. The quantum defect difference curve is obtained from spectroscopic measurements (Newsom 1971). After Becker (1998)

with similar results for the Ar 2p photoionization. The latter are obtained by similar experimental methods (Becker 1990). The absolute values and the relative phase of the photoionization εs and εd amplitudes in a broad energy range are extracted and compared with theoretical calculations. There is a close resemblance between the Ar and the Mg data, except for Mg there is a stronger suppression of the εd wave due to a centrifugal barrier in the potential (Berezhko et al. 1978a) giving rise to a small resonance just above the threshold. Although the phase shift differences are given with relatively large uncertainties, the smooth transition to the quantum defect difference at threshold is clearly demonstrated. Note that the Hartree–Fock calculations better describe the εs ionization channel than the εd channel in Ar, while the situation is opposite in Mg. From a qualitative point of view these results are quite convincing, but from a quantitative point of view there still have to be made significant improvements in theory.

Figure 4.10 illustrates the results of complete photoionization experiment for photoionization of the 3p subshell in Ca (Lörch et al. 1999). The reaction is similar to (4.42a) and (4.42b):

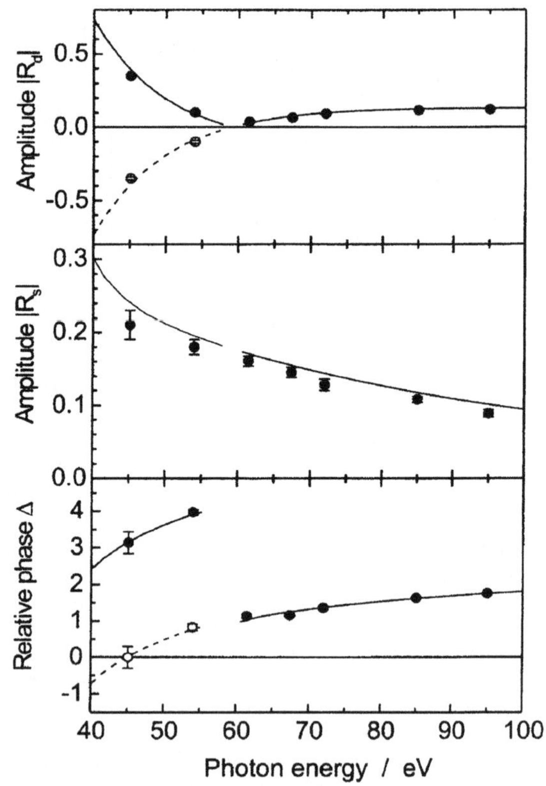

Fig. 4.10 Graphical representation of dipole amplitudes R_d (*top*) and R_s (*middle*) and their relative phase $\Delta \equiv \Delta_{ds}$ (*bottom*) for 3p photoionization in atomic calcium in the region around the Cooper minimum at 59.0 eV photon energy as a function of photon energy. (•) and (∘): experimental data with error bars; (——) and (– – –): RRPA results transferred to the LS coupling limit; the R_d and R_s amplitudes of the calculation have been multiplied with the factor 0.8 in order to take into account satellite processes as observed experimentally at 85 eV photon energy. For photon energies below 59.0 eV the additional parameters $R-d^{\text{mod}} = -|R_d|$ and $\Delta^{\text{mod}} = \Delta - \pi$ are introduced. After Lörch et al. (1999)

$$h\nu + \mathrm{Ca}(2\mathrm{p}^6 3\mathrm{s}^2\,{}^1\mathrm{S}_0) \longrightarrow \mathrm{Ca}^{+*}(2\mathrm{p}^5 3\mathrm{s}^2\,{}^2\mathrm{P}_{1/2}) + e^-_{ph}(\varepsilon \mathrm{s}_{1/2}, \varepsilon \mathrm{d}_{3/2})$$
$$\hookrightarrow \mathrm{Ca}^{++}(2\mathrm{p}^6\,{}^1\mathrm{S}_0) + e^-_{Auger}(\varepsilon \mathrm{p}_{1/2}) \,, \qquad (4.47a)$$

and

$$h\nu + \mathrm{Ca}(2\mathrm{p}^6 3\mathrm{s}^2\,{}^1\mathrm{S}_0) \longrightarrow \mathrm{Ca}^{+*}(2\mathrm{p}^5 3\mathrm{s}^2\,{}^2\mathrm{P}_{3/2}) + e^-_{ph}(\varepsilon \mathrm{s}_{1/2}, \varepsilon \mathrm{d}_{3/2}, \varepsilon \mathrm{d}_{5/2})$$
$$\hookrightarrow \mathrm{Ca}^{++}(2\mathrm{p}^6\,{}^1\mathrm{S}_0) + e^-_{Auger}(\varepsilon \mathrm{p}_{3/2}) \,. \qquad (4.47b)$$

The high accuracy of the data did allow an extensive comparison between experiment and theory in the photon energy range surrounding the Cooper minimum, which provides a particular sensitive case. The Cooper minimum is due to crossing zero of the $3\mathrm{p} \to \varepsilon \mathrm{d}$ ionization channel amplitude. The relativistic random phase approximation (RRPA) results, which include interchannel couplings between the $4\mathrm{s}_{1/2}$, $3\mathrm{p}_{3/2}$, $3\mathrm{p}_{1/2}$ and $3\mathrm{s}_{1/2}$ ionization channels give in their non-relativistic LSJ coupling limit an excellent description for the experimentally derived 3p photoionization amplitudes with their relative phase. This agreement holds for the absolute magnitude of these amplitudes if the considerable satellite intensity is taken into account by a constant and channel-independent spectroscopic factor. The results clearly demonstrate the modification of the main photoprocess by satellite transitions and the applicability of the RRPA for a quantitative analysis of the observed photoionization phenomena.

A detailed analysis of the 4d-subshell photoionization in Xe,

$$h\nu + \mathrm{Xe}(5\mathrm{p}^6) \longrightarrow \mathrm{Xe}^+(4\mathrm{d}^{-1}_{5/2}) + e^-_{ph}(\varepsilon \mathrm{p}_{3/2}, \varepsilon \mathrm{f}_{5/2}, \varepsilon \mathrm{f}_{7/2}), \qquad (4.48a)$$

and

$$h\nu + \mathrm{Xe}(5\mathrm{p}^6) \longrightarrow \mathrm{Xe}^+(4\mathrm{d}^{-1}_{3/2}) + e^-_{ph}(\varepsilon \mathrm{p}_{1/2}, \varepsilon p_{3/2}, \varepsilon \mathrm{f}_{5/2}), \qquad (4.48b)$$

was performed by Becker (1998). The three-parameter model is not sufficient to consistently describe all the available photoionization data for the above process. Nevertheless, adding only one more parameter, the ratio of the integral photoionization cross section to the relativistically splitted ionic states $\mathrm{Xe}^+(4\mathrm{d}^{-1}_{5/2})$ and $\mathrm{Xe}^+(4\mathrm{d}^{-1}_{3/2})$, brings together all the data obtained for the processes (4.48). This 3 + 1 *parameter* model seems to describe satisfactory the dynamics of the photoionization of a large number of elements outside the Cooper minima (Becker 1998).

Finally, we note that the secondary electron measurements have an important advantage: normally the Auger decay proceeds fast enough to neglect the depolarization of the Auger decaying state, which occurs during the time between photoionization and the Auger decay. Therefore the Auger electrons carry information related directly to the polarization of the residual ion due to the photoionization. As we will see below, in the case of secondary fluorescence, the depolarization effects may become a very important factor, which is difficult to take into account.

4.3.3 Secondary Fluorescence Polarimetry

Fluorescence polarimetry is a well developed method in physics of laboratory and astrophysical plasma, which provides information on dynamics of formation of the excited atomic and ionic states. In particular, polarization of the fluorescence lines indicates an anisotropy, induced by particle and radiation flows or external fields in the radiating media (Kazantsev and Hénoux 1995). Observing the fluorescence polarization from excited atoms or ions can be considered as a tool to probe the induced polarization of the collision-excited state. With regard to complete photoionization experiments, we now concentrate on measurements of the alignment and orientation parameters of the residual ion by means of non-coincident fluorescence polarimetry, or the fluorescence angular distribution in the reaction

$$\begin{aligned} h\nu + A(\alpha_i J_i) \longrightarrow\ & A^{+*}(\alpha_f J_f) + e^-_{ph} \\ & \hookrightarrow A^{+}(\alpha_g J_g) + h\nu_{fluo}\ . \end{aligned} \tag{4.49}$$

The first application of the fluorescence method to complete experiment in photoionization was by Jiménez-Mier et al. (1986), who studied photoionization-excitation of He into the $He^{+*}(n = 2)$ states,

$$\begin{aligned} h\nu + \mathrm{He}(1s^2) \longrightarrow\ & \mathrm{He}^{+*}(n=2) + e^-_{ph}(\varepsilon s, \varepsilon d) \\ & \hookrightarrow \mathrm{He}^{+}(1s) + h\nu_{fluo}\ . \end{aligned} \tag{4.50}$$

The data for the asymmetry parameter of the angular distribution of photoelectrons β (Morin et al. 1983) were combined with the data on the alignment of the residual He ion extracted from the angular distribution of the 304 Å $He^{+*}(n = 2)$ fluorescence. With these data the ratio of the partial $\sigma(2p\varepsilon s)$ and $\sigma(2p\varepsilon d)$ photoionization cross sections and the phase difference between the photoionization amplitudes for the two channels, $2p\varepsilon s$ and $2p\varepsilon d$, were deduced.

In the optical region with easy available polarimeters, fluorescence polarization measurements at fixed photon detector position are more convenient. Figure 4.11 shows one of the possible experimental schemes (O'Keeffe et al. 2003, 2004). The alignment parameter $\mathcal{A}_{20}(J_f)$ determines the linear polarization, P_L, of fluorescence from the residual ion state $A^{+*}(\alpha_f J_f)$ in (4.49). General relations between the degree of the fluorescence polarization and polarization of the radiating ion $A^{+*}(\alpha_f J_f)$ for arbitrary fluorescence detector positions for coincidence and non-coincidence experimental set-ups may be found in Lohmann et al. (2003a) and Balashov et al. (2000). In the geometry of the set-up and the coordinate system shown in Fig. 4.11, $\mathcal{A}_{20}(J_f)$ can be extracted for 100 % linearly polarized incoming radiation from the relation

$$P_L = \frac{I_{\parallel} - I_{\perp}}{I_{\parallel} + I_{\perp}} = \frac{3\,\kappa_2\,\mathcal{A}_{20}(J_f)}{\kappa_2\,\mathcal{A}_{20}(J_f) - 2}\ , \tag{4.51}$$

where $I_{\parallel}$ ($I_{\perp}$) is the intensity of the outgoing fluorescence radiation with the axis of the polarimeter parallel (perpendicular) to the direction of the linear polarization of

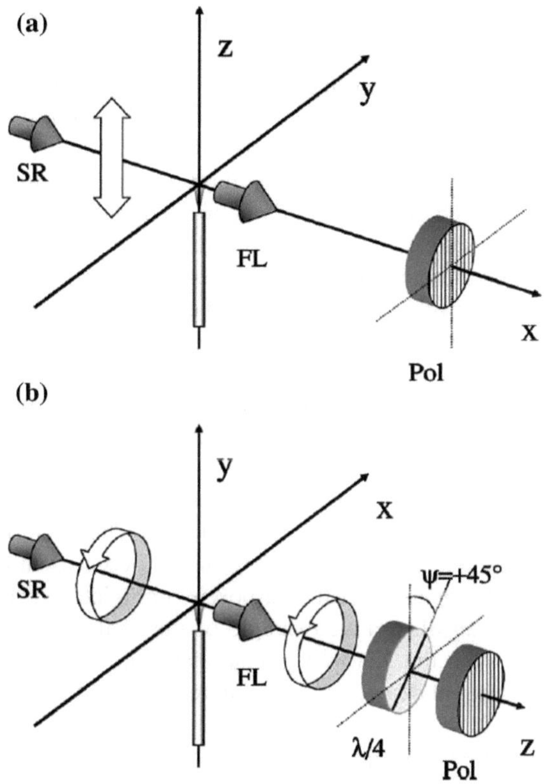

Fig. 4.11 The geometries used to analyze **a** the degree of linear polarization of the fluorescence (FL) following excitation with linearly polarized synchrotron radiation (SR); **b** the degree of circular polarization of the fluorescence following excitation with circularly polarized SR. Note that different coordinate systems are used in (**a**) and (**b**) as the z axis in **a** is defined by the polarization axis of the SR, while in **b** it is defined by the direction of propagation of the SR. The polarimeter (Pol) and the λ/4 plate are indicated. After O'Keeffe et al. (2004)

the incoming radiation. The circular polarization, P_C, of the fluorescence radiation induced by circularly polarized incoming radiation is proportional to the orientation parameter $\mathcal{A}_{10}(J_f)$:

$$P_C = \frac{I_+ - I_-}{I_+ + I_-} = \frac{\sqrt{3}\,\kappa_1\,\mathcal{A}_{10}(J_f)}{\kappa_2\,\mathcal{A}_{20}(J_f) + 1}\,, \tag{4.52}$$

where I_+ (I_-) is the intensity of the right- (left-) handed circularly polarized component of the outgoing fluorescence radiation, provided the incoming radiation is 100 % right-handed polarized. The coefficients κ_k in (4.51) and (4.52),

$$\kappa_k = \sqrt{\frac{3\,(2J_f+1)}{2}}(-1)^{J_f+J_g+k+1}\begin{Bmatrix} 1 & 1 & k \\ J_f & J_f & J_g \end{Bmatrix} D_k(J_f), \tag{4.53}$$

depend on the angular momenta of the ion state before (J_f) and after (J_g) the radiative transition, while the factors $D_k(J_f)$ account for the depolarization of the electronic angular momentum J_f before the radiation takes place.

Note that a cross check of the fluorescence polarization data is possible: the parameters $\mathcal{A}_{k0}(J_f)$ extracted from different fluorescence lines, originating from the same photoion states $A^{+*}(\alpha_f J_f)$, must be equal.

High spectral resolution, which in general allows to resolve the fine-structure final ion states after the photoionization and the possibility of the cross checking are important advantages of the fluorescence polarimetry method. The major drawback of the method is the poorly controlled variation of the alignment and orientation of the residual ion state due to additional population and depolarization of this state before the radiative decay. For example, it is difficult, if possible at all, to measure quantitatively the depolarization effects of the radiative cascades to the $A^{+*}(\alpha_f J_f)$ state. Accurate calculation of the cascade is also problematic, since one generally needs to know the initial relative population of many ionic levels after the photoionization, their polarization, and the probabilities of radiative transitions involved in the cascades. The inherent difficulties in separating the cascade effects (Hamdy et al. 1991; Jiménez-Mier et al. 1993) stimulated correlation measurements between the photoelectron and subsequent polarized fluorescence photon as a method of the complete photoionization experiment. This method will be addressed in Sect. 4.5. Nevertheless, as shown below, the non-coincident fluorescence polarization method can also be used for the purpose of complete experiment.

As an illustrative example of the analysis (see Fig. 4.12), we consider ionization from the Xe(5p^6) ground state into the excited Xe^{+*}(5p^{4}6p) ion states with subsequent detection of the fluorescence radiation

$$\begin{aligned} h\nu + \mathrm{Xe}(5\mathrm{p}^6\,{}^1\mathrm{S}_0) &\longrightarrow \mathrm{Xe}^{+*}(5\mathrm{p}^4(L_cS_cJ_c)\,6\mathrm{p}[K]_{J_f}) + e^-_{ph}(\ell j) \\ &\qquad \hookrightarrow \mathrm{Xe}^{+*}(5\mathrm{p}^4(L_cS_cJ_c)\,6\mathrm{s},5\mathrm{d}) + h\nu_{fluo}\,. \end{aligned} \tag{4.54}$$

Here, the jK-coupling scheme will be used for the assignment of the excited Xe ion levels and the index c indicates the quantum numbers of the Xe^{++}(5p^4) core. The experiments were carried out at the ELETTRA storage ring in Trieste (Italy) at the *Circular Polarization* beamline with a high degree of linear (>0.99) and circular (>0.98) polarization of the XUV photons (O'Keeffe et al. 2003, 2004; Meyer et al. 2001). The photoionization amplitudes (4.15) are specific for each of the 21 final ionic fine-structure states $\left|5\mathrm{p}^4(L_cS_cJ_c)\,6\mathrm{p}[K]_{J_f}\right\rangle$.

Depending on J_f, the contributing photoelectron partial waves are given in Table 4.3 where the crosses mark the contributing channels. Values of the asymmetry parameter β in the angular distribution of photoelectrons (4.21) for all residual ion states of the Xe^{+*}(5p^{4}6p) configuration were experimentally determined by Aksela et al. (1996) and Langer et al. (1996). The dynamical spin polarization P_y (4.28) was measured by Hergenhahn et al. (1999) for some of the final ionic states, but only for one of them, Xe^{+*}(5p^4(^{1}D$_2$)6p[1]$_{1/2}$) with $J_f = 1/2$, it was possible to obtain the parameter ξ in (4.28) and to extract the relative magnitude of the photoionization amplitudes of the $\varepsilon s_{1/2}$ and $\varepsilon d_{3/2}$ channels with reasonable accuracy.

To extract the parameters $\mathcal{A}_{20}(J_f)$, $\mathcal{A}_{10}(J_f)$ from (4.51) and (4.52), respectively, and to further obtain the relative partial cross sections from (4.35) and (4.40), one

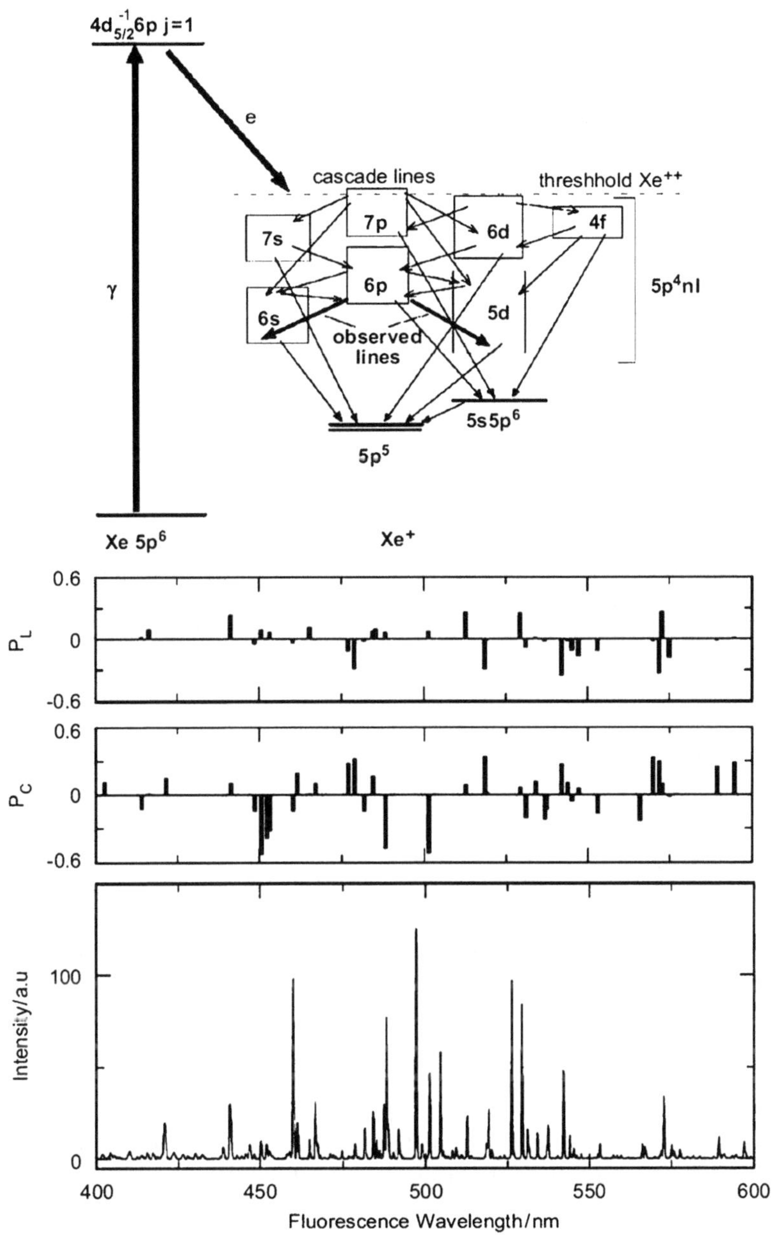

Fig. 4.12 Scheme of transitions for the resonant Auger process in Xe, including radiative cascades in Xe^+ (*top*) and dispersed fluorescence spectrum obtained following excitation of the $Xe^*4d^{-1}_{5/2}6p$ resonance (*bottom*). The degree of circular P_C and linear P_L polarization for each line is shown as gistogram. Adapted from Grum-Grzhimailo and Meyer (2005)

Table 4.3 Contributing partial waves of the photoelectrons for the Xe $^1S_0$ photoionization into the excited $Xe^{+*}(5p^46p)$ ion state

J_f	$\varepsilon s_{1/2}$	$\varepsilon d_{3/2}$	$\varepsilon d_{5/2}$	$\varepsilon g_{7/2}$	$\varepsilon g_{9/2}$
$\frac{1}{2}$	+	+			
$\frac{3}{2}$	+	+	+		
$\frac{5}{2}$		+	+	+	
$\frac{7}{2}$			+	+	+

needs to find the depolarization factors $D_k(J_f)$ in (4.53). With respect to the experimental conditions, there are two depolarization effects that had to be taken into account: the hyperfine interaction, leading to the precession of the electronic angular momentum J_f around the total ionic angular momentum $\mathbf{F}_f = \mathbf{J}_f + \mathbf{I}$, where $\mathbf{I}$ is the nuclear spin, and the fluorescence cascades. Especially the treatment of the developed radiative cascades is very difficult. The complex radiative cascade for the reaction (4.54) induced by photons of the energy 65.1 eV is schematically displayed in Fig. 4.12 (top) with the corresponding measured fluorescence spectrum in the range 400–600 nm and polarization of the fluorescence lines. The geometry for the polarization measurement corresponds to Fig. 4.11. The key point allowing to simplify the task of treating the cascade is the assumption of an *isotropic* radiative cascade population of the $Xe^{+*}(5p^46p)$ ionic levels. The isotropic model, confirmed in this case experimentally, implies that the last line of the radiative cascades is unpolarized and therefore all magnetic substates of the final state in the cascade are equally populated by the cascade. In the isotropic model, the depolarization factor can be expressed in a product form (Meyer et al. 2001): $D_k(J_f) = D(J_f)\, G_k(J_f)$. Here $G_k(J_f)$ is the depolarization factor due to the hyperfine interaction, calculated by standard methods (Fano and Macek 1973; Greene and Zare 1982) under the assumption that the photoionization with excitation populates coherently the hyperfine levels of the final ionic state. The factor $D(J_f)$ is the relative probability to populate the final ionic state $A^{+*}(\alpha_f J_f)$ by the photoionization with excitation process, provided the radiative cascades also populate this state. A way to find $D(J_f)$ is to use the relative line intensities from the experimental photoelectron spectrum as input in the theoretical calculation of the fluorescence cascades. Necessary conditions for using this method are the availability of the photoelectron spectrum in a broad energy range with sufficient energy resolution, the reliable identification of the photoelectron lines, an accurate theoretical model to calculate the relative probabilities of the radiative transitions in the ion, and the possibility to account for all relevant pathways of the cascades. Such an analysis has been realized for the reaction (4.54) in O'Keeffe et al. (2003), Meyer et al. (2001) by utilizing photoelectron spectroscopic data from Aksela et al. (1995) and extensive multiconfigurational intermediate coupling calculations of the optical transition probabilities and the radiative cascades. Generally, at least a simplified empirical estimate of the influence of the cascade, like in the analysis of Schartner et al. (2005), is necessary to extract meaningful values of the alignment and orientation from the fluorescence data.

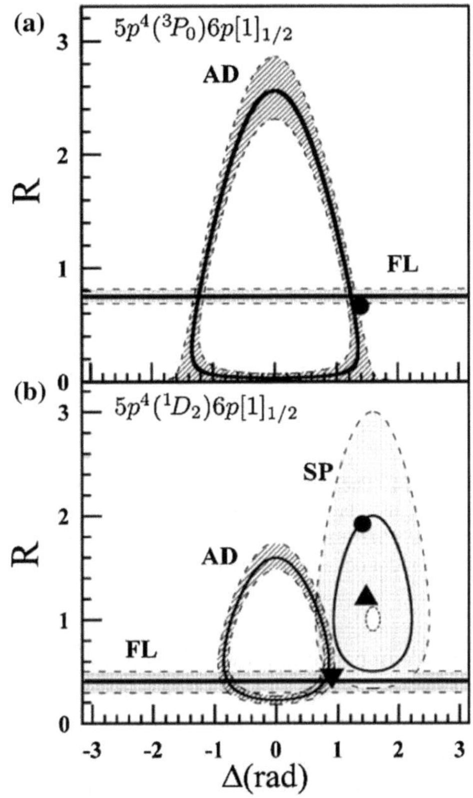

Fig. 4.13 **a** Parametric plot $R(\Delta)$ for the electron angular distribution (AD) data (Aksela et al. 1996; Langer et al. 1996) together with the value of R from the fluorescence polarization measurements (O'Keeffe et al. 2003) for the $Xe^{+*}(5p^4(^3P_0)6p[1]_{1/2})$ final state. **b** the equivalent data for the $Xe^{+*}(5p^4(^1D_2)6p[1]_{1/2})$ final state along with the parametric plot for the spin polarization (SP) data (Hergenhahn et al. 1999). The *shaded areas* show the error bars. Theoretical results: (•) Lohmann and Kleiman (2001); (*L*) Lagutin et al. (2000); (*K*) Hergenhahn et al. (1999); Hergenhahn and Becker (1995). Calculations by Lagutin et al. (2000) yield a negligible value of R in **a**. After O'Keeffe et al. (2003)

Combined data on $\mathcal{A}_{10}(J_f)$ and β are sufficient to determine the absolute ratio $R = d_{s1/2}/d_{d3/2}$ and $\cos(\delta_{s1/2} - \delta_{d3/2})$, as demonstrated in Fig. 4.13 for two $Xe^{+*}(5p^46p)$ ion states with $J_f = 1/2$. To fix the sign of $\Delta = \delta_{s1/2} - \delta_{d3/2}$, an additional measurement of the spin polarization of the emitted photoelectrons is needed, as has been performed for the case of the $Xe^{+*}(5p^4(^1D_2)\,6p[1]_{1/2})$ state, or an appropriate theoretical model can be used, as in the case of the $Xe^{+*}(5p^4(^3P_0)\,6p[1]_{1/2})$ state.

Combined data on $\mathcal{A}_{10}(J_f)$, $\mathcal{A}_{20}(J_f)$, β and ξ provide four independent quantities to find two absolute ratios and two phase differences of the three amplitudes, i.e. to resolve the problem of a complete experiment for the photoionization into the ionic states with $J_f > 1/2$. The statement that the four equations for the above parameters are independent is based on the fact that the corresponding Jacobian does not vanish (Cherepkov and Semenov 2004; O'Keeffe et al. 2004). Only the three parameters $\mathcal{A}_{10}(J_f)$, $\mathcal{A}_{20}(J_f)$, and β of the four quantities are currently available from experiments. Therefore, the particular solution for the amplitudes cannot be extracted, though the allowed amplitude space can be strongly reduced. This procedure is exemplified in Fig. 4.14 for two states of the residual Xe ion in the process

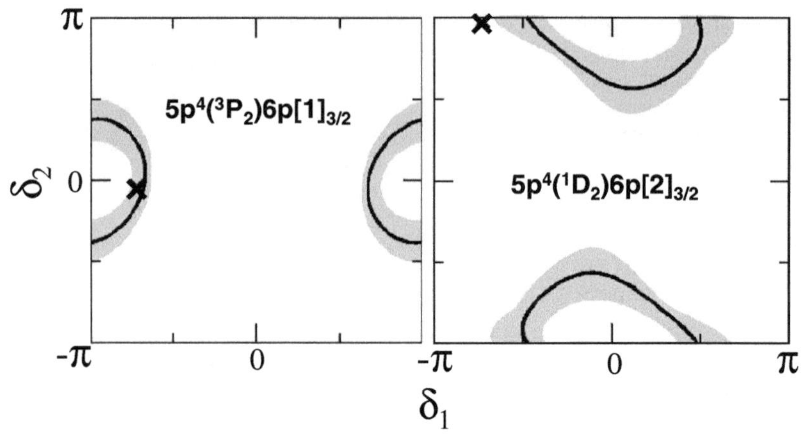

Fig. 4.14 The reduced parameter space δ_2 versus δ_1 found from a combination of the fluorescence polarization data and the Auger electron angular distribution parameter (Aksela et al. 1996; Langer et al. 1996) for the two final ionic $Xe^{+*}(5p^46p)$ states with $J = 3/2$. Results of the calculations in the Dirac–Fock approximation are indicated by crosses. Adapted from O'Keeffe et al. (2004)

(4.54). In particular, the values of $\mathcal{A}_{10}(J_f)$ and $\mathcal{A}_{20}(J_f)$ allow to find absolute ratio of the three amplitudes, while equation for β gives a further relation between the two phase differences, δ_1 and δ_2, between the amplitudes in terms of the above ratios. The latter relation defines, within experimental accuracy, a curve δ_2 versus δ_1 displayed in Fig. 4.14. This curve gives the reduced amplitude space: while the absolute ratios of the amplitudes are defined, for fixing the phase difference one equation is missing. Nevertheless, the values of the relative amplitudes can be fixed, up to ambiguity in sign, with the data available under further realistic assumption, namely assuming that the relativistic splitting of the phase shifts in the $\varepsilon d_{3/2}$ and $\varepsilon d_{5/2}$ channels is negligible. This assumption is confirmed by the multiconfigurational Dirac–Fock calculations (O'Keeffe et al. 2004), which show that this difference is less than 5° for all final ionic states.

Although for the ionic states with $J_f = 3/2$ the calculations give satisfactory results, the data for the $J_f = 7/2$ state (see Table 4.4) on the line intensity and the partial cross sections indicate a strong overestimate of the contribution from the $\varepsilon g_{9/2}$ channel, which is missing in ionization to the $J_f = 3/2$ ionic states.

Table 4.4 Dynamic parameters for the process (4.54) for a final ionic state with $J_f = 7/2$. The relative partial cross sections σ_j/σ are given in %; the relative photoline strengths I_{rel} are normalized to 100 when summed over the $5p^46p$ manifold.

Final ionic state		$\sigma_{5/2}/\sigma$	$\sigma_{7/2}/\sigma$	$\sigma_{9/2}/\sigma$	$\cos(\delta_{\varepsilon d} - \delta_{\varepsilon g})$	I_{rel}
$5p^4(^1D_2)6p[3]_{7/2}$	Exp.	12(9)	13(10)	75(10)	0.50(33)	2.44[a]; 3.66[b]
	Theo.	2.3	1.7	96.0	0.88	17.07

[a]Aksela et al. (1995), [b]Langer et al. (1996)

This points to an effect not included in the theoretical calculations and demonstrates that only the complete data set can assess the validity of a theoretical model. The most important improvement of the theoretical model is expected to arise from a more precise description of the photoelectron continuum beyond the one-electron approximation.

Note that the experimental data for the reaction (4.54) are obtained at the energy of the incoming photon $h\nu\,(SR) = 65.1\,\text{eV}$, corresponding to the strong $\text{Xe}^*\,4\text{d}^{-1}_{5/2}6\text{p}\,(J_r = 1)$ resonance. Although from the point of view of the complete photoionization experiment the ionization mechanism does not matter, theoretical results for the photoionization amplitudes shown in Figs. 4.13 and 4.14 were obtained in the two-step model, peculiar to the photoinduced resonant Auger process. More detailed discussion of the resonant Auger effect is presented in a review by Kabachnik et al. (2007) and in Chap. 3 of this book.

The fluorescence polarimetry method alone, without supplement of other methods or without measurements of the fluorescence in coincidence with the photoelectron (see Sect. 4.5) can provide only partial cross sections without the relative phases of the corresponding amplitudes. Nevertheless, measurements of the partial cross sections are not only an important step towards the complete experiment, but also very informative by themselves, providing a physically transparent quantities. Especially interesting results can be obtained for non-flat continuum: the regions of overlapping autoionizing resonances, ionization thresholds, Cooper minima. Such data can be used, for example, to identify resonances and autoionizing states with respect to their angular momentum. In experiments by Yenen et al. (2001) and McLaughlin et al. (2002) on photoionization of the argon atoms in the region of Rydberg autoionizing states, the incoming circularly polarized VUV radiation produced the Ar^+ ions in the reaction

$$\begin{aligned} h\nu + \text{Ar}(3\text{p}^6) \longrightarrow\ & \text{Ar}^{+*}(3\text{p}^4[^3\text{P}]\,4\text{p}\,^2\text{P}_{3/2}) + \text{e}^-_{\text{ph}}(\varepsilon\text{s}_{1/2}, \varepsilon\text{d}_{3/2}, \varepsilon\text{d}_{5/2}) \\ & \hookrightarrow \text{Ar}^+(3\text{p}^4[^3\text{P}]4\text{s}\,^2\text{P}_{1/2}) + \text{h}\nu_{\text{fluo}}(476.5\,nm). \end{aligned} \tag{4.55}$$

Degree of linear and circular polarization of the secondary fluorescence was measured under fixed angles, as shown in Fig. 4.15, and thus the alignment and orientation of the residual ion were determined. Note that the radiative cascades are negligible in the VUV photon energy range under consideration and the depolarization of the ionic state due to the hyperfine interactions is not relevant because of zero nuclear spin. Displayed in Fig. 4.16 are the results of the partial cross section separation, showing, with high resolution, a complicated dynamics of photoionization with overlapping Rydberg series of resonances of different symmetry. Relativistic effects are very pronounced, which is seen from the lower panel, where the statistical ratio of the $\varepsilon\text{d}_{3/2}$ and $\varepsilon\text{d}_{5/2}$ channels is indicated by dashed line. This example additionally shows the high potential of the fluorescence polarimetry method, provided the depolarization of the residual ionic state does not disturb the results.

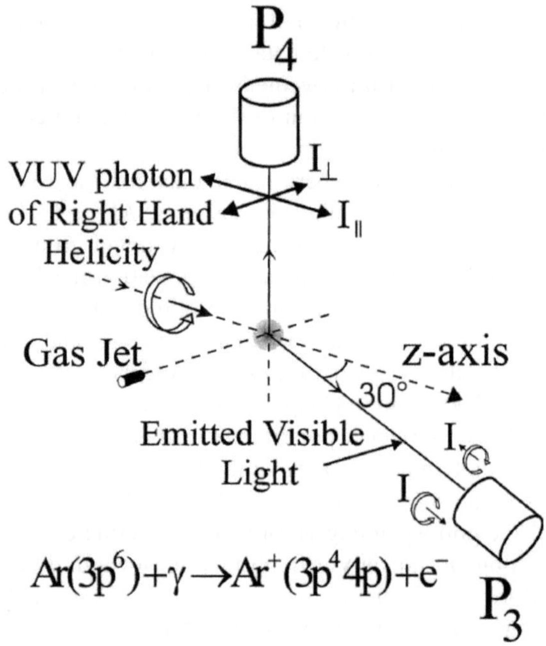

Fig. 4.15 Schematic of experimental setup. Circular polarization and linear polarization of fluorescence are measured for the photons ejected perpendicular and parallel to the incoming radiation, respectively. Here the linear polarization of the fluorescence is defined as $P_4 = (I_{\parallel} - I_{\perp})/(I_{\parallel} + I_{\perp})$. After McLaughlin et al. (2002)

4.4 Coincidence Photoelectron–Auger Electron Spectrometry

It was first predicted theoretically by McFarlane (1975) that the photoelectron and the subsequently ejected Auger electron can be strongly correlated in direction and that the detailed form of the correlation will depend on both photoionization and Auger transition amplitudes. However, for the important special case where the Auger electron can be represented by a single partial wave, the form of the correlation depends only on parameters governing the photoionization. Thus, coincidence photoelectron–Auger electron experiment were proposed. Later, general expressions for the angular correlations between the photoelectron and the Auger electron within the two-step model were derived (Berezhko and Kabachnik, 1979; Kabachnik 1992), which became a theoretical basis for complete photoelectron–Auger electron coincidence experiments. Alternative formulations were developed by Kupliauskiene and Tutlis (2003) in the graphical approach and by da Pieve et al. (2007a) in a single-particle scattering approach, more convenient for extension to the solid state case.

Complete photoionization experiments based on photoelectron–Auger electron coincidence measurements have been pioneered by Kämmerling and Schmidt (1991), preceded by a thorough analysis (Schmidt 1990). In particular, they demonstrated that the combination of the measurements of the angular correlation between the photoelectron and the Auger electron, the angular distribution of the photoelectrons and the angular distribution of the Auger electrons could constitute a complete photoion-

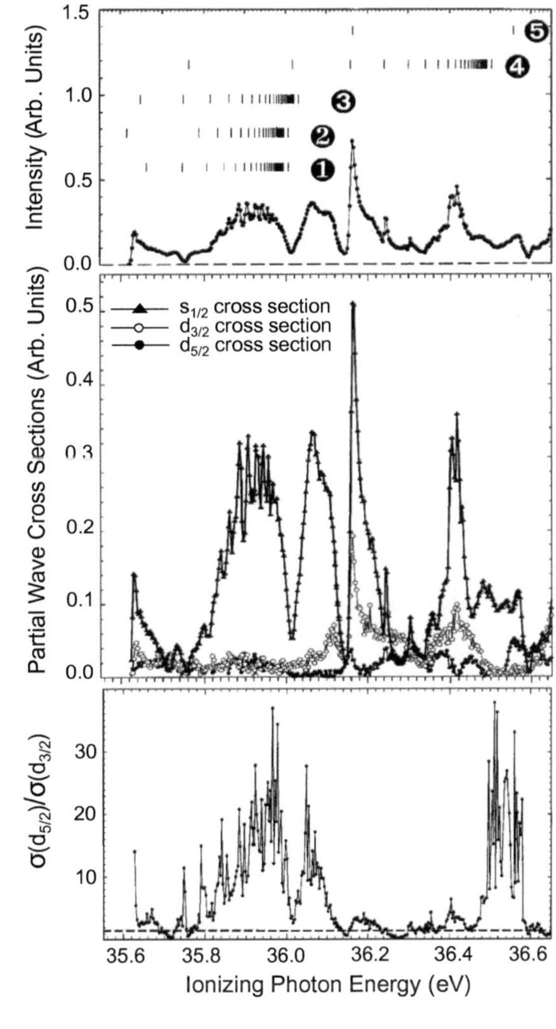

Fig. 4.16 (*Upper panel*) the relative total intensity of fluorescence observed perpendicular to the incoming radiation. The *vertical lines* show the known doubly excited states of Ar. (❶): $3p^4[^1D]3d\,^2F_{5/2}np$ series starting with $n = 8$; (❷): $3p^4[^1D]3d\,^2F_{5/2}nf$ series starting with $n = 6$; (❸): $3p^4[^1D]3d\,^2F_{7/2}nf$ series starting with $n = 6$; (❹): $3p^4[^1S]4s\,^2S_{1/2}np$ series starting with $n = 6$; (❺): $3p^4[^1D]4p\,^2P_{1/2}ns$ series starting with $n = 6$. (*Middle panel*) relative cross sections for $\varepsilon s_{1/2}$ (●), $\varepsilon d_{3/2}$ (○), and $\varepsilon d_{5/2}$ (K) partial waves of the photoelectron. The lines are to guide the eye. (*Lower panel*) the ratio of $\varepsilon d_{5/2}$ partial waves to $\varepsilon d_{3/2}$ partial waves. In absence of relativistic interactions this ratio would have a constant value of 3/2 (− − −) as given by the multiplicities of the partial waves. Adapted from McLaughlin et al. (2002)

ization experiment applicable to the non-LS coupling case, i.e. the case where there are three photoelectron channels. It is important that this approach avoids the spin resolved analysis of the outgoing electrons as well as applying circularly polarized incoming radiation.

The scenario of the complete photoionization experiment for three photoionization amplitudes by means of coincidence electron spectrometry was further discussed in detail and realized by Schaphorst et al. (1997) for the reaction

$$\begin{aligned} h\nu + \mathrm{Xe}(5p^6)\,^1S_0 &\longrightarrow \mathrm{Xe}^{+*}(4d^9 5s^2 5p^6)\,^2D_{5/2} + e^-_{\mathrm{ph}}(\varepsilon p_{3/2}, \varepsilon f_{5/2}, \varepsilon f_{7/2}) \\ &\quad \hookrightarrow \mathrm{Xe}^{++}(5p^4)\,^1S_0 + e^-_{\mathrm{Auger}}(\varepsilon d_{5/2}), \end{aligned} \tag{4.56}$$

where the primary photoionization at the photon energy of 132.2 eV, producing the $4d_{5/2}$ hole, is followed by the $N_5O_{2,3}O_{2,3}$ Auger decay into the $^1S_0$ state. Three amplitudes (4.15) for the three ionization channels, corresponding to the $\varepsilon p_{3/2}$, $\varepsilon f_{5/2}$, and $\varepsilon f_{7/2}$ partial waves of the photoelectrons is the goal of this complete experiment. Note that the total angular momentum in the channels is restricted to $J = 1$ by the dipole selection rules and the residual ionic state is fixed by energy conservation. The electrons are detected in the plane perpendicular to the incoming radiation, which contains also the direction of its linear polarization (reaction plane). Thus, the possible first order non-dipole effects vanish (see Sect. 4.9) and cannot disturb the angular distributions and the angular correlation patterns. The emission angle of the Auger electron was fixed at one position. Rotation of the analyzer for the photoelectrons around the photon beam direction has an advantage that it always views the same source volume. Denoting by d_j the absolute value of the ionization amplitude into the channel with the $\varepsilon\ell j$ photoelectron and by $\Delta_{j,j'}$ the phase difference between the corresponding amplitudes, the alignment of the $Xe^{+*}(4d^9 5s^2 5p^6)\,^2D_{5/2}$ state and the asymmetry parameter of the photoelectrons in the process (4.56) are expressed as (Schaphorst et al. 1997)

$$\mathcal{A}_{20}(^2D_{5/2}) = -\left[5\sqrt{14}\,S\right]^{-1} \times \left(5d_{7/2}^2 - 16d_{5/2}^2 + 14d_{3/2}^2\right), \tag{4.57}$$

and

$$\begin{aligned}\beta(^2D_{5/2}) = \left[35\,S\right]^{-1} \times \Big(&7d_{3/2}^2 - 6\sqrt{14}d_{3/2}d_{5/2}\cos\Delta_{3/2,5/2} \\ &- 32d_{5/2}^2 + 12\sqrt{5}d_{5/2}d_{7/2}\cos\Delta_{5/2,7/2} \\ &+ 25d_{7/2}^2 - 12\sqrt{2}d_{3/2}d_{7/2}\cos\Delta_{7/2,3/2}\Big),\end{aligned} \tag{4.58}$$

where

$$S = d_{7/2}^2 + d_{5/2}^2 + d_{3/2}^2\,. \tag{4.59}$$

The angular correlation function between photoelectron and Auger electron in the reaction plane is a function of the two angles ϕ_p and ϕ_A, respectively, counted from the direction of the linear polarization of the radiation beam.

It is parameterized in the form

$$P(\phi_p, \phi_A) = \sum_{q_p q_A} C_{q_p q_A} \cos\left(q_p\phi_p - q_A\phi_A\right), \tag{4.60}$$

where $q_p = 0, 2, 4, 6$, $q_A = 0, 2, 4$ and $\left|q_p - q_A\right| \leq 2$. The dynamical coefficients $C_{q_p q_A}$ are bilinear combinations of the partial amplitudes, including interference terms with $\cos\Delta_{j,j'}$ as in (4.58). Since the phase differences appear only via the cosine function, the signs of the phase differences remain experimentally undetermined. The photoionization amplitudes are obtained by fitting to the measured parameters (4.57), (4.58) and correlation functions (4.60). Figure 4.17 shows one of the measured angular correlation patterns together with fitted and theoretically calculated curves and Table 4.5 contains final results of Schaphorst et al. (1997) for the amplitudes. They are compared to the two versions of relativistic random phase calculations: frozen (Johnson and Cheng 1992a) and relaxed (Johnson and Cheng 1992b). The extracted matrix elements differ significantly from the theoretical predictions. It indicates the need for further refinement of these rather sophisticated calculations.

In the example of the 4d-photoionization in Xe, the complete experiment (up to the sign of the relative phases of the partial amplitudes) incorporated data on the asymmetry parameter in the angular distributions of photoelectrons β, the alignment

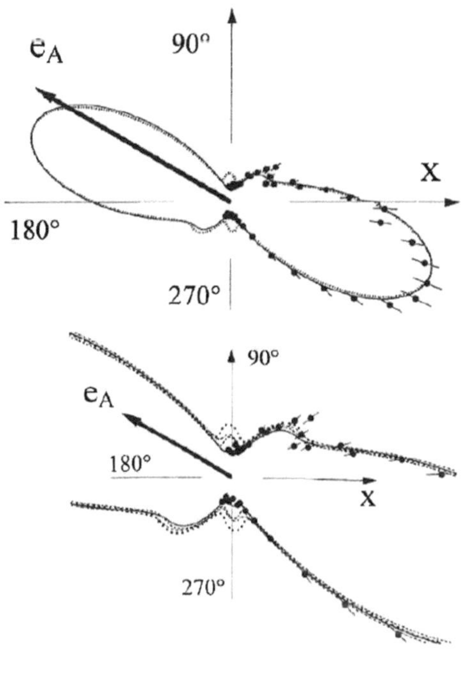

Fig. 4.17 (*Upper panel*) angular correlation patterns between $4d_{5/2}$ photoelectrons and $N_5O_{2,3}O_{2,3}\,{}^1S_0$ Auger electrons in xenon for completely linearly polarized light at 132.2 eV photon energy. Observation is made in a plane perpendicular to the incident photon beam. The direction of the electric field vector is indicated by X, the average direction for the acceptance of Auger electrons at the angle 149° by e_A. The angle dependent intensity of the coincident photoelectron is shown as a polar plot. (•): Experimental data with error bars; (⋯⋯) and (——): Results from fitting the theoretical expression for the correlation pattern (finite acceptance angles of the electron analyzers taken into account) to the experimental values, using the matrix elements and phases from theoretical predictions (relaxed relativistic random phase approximation (RRPA) (Johnson and Cheng 1992b)) and best fit to experimental data in Table 4.5, respectively. (*Lower panel*) details of the angular correlation pattern, demonstrating its sensitivity to the values of the photoionization matrix elements and their relative phases. After Schaphorst et al. (1997)

Table 4.5 Compilation of parameters needed to describe $4d_{5/2}$ photoionization in xenon at 132.2 eV photon energy

	$d_{7/2}$ [au]	$d_{5/2}$ [au]	$d_{3/2}$ [au]	$\Delta_{5/2,3/2}$ [rad]	$\Delta_{7/2,3/2}$ [rad]	$\Delta_{5/2,7/2}$ [rad]	σ [Mb]	β_{ph}	$\mathcal{A}_{20}(^2D_{5/2})$
Exp[a]							3.0	1.54	−0.274
							±0.3	±0.02	±0.019
Exp[a]	0.44	0.10	0.16	2.6	2.6	0.0	3.0	1.54	−0.271
Theory[b]	0.654	0.145	0.194	2.27	2.27	0.0	6.35	1.26	−0.256
Theory[c]	0.703	0.159	0.205	2.27	2.27	0.0	7.33	1.25	−0.252

Note that for the values obtained from fitting the experimental observables there exists an ambiguity in the sign of the phase differences $\Delta_{i,j}$, i.e. the calculated angular correlation pattern is the same for positive and negative values; guided by theory, the positive values are presented. (Fragment of Table 4.1 from Schaphorst et al. (1997)). [a]Schaphorst et al. (1997); [b]Johnson and Cheng (1992a); [c]Johnson and Cheng (1992b)

of the residual ion $\mathcal{A}_{20}(J_f)$, and one electron angular correlation pattern for linearly polarized radiation. As demonstrated by Bolognesi et al. (2004), using larger number of the angular correlation functions, it is possible to extract the three relative amplitudes (again up to the sign of the relative phases) solely by the coincidence spectrometry with linearly polarized radiation and detecting the both electrons in the reaction plane. Then, the values of β and $\mathcal{A}_{20}(J_f)$, as well as the photoelectron spin polarization parameters could be predicted from the extracted amplitudes. Figure 4.18 shows the photoelectron–Auger electron angular correlations in the reaction

$$
\begin{aligned}
h\nu + \mathrm{Ar}(3\mathrm{p}^6)\,^1\mathrm{S}_0 \longrightarrow\; & \mathrm{Ar}^{+*}(2\mathrm{p}^5 3\mathrm{s}^2 3\mathrm{p}^6)\,^2\mathrm{P}_{3/2} + e^-_{ph}(\varepsilon \mathrm{s}_{1/2}, \varepsilon \mathrm{d}_{3/2}, \varepsilon \mathrm{d}_{5/2}) \\
& \hookrightarrow \mathrm{Ar}^{++}(3\mathrm{p}^4)\,^1\mathrm{S}_0 + e^-_{Auger}(\varepsilon \mathrm{p}_{3/2}),
\end{aligned}
\tag{4.61}
$$

at the photon energy $h\nu = 253.6$ eV and three directions of the photoelectron emission. The fitting procedure with three amplitudes (jj coupling) and with two nonrelativistic amplitudes (LS coupling) gives very similar results, showing excellent quality of the Cooper–Zare model for this process. With the extracted partial photoionization amplitudes, the predicted parameters β, $\mathcal{A}_{10}(^2P_{3/2})$, and $\mathcal{A}_{20}(^2P_{3/2})$ are in good agreement with earlier independent measurements (Lindle et al. 1985; Avaldi et al. 1994; Becker 1998; Becker and Langer 1998), but diverge from theoretical predictions available (Lindle et al. 1985; Becker 1998; Becker and Langer 1998; Kleiman and Lohmann 2003; Kutzner et al. 1997). Furthermore, the complete experiment of Bolognesi et al. (2004) allowed to predict the unknown dynamic spin polarization parameter and polarization transfer parameters (see (4.28), (4.30), (4.31)), as well as the angular correlation patterns for fixed direction of the Auger emission. Eventually, the anisotropy parameters of the Auger decay of the state $\mathrm{Ar}^{+*}(2p^5 3s^2 3p^6)^2 P_{3/2}$ to the $\mathrm{Ar}^{++}(3p^4)^1D_2$ and $\mathrm{Ar}^{++}(3p^4)^3P_{0,1,2}$ states was predicted in very good agreement with earlier measurements (Sarkadi et al. 1990) and calculations (Kabachnik et al. 1991). The latter was possible after measuring the angular correlation patterns for these Auger transitions, since the photoionization step is identical to one in (4.61).

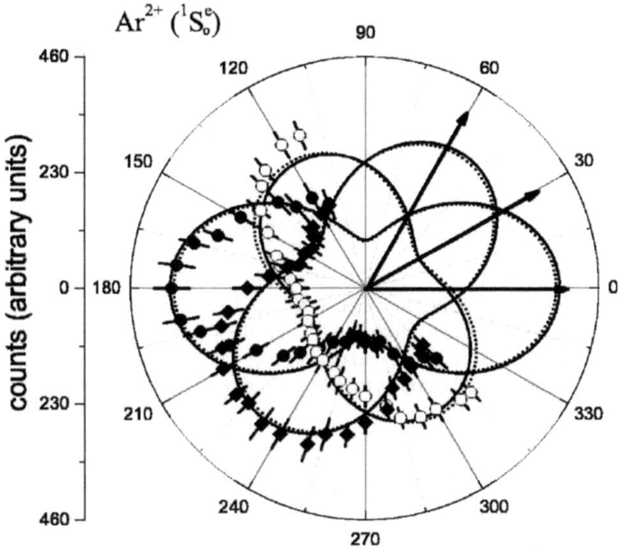

Fig. 4.18 Photoelectron–Auger electron coincidence angular correlations for the $Ar^{++}(3p^4\,{}^1S_0)$ state. The photoelectron was detected at 0° (●), 30° (♦), and 60° (○), respectively. The *curves* are the result of a simultaneous fit to the three sets of data with the LSJ constraints (——) or without any constraints (– – –). After Bolognesi et al. (2004)

The analysis of Bolognesi et al. (2004) can serve as a showcase of a complete photoionization experiment with all the main constituents involved: extraction of the photoionization amplitudes, verifying theory, prediction of unknown photoionization parameters, and even gaining meaningful information on a subsequent process (Auger decay) due to the knowledge of the primary photoionization amplitudes. Thus, the coincidence photoelectron–Auger electron spectrometry showed itself as a very powerful method of complete experiment for photoionization in the inner atomic shells.

Remind, that the theoretical description of the photoelectron–Auger electron correlations used in the above analysis is based on the two-step model, when the Auger emission is connected to the primary photoionization only via an intermediate hole state (with possible evolution of the polarization of this state during the time between photoionization and Auger decay). Thus the direct interaction between the photoelectron and the Auger electron in the final tree-body system, a so-called *post-collision interaction* (PCI), is neglected. The post-collision interaction is known to strongly affect the double photoionization process in some kinematical domains. In particular, due to the Coulomb repulsion, the post-collision interaction suppresses ejection of the photoelectron and the Auger electron at small relative angles, especially when the speeds of the photo- and Auger electrons are comparable. Additionally, interference effects due to the indistinguishability of the two electrons show up if the two electrons have almost equal energy and are ejected at small mutual angle (Rioual et al. 2001).

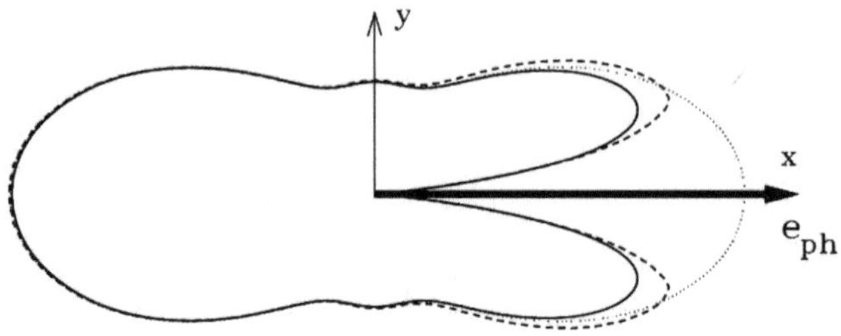

Fig. 4.19 Experimental data (– – –) for the photoelectron–Auger electron angular correlation pattern in the process (4.56) at the photon energy of 84.52 eV. For comparison, the theoretical predictions from the quantummechanical model (——) and the shape without post-collision interaction (· · · · · ·) have been included. Both calculated *curves* take into account the partial light polarization as well as the finite solid angle acceptances of the electron spectrometers. After Scherer et al. (2004)

In the example of Xe (4.56) considered by Schaphorst et al. (1997), the photo- and Auger electron energies were 64.65 and 29.97 eV, respectively, and the relative emission angle of the two electrons was large (see Fig. 4.17). At smaller photoelectron energy of 16.97 eV and smaller relative angles in the same reaction, the post-collision interaction dominates the angular correlation pattern. When the difference between the kinetic energies of the photoelectron and the Auger electron is large compared to the Auger decay width, the post-collision interaction can be included in the two-step model through a correlation factor which modifies the known angle-dependence of two-electron emission (Kämmerling et al. 1993; Sheinerman and Schmidt 1997; Wiedenhoeft et al. 2008). In the latter paper the complete experiment on photoionization of the $4d_{5/2}$ subshell of xenon was performed with accounting for the post-collision interaction effects. Figure 4.19 demonstrates observations of the post-collision interaction in the angular correlations between photoelectrons and Auger electrons by Scherer et al. (2004). We refer to the above and other original papers by Kuchiev and Sheinerman (1989), Vegh and Macek (1994), Briggs and Schmidt (2000), Rouvellou et al. (2003), which contain further references, for discussions on the phenomenon of post-collision interaction and interference in double photoionization and its theoretical description.

4.5 Coincidence Photoelectron–Fluorescence Spectrometry

Studies of the secondary fluorescence by coincidences between the fluorescence and the photoelectron have an important advantages in comparison with recording only the fluorescence, because the influence of the cascading population of the residual ionic state is suppressed. The post-collision interaction, inherent to the photoelectron–Auger electron coincidence method, is absent in this type of experiment. In certain cases, when the residual ion is left in the discrete excited state, only observations of fluorescence in coincidence with the photoelectron can constitute a complete experiment. A theoretical description of the angular correlations between the photoelectron and polarized fluorescence photons has been developed by Berezhko and Kabachnik (1979), Klar (1980a), Kabachnik (1992), Kabachnik and Ueda (1995). The coherence analysis of these, so called, $(\gamma, e\gamma)$ processes has been reviewed by Lohmann et al. (2003a) who further extended the theory including polarized beams and/or targets, as well as analyzing the photoelectron and fluoresence polarization by means of tensor polarization s. In comparison with the coincidence photoelectron–Auger electron studies, the angular distribution or polarization of the secondary fluorescence is determined solely by the polarization of the residual ion $A^{+*}(J_f)$ in (4.49) before the radiative decay and the values of J_f and J_g. Therefore the second step does not add new independent dynamical parameters, which is an advantage over the Auger electron emission. As has been discussed in Sect. 4.3.2 in the case of the Auger decay this happens only in a particular situation with one Auger decay channel.

The electron–photon coincidence technique has been used for studying in metallic vapors, starting with the pioneering work by Beyer et al. (1995) in calcium for *photoionization with simultaneous excitation*, so-called 'conjugate shake up',

$$\begin{aligned} h\nu + \mathrm{Ca}(4\mathrm{s}^2\,{}^1\mathrm{S}_0) \longrightarrow\ & \mathrm{Ca}^{+*}(4\mathrm{p}\,{}^2\mathrm{P}_{3/2}) + \mathrm{e}^-_{\mathrm{ph}}(\varepsilon\mathrm{s}_{1/2}, \varepsilon\mathrm{d}_{3/2}, \varepsilon\mathrm{d}_{5/2}) \\ & \hookrightarrow \mathrm{Ca}^+(4\mathrm{s}\,{}^2\mathrm{S}_{1/2}) + h\nu_{fluo}(393.3\,\mathrm{nm}), \end{aligned} \tag{4.62}$$

in the region of the $3\mathrm{p} \rightarrow 3\mathrm{d}$ photoabsorption resonance around 31.5 eV.

Out of the resonance, the conjugate shake up in the direct photoionization with excitation has small cross section. The principle motivation for this experiment was the fact that the alignment of the $\mathrm{Ca}^{+*}(4\mathrm{p}\,{}^2\mathrm{P}_{3/2})$ level previously measured by Hamdy et al. (1991) in reaction (4.62) by the non-coincidence method was not adequate due to the presence of radiative cascade processes and could not be directly related to the photoionization amplitudes. For the coincidence electron–fluorescence measurements, the experimental set-up is shown in Fig. 4.20. The asymmetry parameter of the angular distribution of the photoelectrons was calculated from the two measurements of the electron analyzer at angles 0° and 90°, with respect to the polarization of thc incoming photon. Variation of the linear polarization of the fluorescence emitted under a fixed angle of 90° to the synchrotron radiation beam was analyzed as a function of the electron emission angle. The LSJ coupling scheme was assumed to be valid, leading to the Cooper–Zare model. Experiments with circular polarized incoming photons and observation of circular dichroism in the coincident fluorescence polarization (West et al. 1996) allowed to fix the sign of the phase difference between the outgoing εs and εd waves of the ejected electrons. To our knowledge, so far this experiment on Ca is the only coincident photoelectron–fluorescence experiment with circularly polarized synchrotron radiation.

Similar measurements were performed with Sr (West et al. 1998; Ueda et al. 1998) for the reaction

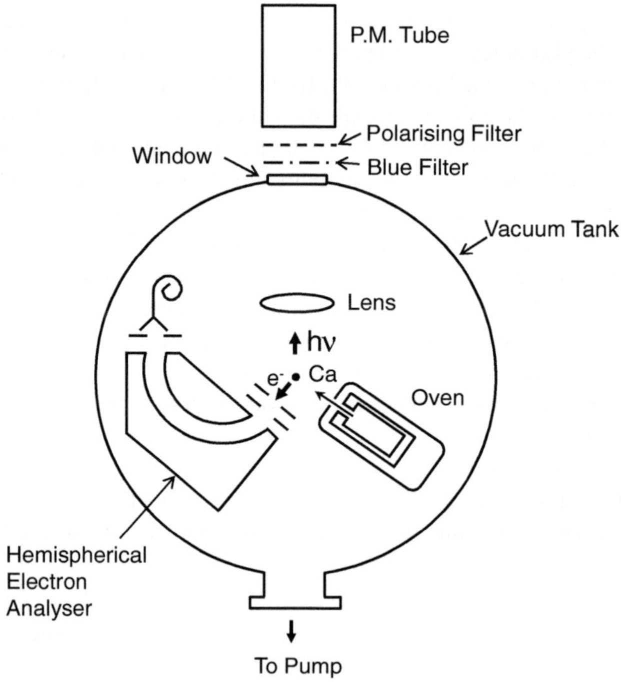

Fig. 4.20 The experimental equipment for the $e-\gamma$ coincidence measurements of photoelectrons and subsequent fluorescence photons. The primary photon beam of linearly polarized radiation from the Daresbury Synchrotron Radiation Source is perpendicular to the plane of the figure. After Beyer et al. (1996)

$$
\begin{aligned}
h\nu + \mathrm{Sr}(5\mathrm{s}^2\,{}^1\mathrm{S}_0) \longrightarrow\ & \mathrm{Sr}^{+*}(5\mathrm{p}\,{}^2\mathrm{P}_{J_f}) + \mathrm{c}^-_{ph}(\varepsilon\mathrm{s}_{1/2}, \varepsilon\mathrm{d}_{3/2}, \varepsilon\mathrm{d}_{5/2}) \\
& \hookrightarrow \mathrm{Sr}^{+}(5\mathrm{s}\,{}^2\mathrm{S}_{1/2}) + h\nu_{fluo}(407.8\,\mathrm{nm}),
\end{aligned}
\tag{4.63}
$$

in the region of 4p excitations 25.1–26.9 eV. For Sr, the resonant model of ionization via excitation of autoionizing states, located in this region, was implied. The combination of the coincidence data with the data on the angular distributions of photoelectrons provided enough parameters to perform a complete experiment in the relativistic case of three photoionization amplitudes for the residual ionic state with $J_f = 3/2$, but neglecting the relativistic phase difference between $\varepsilon d_{3/2}$ and $\varepsilon d_{5/2}$ photoelectron waves. The analysis of the amplitudes and phase difference between εs and εd photoelectron waves showed that this technique can identify the presence of components of different symmetries within a resonance line, giving quantitative information difficult to obtain from other experimental methods. Quantitative theoretical predictions for the extracted amplitudes are missing so far.

The great potential of the coincidence photoelectron–fluorescence studies is, unfortunately, up to the moment still kept on the sidelines.

4.6 Photoionization of Polarized Atoms

4.6.1 Angular Distribution of Photoelectrons from Polarized Atoms and Dichroism

It has been established by Klar and Kleinpoppen (1982) that a set of measurements of the angular distribution of photoelectrons from a polarized atom without spin-resolved analysis of the photoelectron can constitute a complete experiment for polarizable one-electron atom and many other atoms. In practice of the complete photoionization experiments, the initial state was polarized by the hexapole magnet, by optical laser pumping, or by a single photon absorption. A principle restriction of the method is its applicability to only atoms with non-vanishing angular momentum of the initial state of the photoionization process. Another weak point of the method is that sometimes the values of the initial polarization parameters, such as the alignment and the orientation of the initial atomic state, generally needed for the data analysis, are not well known.

Photoionization of polarized atoms is now a well established field. General theoretical description of the angular distribution of photoelectrons from polarized atoms within the dipole approximation was developed by Jacobs (1972), Klar and Kleinpoppen (1982), Cherepkov and Kuznetsov (1989), Baier et al. (1994a), Manakov et al. (1996), using different formulations. Using the results of Baier et al. (1994a), the angular distribution of the photoelectrons for arbitrary polarization of the incoming radiation takes the form

$$\frac{d\sigma}{d\Omega} = \frac{\sigma}{4\pi}\left(1 + \sum_{k_0 k k_\gamma} \mathcal{A}_{k_0 0}\, \beta_{k_0 k k_\gamma}\, F_{k_0 k k_\gamma}\right), \tag{4.64}$$

where σ is the angle integrated photoionization cross section of unpolarized atom in the dipole approximation, $F_{k_0 k k_\gamma}$ are the geometrical factors containing the Stokes parameters of the incoming radiation, directions of the photoelectron emission and the polarization axis of the target atom (axial symmetry is implied for the polarized target). General expressions for $F_{k_0 k k_\gamma}$ are given by Baier et al. (1994a). The reduced statistical tensors (state multipoles) $\mathcal{A}_{k_0 0}$ with rank $k_0 = 1, 2, \ldots, 2J_i$ characterize the polarization of the initial atomic state with angular momentum J_i. They are expressed in terms of relative populations of magnetic substates:

$$\mathcal{A}_{k_0 0} = \sqrt{2J_i + 1}\left(\sum_{M_i=-J_i}^{J_i} N_{M_i}\right)^{-1} \times \sum_{M_i=-J_i}^{J_i} (-1)^{J_i - M_i} \left(J_i M_i, J_i - M_i | k_0 0\right) N_{M_i}, \tag{4.65}$$

where N_{M_i} is the number of atoms in the magnetic substate $|J_i M_i\rangle$. Summation in (4.64) is over the values of k_0 ($1 \leq k_0 \leq 2J_i$), $k_\gamma = 0, 1, 2$, and $k =$ even; the set $\{k_0, k, k_\gamma\}$ obeys the triangle condition. The generalized anisotropy coefficients $\beta_{k_0 k k_\gamma}$ are bilinear combinations of the dipole ionization amplitudes and describe the dynamics of the photoionization. Expressions for $\beta_{k_0 k k_\gamma}$ may be found in the literature (Baier et al. 1994a; Verweyen et al. 1999; Wernet et al. 2001). In the particular case of an unpolarized target atom, i.e. $k_0 = 0$, (4.64) turns to (4.21) with the asymmetry parameter

$$\beta = -\sqrt{\frac{10}{3}}\,\beta_{022}\,. \tag{4.66}$$

The angular distribution of photoelectrons (4.64) is shaped by the spherical harmonics up to rank k.

For polarized atoms, it is often not of advantage to analyze the angular distribution of the photoelectrons itself but various kinds of dichroism (Cherepkov et al. 1995). The dichroism in the photoelectron spectra of polarized atoms is defined as the intensity difference for two different directions either of the target atom polarization (*magnetic dichroism*) or of the polarization of the ionizing photons. For example, the magnetic dichroism in the angular distribution of photoelectrons (MDAD) is defined as the difference of the differential photoionization cross sections (4.64)

$$\mathrm{MDAD} = \left(\frac{\mathrm{d}\sigma}{\mathrm{d}\Omega}\right)_{\mathbf{a}} - \left(\frac{\mathrm{d}\sigma}{\mathrm{d}\Omega}\right)_{\mathbf{a}'}, \tag{4.67}$$

where $\mathbf{a}$, $\mathbf{a}'$ indicate two directions of the atomic axis with all other parameters of the experiment fixed. The advantage of the dichroism is two-fold. First, when taking the difference, many terms in (4.64) cancel out, leading to a substantial simplification of expressions. Secondly, experimentally it is easier to change the direction of the atomic axis (e.g. by rotating the linear polarization or changing the light helicity of the pumping laser), then to rotate the electron spectrometer, correcting an instability of signal due to change in geometry of the effective reaction volume. Expressions for particular kind of dichroism can be obtained from (4.64).

Some dichroisms have been given special names, for example:

CDAD: Circular dichroism in the angular distribution

$$\mathrm{CDAD} = \left(\frac{\mathrm{d}\sigma}{\mathrm{d}\Omega}\right)_{+} - \left(\frac{\mathrm{d}\sigma}{\mathrm{d}\Omega}\right)_{-}, \tag{4.68}$$

where $+$ and $-$ denote right and left circularly polarized incoming ionizing radiation;

LDAD: Linear dichroism in the angular distribution

$$\mathrm{LDAD} = \left(\frac{\mathrm{d}\sigma}{\mathrm{d}\Omega}\right)_{\parallel} - \left(\frac{\mathrm{d}\sigma}{\mathrm{d}\Omega}\right)_{\perp}, \tag{4.69}$$

where $\parallel$ and $\perp$ denote two perpendicular directions of the linear polarization of the ionizing radiation;

CMDAD: Circular magnetic dichroism in the angular distribution

$$\text{CMDAD} = \left(\frac{\mathrm{d}\sigma}{\mathrm{d}\Omega}\right)_{\mathbf{a}} - \left(\frac{\mathrm{d}\sigma}{\mathrm{d}\Omega}\right)_{-\mathbf{a}}, \tag{4.70}$$

where the ionizing radiation is implied to be circularly polarized and the indices **a** and −**a** denote opposite directions of the target atom orientation;

LMDAD: Linear magnetic dichroism in the angular distribution is even to be defined by (4.70), but the ionizing radiation is implied to be linearly polarized;

LADAD: Linear alignment dichroism in the angular distribution

$$\text{LADAD}(\eta/\eta - 90^\circ) = \left(\frac{\mathrm{d}\sigma}{\mathrm{d}\Omega}\right)_{\eta} - \left(\frac{\mathrm{d}\sigma}{\mathrm{d}\Omega}\right)_{\eta\,-\,90^\circ}, \tag{4.71}$$

where the ionizing radiation is implied to be linearly polarized and η and $\eta - 90^\circ$ denote two perpendicular directions of the initial atomic alignment. The LADAD depends on the angle η between the atomic axis and the polarization of the ionizing radiation. The LADAD for different angles η are generally dependent. For the two independent quantities one can take $\eta = 0^\circ$ and $\eta = 90^\circ$. Other names are also used for this kind of dichroism in the literature.

Experimentally, to avoid the normalization problems it is convenient to study a dichroism normalized to the cross section. In this case the same name for the dichroism is used.

Integrating over the angles of the photoelectron emission, the CMDAD transforms into the circular magnetic dichroism (CMD), the LMDAD transforms into the linear magnetic dichroism (LMD), the CDAD transforms into the circular dichroism (CD), the LDAD transforms into the linear dichroism (LD) and so on. These photoelectron–angle integrated quantities depend only on the absolute squares of the partial photoionization amplitudes in the ℓj representation in accordance with a general quantum mechanical prescription: in order to find an observable the trace of the density matrix of the final state over the quantum numbers of an unobserved subsystem must be taken.

4.6.2 *Photoionization of Atoms Polarized by Laser Optical Pumping*

Figure 4.21 shows a principle of polarization of an atomic target by the optical laser pumping (see Sect. 2.5.5). Linearly polarized light leads to the excited or/and ground atomic state aligned along the direction of the linear polarization. Circularly polarized light leads to the states oriented along the direction of the laser beam.

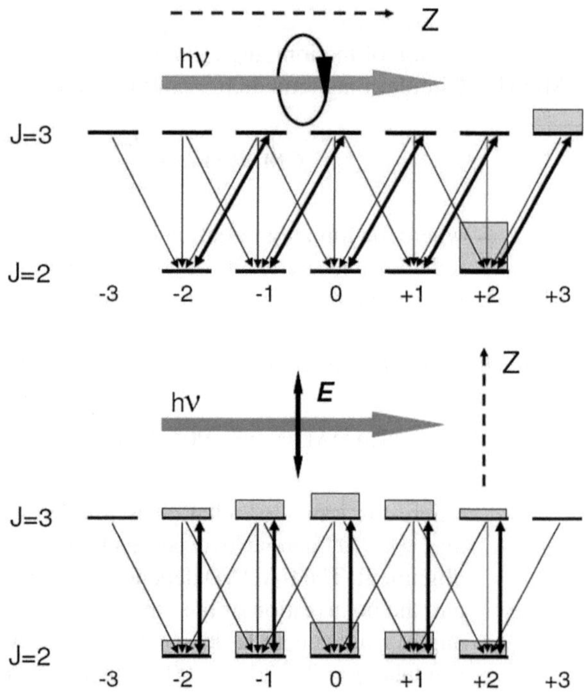

Fig. 4.21 A scheme of the pump process for levels with the total angular momentum $J = 2$ and $J = 3$ by right circularly polarized (*top*) and linearly polarized (*bottom*) laser light. The quantization axis z is chosen along the laser beam and along its polarization, respectively. (➔): laser induced transitions; (→): spontaneous radiative decay

As example, Fig. 4.22 shows a scheme of transitions in the laser pumping of the Cr atom by linearly polarized light, which leads to the aligned atomic ground state $^7S_3$. For isotopes with non-zero nuclear spin, the laser pumping usually proceeds between the magnetic substates of the hyperfine structure levels. The nuclear spin is not influenced by the subsequent photoionization process, where the polarization of the angular momentum of the electronic shell is only important, except some very special cases.

Combining laser optical pumping with the vacuum ultraviolet radiation for photoionization was pioneered by Kerling et al. (1990). They used the *CW*-laser to produce large fractions of aligned Yb atoms in the excited 6s6p $^3P_1$ state and linearly polarized Ar line radiation with 11.63 and 11.83 eV photon energies for the photoionization. The measured angular distribution of photoelectrons at different relative angles between the polarization vectors of the two radiation beams allowed to extract, within the Cooper–Zare model, the absolute ratios of the two ionization amplitudes and the cosine of their relative phase.

The combination of optical laser pumping and intense synchrotron radiation resulted in a breakthrough in investigations of photoionization of polarized atoms in gaseous atomic targets (Meyer et al. 1987; Pahler et al. 1992; Baier et al. 1994b; Wedowski et al. 1995, 1997). High intensity of the synchrotron radiation permits to decrease the atomic density in the reaction volume and therefore to avoid collisions and radiation trapping, which destroy the target polarization. Furthermore, the energy

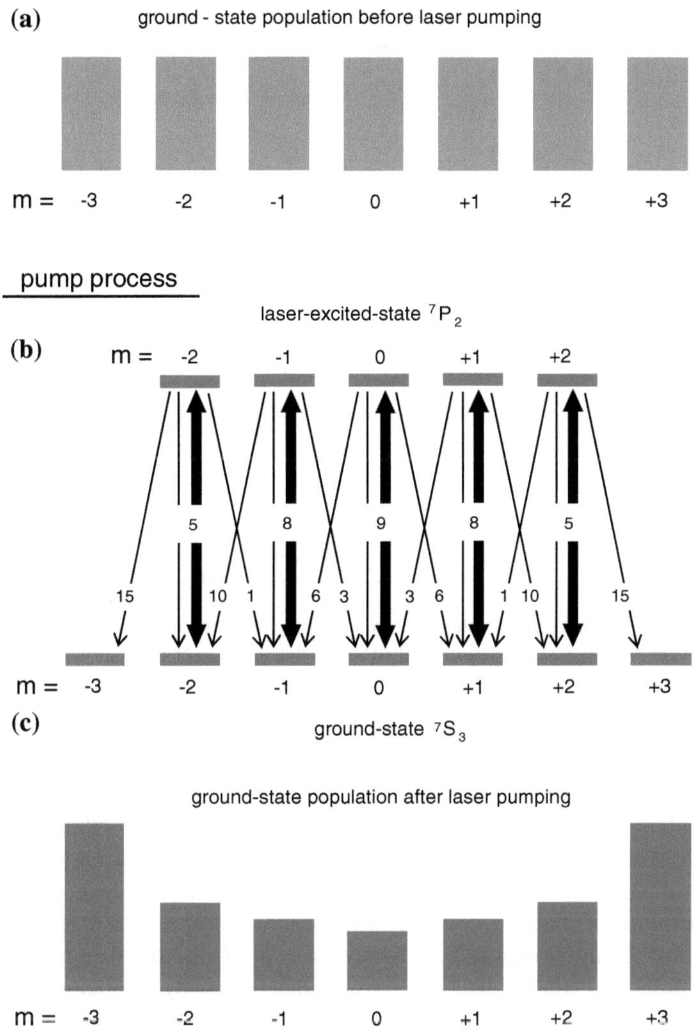

Fig. 4.22 Ground state population (**a**) and scheme of the pump process (**b**) used to align the Cr atoms in the ground state (**c**). The numbers related to the arrows give the relative transition probabilities. After Dohrmann et al. (1996)

resolution improved dramatically, leading e.g. to fine structure resolved photoelectron spectra in the XUV (Cubaynes et al. 2004, 2009; Meyer et al. 2006; Schulz et al. 2005, 2006). Figure 4.23 illustrates a progress in photoionization experiments with laser excited atoms and dramatic improvement in the energy resolution by comparing photoelectron spectra from laser-excited sodium atoms measured within a 15-year interval. Compared to the first studies on laser-polarized Li and Na atoms (Meyer et al. 1987, 1990), an essential experimental improvement was the extension

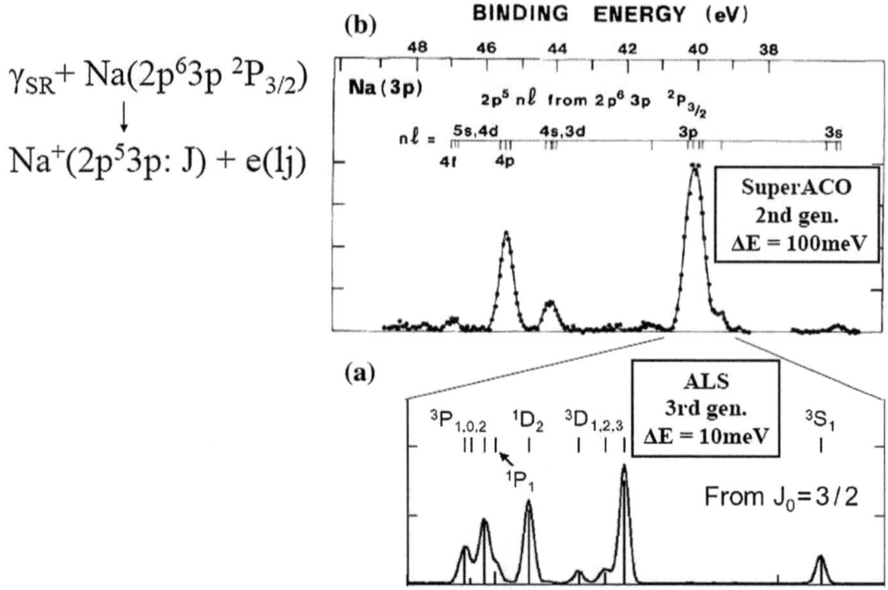

Fig. 4.23 Region of $Na^+(2p^5 3p)$ photoelectron spectra produced by photoionization from laser-excited $Na^*(2p^6 3p)\ ^2P_{3/2}$ state measured at (**a**) the Advanced Light Source synchrotron radiation facility at the Lawrence Berkeley National Laboratory by Cubaynes et al. (2004) a and (**b**) at the SuperACO synchrotron radiation facility in Orsay by Cubaynes et al. (1989). Adapted from Cubaynes et al. (2004) and Cubaynes et al. (1989)

of the wavelength range of the pumping laser (Dohrmann et al. 1996; von dem Borne et al. 1997; Müller et al. 2007), which allows to cover many new targets with higher excitation energies of first resonant levels. As another challenge, high evaporation temperature of metals necessitates incorporating appropriate heating devices (Sonntag and Zimmermann 1992). Many aspects of photoionization experiments with laser polarized atoms have been covered in a review by Wuilleumier and Meyer (2006).

The direction of the target atom polarization in experiments with laser-excited atoms is controlled by changing the polarization of the pumping laser light. This allows experimental studies of magnetic dichroism. Figure 4.24 illustrates geometries of measurements of different types of magnetic dichroism in photoemission, which were observed in experiments based on laser pumping. A powerful *phase tilt* method was suggested by Wedowski et al. (1995), which consists in determination of the intensity modulation of the photoelectron signal caused by rotation of the direction of linear polarization of the laser radiation. The detected photoelectron flux is modulated as

$$I(\eta) = a + b\cos 2(\eta - \delta), \qquad (4.72)$$

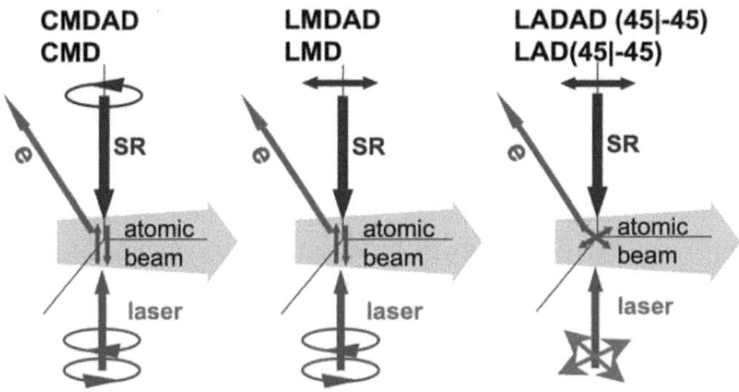

Fig. 4.24 Different kinds of magnetic dichroism in photoemission as studied in processes with collinear laser and synchrotron beams (SR). Circular or linear polarization of the SR is indicated in each panel, as well as two orthogonal polarizations of the laser light with the two corresponding directions of target polarization in the atomic beam. (*Left*) circular magnetic dichroism in the angular distribution of photoelectrons (CMDAD), which after integrating over the photoemission angle gives circular magnetic dichroism (CMD). (*Center*): linear magnetic dichroism in the angular distribution of photoelectrons (LMDAD) and linear magnetic dichroism (LMD). (*Right*) linear alignment dichroism in the angular distribution of photoelectrons (LADAD) and linear alignment dichroism (LAD); the two numbers in parenthesis indicate the angles of the linear laser polarization with respect to the linear SR polarization. After Grum-Grzhimailo and Meyer (2009)

where η is the angle between linear polarizations of the laser and ionizing radiation beams, and the phase tilt δ depends on the geometry of the detector (for example, complete angular resolved measurements or integrating over an entrance slit of the electron analyzer), and on the photoionization amplitudes. Compared with the usual determination of the angular distribution of photoelectrons by rotation of the spectrometer, this method offers the advantage of being insensitive to smaller misalignments of apparatus and independent of the numerical value of the atomic alignment and absolute count rates produced by the laser pumping. The phase tilt method was first successfully applied to photoionization of laser excited Ca atoms (Wedowski et al. 1995, 1997)

$$h\nu + \mathrm{Ca}^*(3\mathrm{p}^6 4\mathrm{s}4\mathrm{p}\,^1\mathrm{P}_1) \hookrightarrow \mathrm{Ca}^+(3\mathrm{p}^6 4\mathrm{s}\,^2\mathrm{S}_{1/2}) + \mathrm{e}^-_{\mathrm{ph}}(\varepsilon\mathrm{s}_{1/2}, \varepsilon\mathrm{d}_{3/2}, \varepsilon\mathrm{d}_{5/2}), \tag{4.73}$$

and a reduced parameter space was found for the ratio of the two relativistic amplitudes, corresponding to ionization into the $\varepsilon d_{3/2}$ and $\varepsilon d_{5/2}$ channels.

As an example of complete photoionization experiment utilizing atoms polarized by the laser optical pumping, consider photoemission from the 5p subshell of the ground state of Eu in the process (Godehusen et al. 1998)

$$h\nu + \mathrm{Eu}(7\mathrm{f}^7 5\mathrm{p}^6 6\mathrm{s}^2\ {}^8\mathrm{S}_{7/2}) \longrightarrow \mathrm{Eu}^+(7\mathrm{f}^7 5\mathrm{p}^5 6\mathrm{s}^2\ {}^9\mathrm{P}_5) + \mathrm{e}^-_{\mathrm{ph}}(\varepsilon \mathrm{s}_{1/2}, \varepsilon \mathrm{d}_{3/2}, \varepsilon \mathrm{d}_{5/2})\ . \tag{4.74}$$

The corresponding line is well separated in the photoelectron spectrum. Because of large values of the total angular momentum of the target atom and the residual ion, as many as 24 photoionization channels (i.e. 24 complex photoionization amplitudes in the dipole approximation) occur with different angular momenta of the photoelectron ℓ and j and total angular momentum of the final system ion + photoelectron J. In the Cooper–Zare model, the number of independent amplitudes reduces to two. These two channels correspond to the outgoing εs and εd waves of the photoelectron. The absolute ratio of the single particle amplitudes, $x = |D_s|/|D_d|$, and the relative phase between the outgoing εs and εd electron waves, $\delta_s - \delta_d$, have been extracted from data on the LMDAD, the LADAD, and the phase tilt, as shown in Fig. 4.25. The agreement between theoretical and experimental values for the absolute ratio of the amplitudes is good, while a clear deviation in the phase difference can be due to neighboring 5s thresholds, which were not included in the calculation. The parameters x and $\delta_s - \delta_d$ were extracted from the experiment with good accuracy. The main source of errors was the uncertainties in the determination of the degree of polarization of the target atom after laser pumping.

Interestingly, the relative intensities of the photoelectron lines and spectra of dichroism corresponding to one or a few allied ionic configurations can be described already to a good approximation even without explicit calculation of the photoionization amplitudes within so-called generalized geometrical model (Cubaynes et al. 2004, 2009). For information about the amplitudes one needs to compare different types of dichroism.

A new method of complete photoionization experiment could be photoelectron–Auger electron coincidence technique with atoms polarized by the laser pumping. A general theory within the two-step model and first numerical predictions by Pieve et al. (2007b) show significant dependence of dichroic effects on the geometry chosen in the experiment.

4.6.3 Photoionization of Atoms Polarized by Magnetic Field

Polarized atoms in the ground state can be efficiently prepared by passing a hexapole magnet, which focuses the atoms with positive magnetic quantum numbers M_i, while those with negative M_i are defocused (see Sect. 2.5.5). Linear magnetic dichroism in the photoelectron angular distribution (LMDAD) is measured by inverting the guiding magnetic field. The intensities of the Stern-Gerlach components give necessary information on the population of the magnetic substates N_{M_i} in (4.65), therefore determining alignment and orientation of the target atom. Accounting for thermal distribution among the fine structure states is needed in the case of high evaporating temperatures and small fine structure splitting (Prümper et al. 2001a, b).

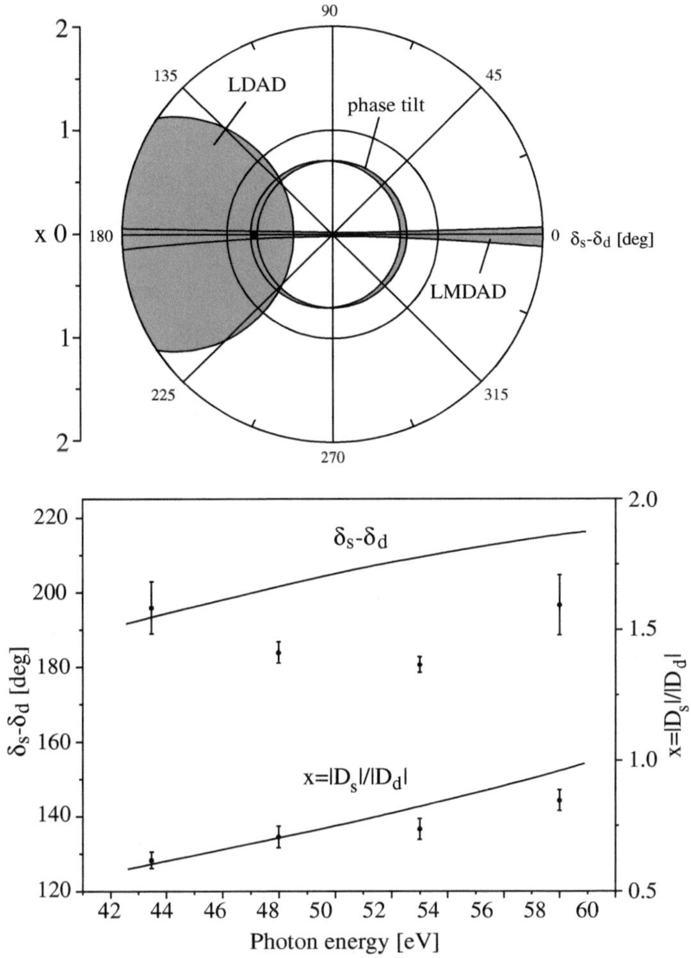

Fig. 4.25 Absolute ratio $x = |D_s|/|D_d|$ of the dipole amplitudes and phase difference $(\delta_s - \delta_d)$ for the outgoing electron waves εs, εd in the 5p photoionization of Eu. (*Top*) polar plot of the different sets of values derived from measurements of LADAD, LMDAD, and phase tilt at $h\nu = 53$ eV. The area where all three sets intersect is the black square. The cross is the theoretical result obtained using the Cowan code. (*Bottom*) energy dependence of the extracted parameters together with theoretical results obtained using the Cowan code. The data were obtained with a cylindrical mirror analyzer, which partly integrates the photoelectron flux over the azimuthal angle. Adapted from Godehusen et al. (1998)

Figure 4.26 shows the scheme of the apparatus applied in complete photoionization experiments of Plotzke et al. (1996), Prümper et al. (1997, 2000, 2001a, b) with oxygen and thallium atoms. Figure 4.27 presents results of the complete experiment in photoionization of oxygen,

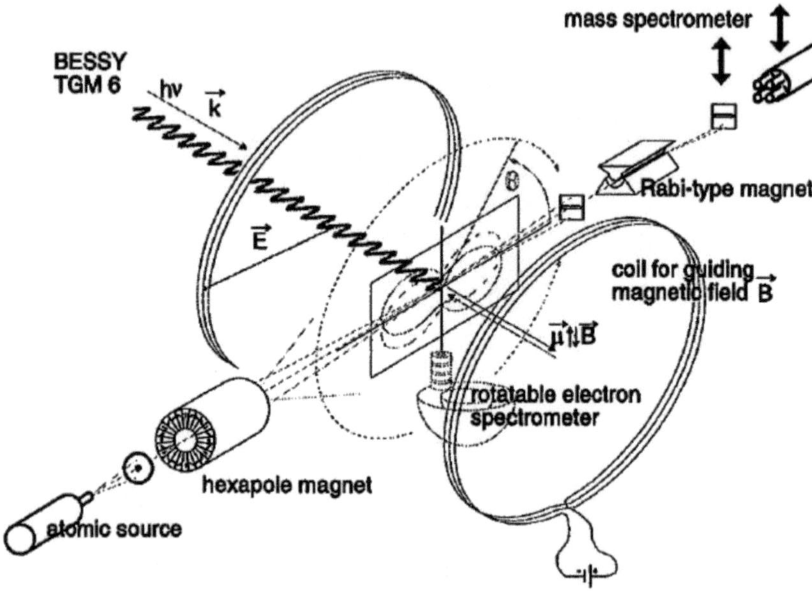

Fig. 4.26 Schematic of the experiment including the atomic source, a collimator, the hexapole magnet, the rotatable electron spectrometer, a Rabi magnet, mass spectrometer, the incoming synchrotron radiation, and the magnetic coil pair for the guiding field. After Plotzke et al. (1996)

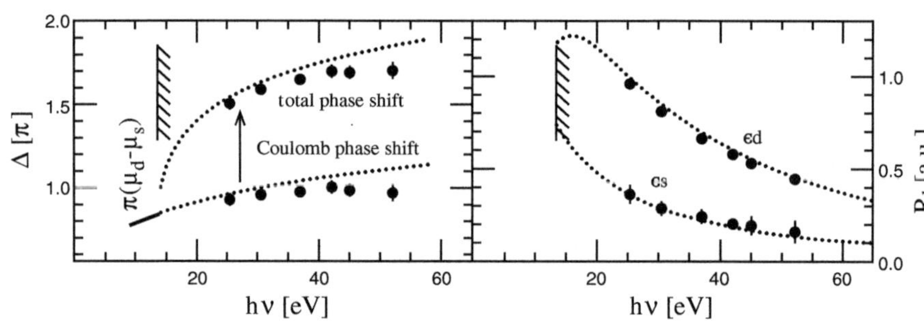

Fig. 4.27 The difference in the phase Δ of the amplitudes, (*left panel*), and absolute values of the amplitudes $R_{\varepsilon\ell}$, (*right panel*), for photoionization of the oxygen ground state; see (4.75). (······): Hartree–Fock calculations. (——): Quantum defect phase difference of the two ns and nd Rydberg series which converge to the ^{4}S threshold. Adapted from Plotzke et al. (1996)

$$h\nu + \mathrm{O}(2\mathrm{p}^4\,{}^3\mathrm{P}_2) \longrightarrow \mathrm{O}^+(2\mathrm{p}^3\,{}^4\mathrm{S}_{3/2}) + \mathrm{e}^-_{\mathrm{ph}}(\varepsilon\mathrm{s}_{1/2}, \varepsilon\mathrm{d}_{3/2}, \varepsilon\mathrm{d}_{5/2}), \qquad (4.75)$$

where the combination of the angular distribution of photoelectrons and magnetic dichroism measurements were used to extract, within the Cooper–Zare model, the ratio R_d/R_s of the two radial components of the matrix elements for the 2p electron photoionized into εs and εd continuum states and the difference in the asymptotic phase shifts $\Delta = \delta_d - \delta_s$ of these states. The transformation to absolute amplitudes

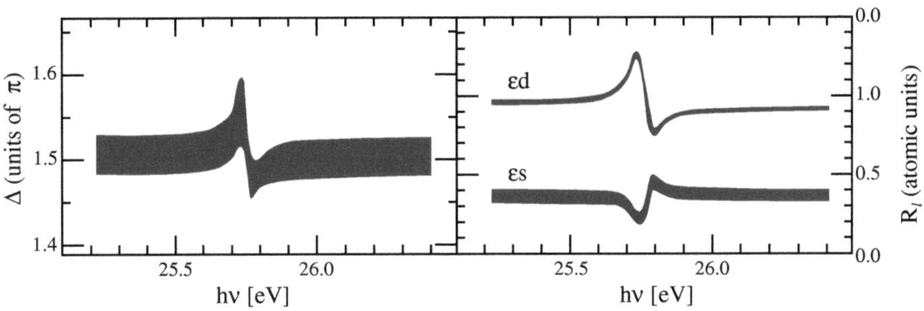

Fig. 4.28 Same as Fig. 4.27 but in the region of the autoionizing $O^*(2s2p^4 3p\,^3D)$ resonance. The error limits are represented by the shadowed zone. Adapted from Plotzke et al. (1996)

is made using the absolute cross section from Schaphorst et al. (1995). To choose between two possible solutions of the set of equations for the amplitudes, an additional bit of information was taken from the fact that far from the threshold the d amplitude is expected to exceed the s amplitude. Hartree–Fock calculations show very good agreement for the values of the amplitudes, while the relative phase shift deviate from the experimental data at the photon energies higher than 40 eV. The Fano profiles in the extracted photoionization amplitudes in the region of the autoionizing state $O^*(2s2p^4 3p\,^3D)$ are shown in Fig. 4.28. Interestingly, the profiles for s and d amplitudes have opposite asymmetry, which indicates a sign difference between the Auger decay matrix elements of the autoionizing state into the two s and d continua.

4.6.4 Resonant Two-Colour Two-Photon Ionization

The excited atomic state can be polarized not only by the laser pumping, but just by absorption of one photon. High brightness of lasers gives the opportunity to achieve intense flux of photoelectrons from a process, when the first laser photon transfers atom into an excited polarized state, while the second laser photon, not necessarily from the same laser, ionizes this state. Varying polarizations of the laser beams and analyzing angular distribution and/or spin polarization of photoelectrons, as well as the photoemission angle-integrated ion yield, provides conditions for performing the complete experiment for photoionization from atomic excited states. First such two-photon laser experiments with the purpose of a complete experiment were realized at the end of the 1970s and the beginning of the 1980s by Duncanson et al. (1976), Hansen et al. (1980), Kaminski et al. (1980) with alkali metal atoms and later with neon initially in a metastable state by Siegel et al. (1983). The absolute ratio of partial amplitudes and their relative phase were extracted in the Cooper–Zare three-parameter model. Since the dominant neon isotope possesses spinless nuclei, this case has a preference due to the lack of the hyperfine interactions, which disturb polarization of the intermediate laser-excited state before being ionized by the second photon. In the general case this depolarization should be carefully taken into consideration during the extraction of the photoionization amplitudes.

Figure 4.29 presents schemes of the set-up and studied transitions in experiments by Siegel et al. (1983). Initially isotropic Ne atoms in the metastable Ne* $(2p^5 3s\,^3P_2)$ state were photoexcited by linearly polarized photons (640.2 nm). Thus, the aligned Ne* $(2p^5 3p\,^3D_3)$ states were produced, subsequently ionized by another counterpropagating linearly polarized laser beam. In the Cooper–Zare model, ionization of the 3p electron gives photoelectrons in the εs and εd channels:

$$h\nu + \mathrm{Ne}^*(2p^5 3p\,^3D) \longrightarrow \mathrm{Ne}^+(2p^5\,^2P) + e^-_{\mathrm{ph}}(\varepsilon s, \varepsilon d)\ . \tag{4.76}$$

The polarization directions of the two beams, $\mathbf{E_P}$ and $\mathbf{E_I}$, respectively, could be independently rotated. The photoelectrons were either extracted by a weak electric field leading to the angle-integrated photoionization cross section (depending on the mutual angle of the polarization vectors α), or detected in the plane perpendicular to the beam by an analyzer in a direction $\hat{\mathbf{p}}$ leading to the photoelectron angular distributions. Figure 4.30 demonstrates part of the experimental results. Theoretical interpretation of the results are based on the formalism of Klar and Kleinpoppen (1982) and leads to the ratio R_d/R_s and the relative phase $\cos(\delta_s - \delta_d)$. Three remarks are appropriate here. First, the distribution shown in Fig. 4.30a represents the angular part of the anisotropic 3p electron density distribution prepared by the

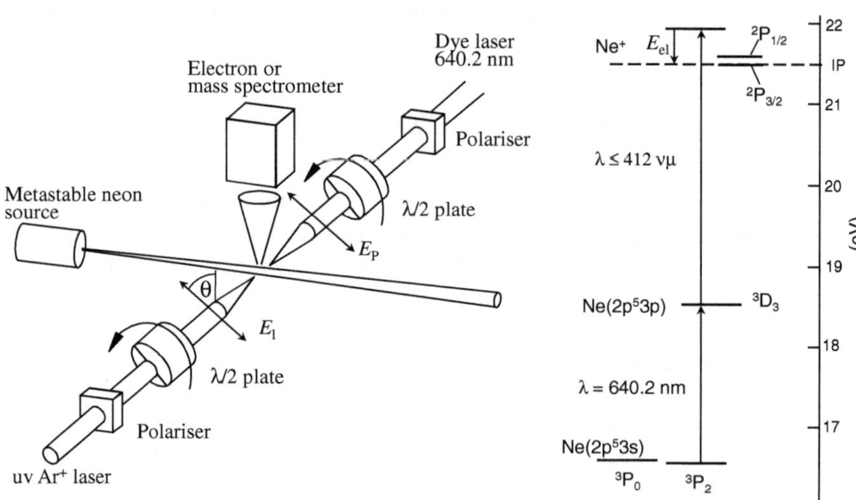

Fig. 4.29 (*Left panel*) schematic diagram of the experimental set-ups. In the photoelectron angular distribution measurements electrons are detected with an electron spectrometer plus a channel electron multiplier. Total cross sections are measured with a quadrupole mass spectrometer as ion detector. (*Right panel*) energy level scheme for the resonant two-photonionization of Ne* $(2p^5 3s\,^3P_2)$. After Siegel et al. (1983)

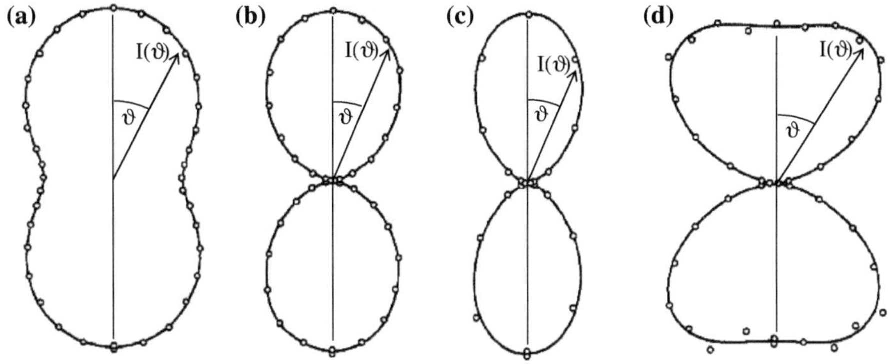

Fig. 4.30 Examples of angular distributions of photoelectrons from ionization of the aligned $\mathrm{Ne}^*(2\mathrm{p}^5 3\mathrm{p}\,{}^3\mathrm{D}_3)$ state. **a** $\mathbf{E_I}$ fixed, parallel to $\hat{\mathbf{p}}$, $\mathbf{E_P}$ rotated; **b** $\mathbf{E_P}$ fixed, parallel to $\hat{\mathbf{p}}$, $\mathbf{E_I}$ rotated; **c** $\mathbf{E_I} \parallel \mathbf{E_P}$, rotated simultaneously; **d** $\mathbf{E_I} \perp \mathbf{E_P}$, rotated simultaneously. The full curves represent least-squares fits by $a_0 + a_2 P_2(\cos\vartheta) + a_2 P_4(\cos\vartheta)$ (with $a_4 = 0$ for panels **a** and **b**). After Siegel et al. (1983)

first laser, thus the alignment of the atomic state before photoionization is fixed experimentally. Second, the set of data is 'overdetermined' in a sense that an extensive cross checking of the results is possible with respect to extracted photoionization amplitudes. Finally, the modulation of the angle-integrated cross section as function of the angle α alone gives the absolute ratio R_d/R_s provided the alignment of the initial state is known. Due to a limited range of the laser wavelength, the complete experiment on photoionization from the $\mathrm{Ne}^*(2\mathrm{p}^5 3\mathrm{p}\,{}^3\mathrm{D}_3)$ state was performed close to its ionization threshold. The results of the complete experiment presented in Fig. 4.31 are in good agreement with many-body perturbation theory of Chang (1982) and muliticonfiguration close-coupling calculations by Luke (1982).

As was used in experiments of Siegel et al. (1983) (although within the three-parameter model), by combining different polarizations of laser beams and detecting the angle-integrated cross section (or the ion yield as an option), the partial photoionization cross sections from isotropic excited states can be separated. Before, by the two-photon resonant ionization Duong et al. (1978) found partial photoionization cross sections of the $\mathrm{Ne}^*(3\mathrm{p})$ state into the εs and εd channels. Separation of the partial cross sections in photoionization of polarized atoms was discussed, for example, by Elizarov and Cherepkov (1986), Balashov et al. (1988, 1994), Schohl et al. (1997). This task has been considered already as an application of the fluorescence polarimetry method in Sect. 4.3.3.

We show now general principles of separating partial cross sections beyond the three-parameter model on an example of the two-photon ionization scheme

$$A(J_0 = 0) \xrightarrow{h\nu_1} A^*(J_1 = 1) \xrightarrow{h\nu_2} (A^+ + e)_{J=0,1,2}\,. \tag{4.77}$$

Assuming parallel radiation beams, a simple angular momentum algebra gives the following expressions for the photoelectron–angle integrated cross sections in case

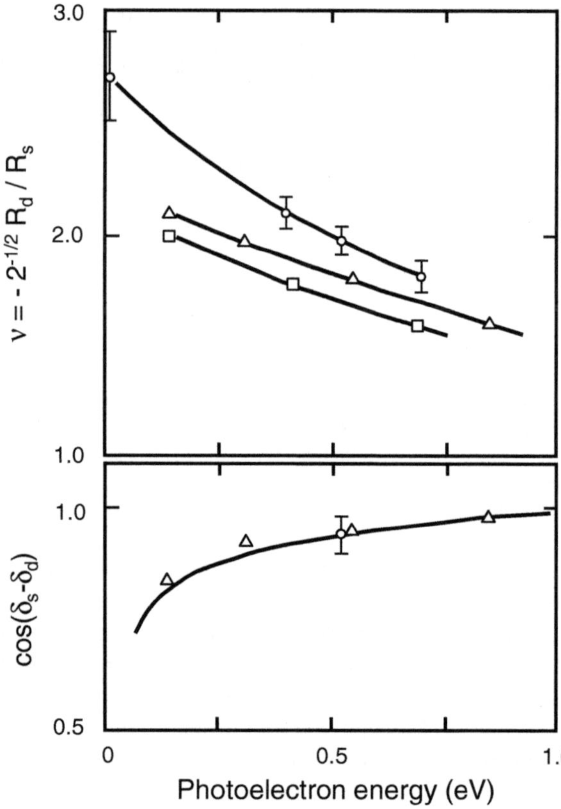

Fig. 4.31 Parameters of photoionization of $Ne^*(2p^5 3p\,^3D_3)$. Experimental results are shown together with calculations by Chang (1982) (△) and Luke (1982) (□). Adapted from Siegel et al. (1983)

of different combinations of polarization of the beams (He et al. 1995):

$$\sigma^{\parallel} = 3\sigma_0 + \frac{6}{5}\sigma_2 \tag{4.78}$$

for parallel linear polarizations,

$$\sigma^{\perp} = \frac{3}{2}\sigma_1 + \frac{9}{10}\sigma_2 \tag{4.79}$$

for perpendicular linear polarizations,

$$\sigma^{++} = \frac{9}{5}\sigma_2 \tag{4.80}$$

for two right (two left) circular polarizations, and

$$\sigma^{+-} = 3\sigma_0 + \frac{3}{2}\sigma_1 + \frac{3}{10}\sigma_2 \tag{4.81}$$

for the opposite sign of circular polarizations, with σ_J being the isotropic cross section for photoionization from the excited atomic state $A^*(J_1 = 1)$ to final angular momentum state J (see (4.77)). The four measurable combinations (4.78)–(4.81) provide the three cross sections σ_J $(J = 0, 1, 2)$ and, additionally, a self checking of the results. Equations (4.78)–(4.81) are valid when depolarization of the intermediate state $A^*(J_1)$ between the excitation and ionization is negligible. The sources of the depolarization may be collisions, radiation trapping, precession of the electronic angular momentum J_1 due to the hyperfine interactions. Accounting for the depolarization changes the numerical coefficients in (4.78)–(4.81) and is generally not trivial. To account for the hyperfine interactions, averaging over time and isotopic abundances is needed (Fano and Dill 1972; Greene and Zare 1982; Wood et al. 1993; Meyer et al. 2001).

Figure 4.32 presents separated cross sections for channels with different total angular momentum J in the reactions with excited barium atoms (He et al. 1995

$$h\nu + \mathrm{Ba}^*(6\mathrm{s}6\mathrm{p}\,^1\mathrm{P}_1) \longrightarrow \left(\mathrm{Ba}^+ + \mathrm{e}^-_{\mathrm{ph}}\right)_J\,. \tag{4.82}$$

in the region of autoionizing states Ba**(5dnd), Ba**(5dng). The extracted partial cross sections are very detailed and give an authentic view of contributions from channels with different total angular momentum with all the resonance structures. These data can serve as an excellent test for advanced theoretical models, describing photoionization of such a complex target as barium. The R-matrix calculations with inclusion of spin-orbit interaction terms in atomic Hamiltonian by Greene and Aymar (1991) give a good qualitative explanation of all the main features in the separated channels, as demonstrated in Fig. 4.33 for the channel with $J = 2$.

The channel-separated cross sections for photoionization of excited argon atoms, studied by Schohl et al. (1997)

$$h\nu + \mathrm{Ar}^*(3\mathrm{p}^5 4\mathrm{p}\,^3\mathrm{D}_3) \longrightarrow \left(\mathrm{Ar}^+ + \mathrm{e}^-_{\mathrm{ph}}\right)_J, \tag{4.83}$$

in the region of the autoionizing Rydberg states Ar**3p^{5}10d$'$ are shown in Fig. 4.34. In this experiment, the initial atomic state is the metastable Ar*(3p^{5}4s ^{3}P$_2$) state, further excited by linearly polarized (811.75 nm) laser light to the Ar*(3p^{5}4p ^{3}D$_3$) state, and then ionized by linearly polarized radiation from a tunable multi-mode dye laser. Within the Cooper–Zare model, the partial photoionization cross sections σ_J into the states with $J = 2, 3, 4$ are expressed in terms of only two absolute squares of the ionization amplitudes, D_s and D_d, corresponding to the photoelectrons in the εs and εd continua. With only two parameters it becomes possible to extract the partial cross sections σ_2, σ_3, and σ_4 utilizing only linearly polarized laser beams. In contrast to the method described above for photoionization of barium, the method

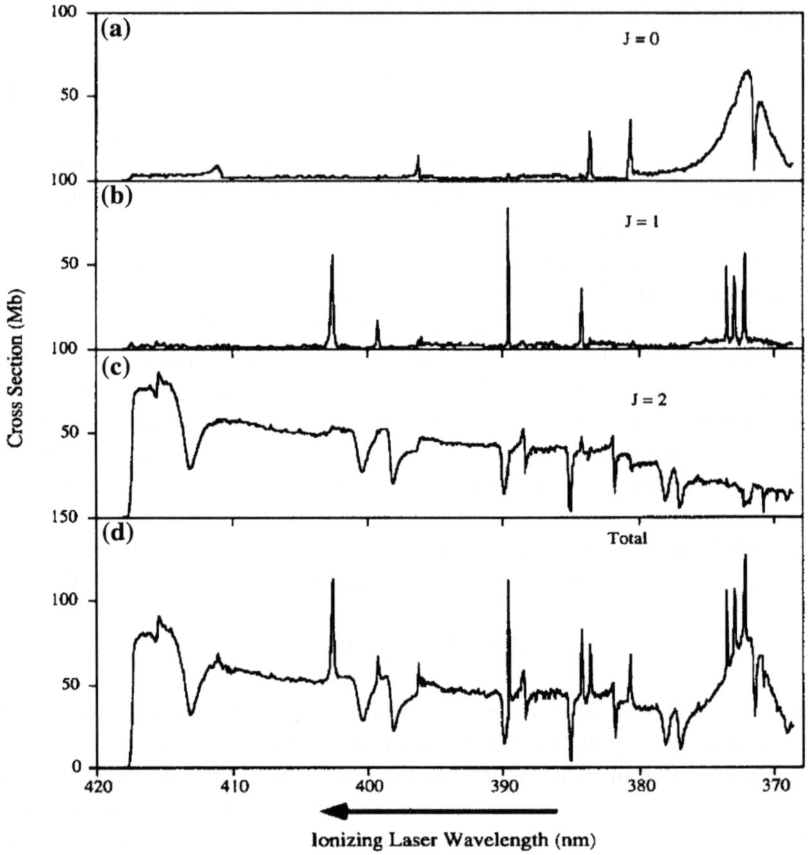

Fig. 4.32 **a–c** Cross sections for photoionization of $Ba^*(6s6p\,^1P_1)$ to specific final J states. **d** total isotropic photoionization cross section of $Ba^*(6s6p\,^1P_1)$. After He et al. (1995)

used by Schohl et al. (1997) is applicable within more restrictive model. Again, the experiment leads to a clear picture of the channel contributions, showing differently shaped and overlapping autoionizing resonances with the total angular momenta $J = 2$ and $J = 3$. Satisfactory overall agreement is observed between the experimental and theoretical partial cross sections, although they deviate quantitatively in the relative contribution of the channels and in the details of the resonance profiles.

Two-photon resonant ionization experiments with lasers to prepare the excited atoms and to ionize them are limited by the narrow range of the laser tunability. On the other hand, intensity of easily tunable synchrotron radiation sources is not enough to perform such experiments. Furthermore, a poorly matching time structure of laser and synchrotron radiation pulses prevents their joint use in the experiments and up to recently a combination of lasers with synchrotron radiation sources was limited by the application of high-intensity *CW*-dye lasers for the optical pumping of

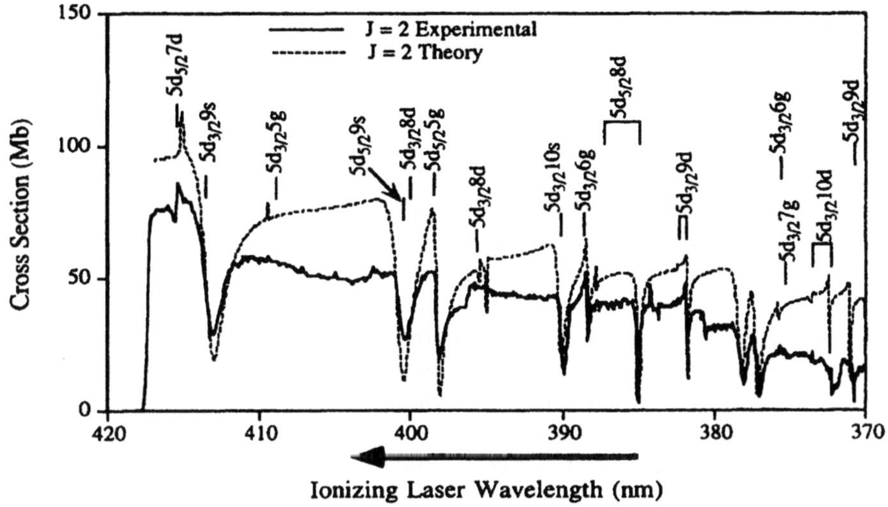

Fig. 4.33 Isotropic cross section, σ_2, for photoionization of $Ba^*(6s6p\,^1P_1)$ to $J = 2$ states versus the ionizing laser beam wavelength. (——): measured spectrum; (– – –): theoretical spectrum of Greene and Aymar (1991) convoluted with the experimentally determined laser beam profile to account for the bandwidth. The positions and designations of some autoionizing resonances taken from Camus et al. (1983) are indicated. After He et al. (1995)

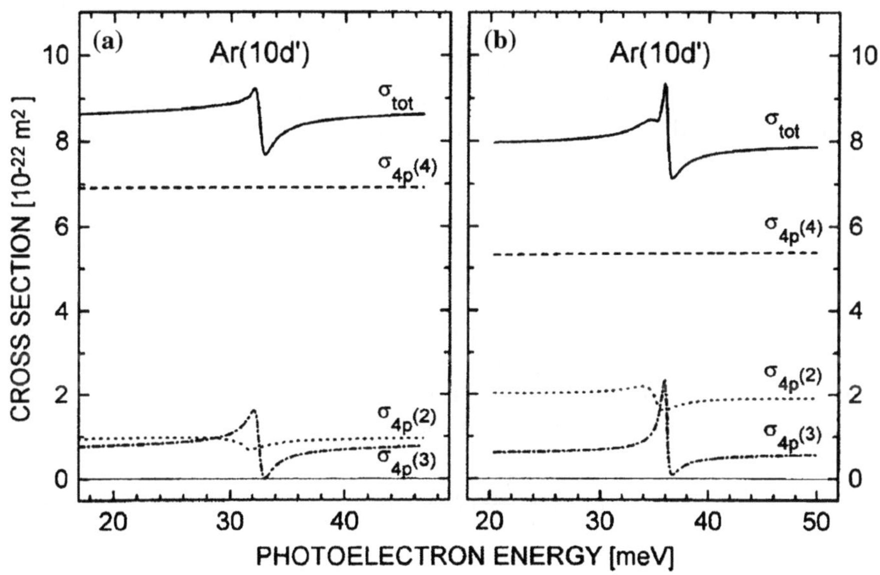

Fig. 4.34 Total and partial $J = 2, 3, 4$ cross sections for photoionization of $Ar^*(3p^5 4p\,^3D_3)$ atoms in the region of the $10d'$ resonances: **a** deduced from experiment; **b** calculated. After Schohl et al. (1997)

the target with its subsequent ionization by the synchrotron radiation, as have been discussed in Sect. 4.6.2. A technique of time synchronization between laser and synchrotron radiation pulses (Pizzoferrato et al. 1986; Mitani 1989; Ederer et al. 1992; Lacoursière et al. 1994; Meyer et al. 1996; Gisselbrecht et al. 1998) allowed to combine these two essentially different photon sources in a *pump-probe* arrangement. By this means photoionization of highly excited and polarized short-lived states can be studied, when the synchrotron radiation is used as a *pump*. Experiments on photoionization from the excited states of He (Lacoursiére et al. 1994) and Xe(Gisselbrecht et al. 1998) with recording of the photoion yield demonstrated potentialities of the method. A possibility of using synchrotron radiation for the preparation of excited atoms with subsequent ionization by absorbing a laser photon even without the time synchronization between the synchrotron radiation and laser pulses was shown by Mitsuke et al. (2000) in experiments with Ar atoms, where a laser with high repetition rate and long pulse duration was used.

A further step was made by Aloïse et al. (2005) in study of photoionization from Xe with the use of synchrotron radiation for atomic excitation of a short-lived Rydberg state and a CW-dye laser for further photoionization of the excited atoms, illustrated in Fig. 4.35. As a result of this study, which combined circular and linear polarization of the synchrotron as well as of the laser photons, photoionization cross sections were separated in the region of overlapping autoionizing resonances of different symmetry and the parameters of the resonances were extracted. Figure 4.36 demonstrates that although the quality of the data is not as good as with two laser beams, two-photon two-colour complete experiments with the first photon from synchrotron source are feasible. Since only the laser beam could be considered as ideally polarized, (4.78)–(4.81) have to be generalized to the case of arbitrary polarized photon beams. For the parallel laser and synchrotron radiation beams the general equation for the cross section reads (Aloïse et al. 2005)

$$\sigma = \sigma_{tot} \left[1 + \xi_1 P_c^{\text{las}} P_c^{\text{SR}} + \xi_2 \big(1 + 3 P_\ell^{\text{las}} P_\ell^{\text{SR}} \cos 2\varphi \big) \right], \tag{4.84}$$

where $\sigma_{tot} = \sigma_0 + \sigma_1 + \sigma_2$. The degree of circular (linear) polarization of the ionizing laser radiation and synchrotron radiation is denoted by P_c^{las} (P_ℓ^{las}) and P_c^{SR} (P_ℓ^{SR}), respectively, while φ is the angle between the principal axes of the polarization ellipse of the laser and synchrotron radiation beams. Information on the channel cross sections σ_J is contained in the two parameters

$$\xi_1 = \frac{3}{4\,\sigma_{tot}} \big(2\sigma_0 + \sigma_1 - \sigma_2 \big), \tag{4.85}$$

and

$$\xi_2 = \frac{1}{20\,\sigma_{tot}} \big(10\sigma_0 - 5\sigma_1 + \sigma_2 \big)\,. \tag{4.86}$$

To find the individual σ_J, the circular dichroism and the linear dichroism were measured:

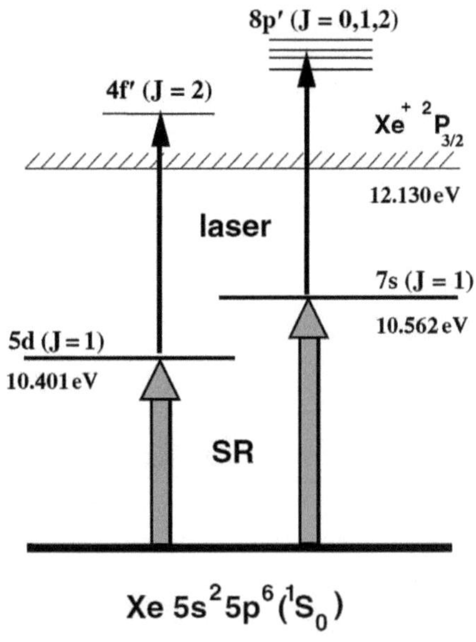

Fig. 4.35 Two-photon excitation scheme of Xe by synchrotron radiation + laser combination. After Aloïse et al. (2005)

$$\frac{\sigma^{++} - \sigma^{+-}}{\sigma^{++} + \sigma^{+-}} = \frac{\xi_1}{1 + \xi_2} P_c^{\text{las}} P_c^{\text{SR}}, \tag{4.87}$$

and

$$\frac{\sigma^{\|} - \sigma^{\perp}}{\sigma^{\|} - \sigma^{\perp}} = \frac{3\xi_2}{1 + \xi_2} P_\ell^{\text{las}} P_\ell^{\text{SR}}, \tag{4.88}$$

together with the isotropic cross section

$$I(\varphi_0) = \text{const} \times \sigma_{tot}, \tag{4.89}$$

where φ_0 is the *magic* angle defined by the relation $\cos 2\varphi_0 = -(3P_\ell^{\text{las}} P_\ell^{\text{SR}})^{-1}$. It is important for experiments with variously polarized synchrotron radiation beams that the described above extraction procedure for σ_J is independent of the intensity of the radiation. Depolarization of the excited Xe Rydberg state due to hyperfine interactions and collisions is incorporated by a modification $\xi_k \rightarrow DG_k\xi_k$ $(k = 1, 2)$, where the factors G_k and D account for depolarization due to the hyperfine interactions and collisions, respectively.

A velocity map imaging analyzer incorporated in experiments in which synchrotron is used as a pump and the laser radiation is used as a probe (O'Keeffe et al. 2010) provides more information on the ionization process than geometries in which fixed one-dimensional detectors are used and only the polarization of the ionizing light is changed.

The presented examples illustrate the great potential of the dichroism and magnetic dichroism studies for the complete description of atomic photoionization. Besides the

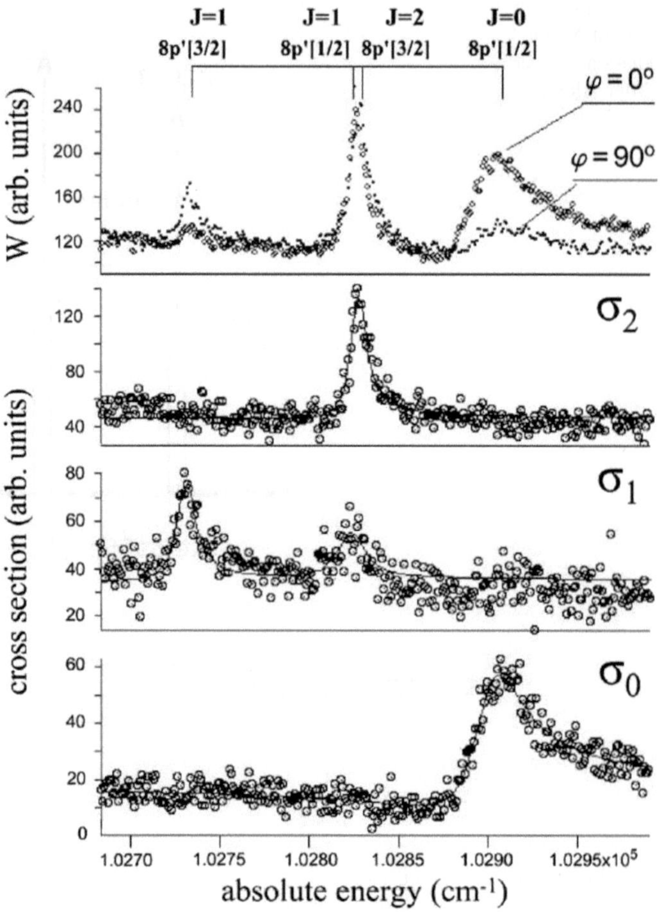

Fig. 4.36 Photoionization spectra in the region of the Xe8p′ resonances excited from the $7s[\frac{3}{2}]$ state using linearly polarized laser and synchrotron radiation of parallel ($\varphi = 0°$) and perpendicular ($\varphi = 90°$) relative orientation of their electric field vector (*upper panel*). Partial photoionization cross sections σ_0, σ_1, and σ_2 in the same energy region (*lower panels*). After Aloïse et al. (2005)

complete experiment, the dichroism in inner-shell photoelectron spectra is a powerful tool for the element-specific determination of magnetic properties from thin films and multilayers. Since the inner-shell photoionization in solids is strongly influenced by local interactions, atomic models can be successfully used in many cases as a first step for a qualitative understanding of the dichroism spectra of atoms bound to a surface. Figure 4.37 illustrates an obvious similarity between the LMDAD in the 2p photoelectron spectra of free Cr atoms and of Cr atoms bound at a surface. The 2p photoelectron spectra of the Cr surface layer was excited at a photon energy of 705 eV and the LMDAD was measured by taking the magnetization in two opposite directions. We refer to original papers and reviews for further discussions of this

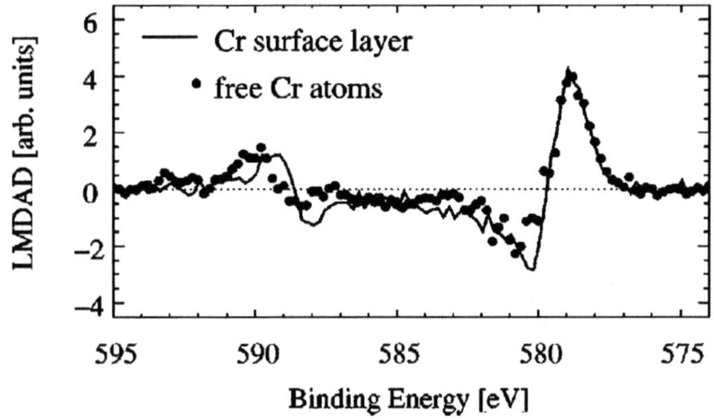

Fig. 4.37 Comparison of the LMDAD of a magnetized Cr surface layer and the LMDAD of free oriented Cr atoms. Both curves are displayed along the binding energy axis of the free atoms and normalized to each other in the maximum. The spectrum of the Cr surface layer has been shifted by 5.5 eV towards higher energies. After Wernet et al. (2000)

interesting field (for example Starke 2000; Henk and Ernst 2002; Bethke et al. 2005 and references therein).

As shown above, photoionization from polarized atoms can lead to complete experiment provided the initial polarization state of the target atom is known. Postulating that the amplitudes of photoionization are known, one can solve a reverse problem, namely, using the observed dependence of the photoelectron flux on the emission angle and polarization of the incoming photon beam to deduce polarization state of the target atom. This method is widely used in solid state physics, where the angular resolved photoelectron spectroscopy (ARPES) is a well established technique to draw conclusions about three-dimensional band structures (Fermi surfaces) and the orbital character of atomic electrons composing the Fermi surface (Schattke and Van Hove 2003; Damascelli 2004).

4.7 Non-Resonant Multiphoton Ionization

Complete experiments in nonresonant (direct) multiphoton ionization of atoms are in their infancy. Assuming a moderate field intensity of not larger than 10^{12}–$10^{13}\,\mathrm{W/cm^2}$, appropriate theoretical grounds to describe the process would be the perturbation theory with respect to interaction of atom with the radiation. The simplest case is two-photon ionization by a one-colour laser beam:

$$2h\nu + A(\alpha_i J_i) \rightarrow A^+(\alpha_f J_f) + e^-_{ph} \,. \tag{4.90}$$

In this case the energy and the polarization of two absorbed photons $h\nu$ are equal. The partial wave second-order photoionization amplitude of the process (4.90) in the dipole approximation is expressed in terms of the second-order reduced matrix elements

$$S_n(J_f\ell j, J; \omega) \equiv \big\langle (\alpha_f J_f, \ell j) J \| D G_n(\omega) D \| \alpha_i J_i \big\rangle, \tag{4.91}$$

where notations for the quantum numbers are similar to (4.13)–(4.15), ω is the photon frequency, n symbolizes symmetry of intermediate virtual states and $G_n(\omega)$ is the corresponding part of the atomic Green's function. More details on the formal theory of the two-photon ionization amplitudes and on expressions of different observable quantities in terms of these amplitudes can be found, for example, in Manakov et al. (1986, 1999). The number of independent two-photon amplitudes S_n is determined by two sets of individual selection rules of the type (4.16)–(4.17) for two dipole momentum operators D in (4.91). Similar to one-photon ionization, this number reduces in case of additional approximations. Thus, an analogue of the Cooper–Zare model can be formulated for the two-photon ionization. In general, the number of independent two-photon amplitudes is larger than the number of the one-photon amplitudes, although the direct comparison is not possible; there are no common ionization channels from the same atomic state since these two mechanisms are governed by transition operators of different parities. The generally larger number of the two-photon amplitudes is due to larger possible interval of the angular momentum of ionization channels and various possible symmetries of the intermediate states. From another side, the presence of the two photons, despite of their identity, gives more flexibility in using polarization properties of the radiation for the needs of complete experiment.

Figure 4.38 shows a scheme of transition pathways in complete two-photon ionization experiment with the rubidium atom by Wang and Elliott (2000)

$$2h\nu + \mathrm{Rb}(5\mathrm{s}\,{}^2\mathrm{S}_{1/2}) \longrightarrow \mathrm{Rb}^+({}^1\mathrm{S}_0) + \mathrm{e}^-_{\mathrm{ph}}(\varepsilon \mathrm{s}_{1/2}, \varepsilon \mathrm{d}_{3/2}, \varepsilon \mathrm{d}_{5/2})\,. \tag{4.92}$$

There are five two-photon ionization amplitudes, contributing to the process; the photoelectron in the $\varepsilon \mathrm{s}_{1/2}$ and $\varepsilon \mathrm{d}_{3/2}$ states can be generated via virtual intermediate states with ${}^2\mathrm{P}_{1/2}$ and ${}^2\mathrm{P}_{3/2}$ symmetries, whereas the $\varepsilon \mathrm{d}_{5/2}$ state can be reached only via the ${}^2\mathrm{P}_{3/2}$ state. Wang and Elliott (2000) used elliptically polarized light and showed its ability to constitute the complete two-photon ionization experiment when measuring only the angular distribution of the photoelectrons. The apparatus of Wang and Elliott (2000) has been able to measure the angular distribution in the entire 4π steradian solid angle by using the photoelectron imaging technique developed by Helm et al. (1993). This technique has advantages of high collection efficiency and stability in comparison with the traditional technique of ionization in the field-free region and counting the photoelectrons emitted towards a single-channel electron detector, while rotating the laser polarization.

Figure 4.39 demonstrates the angular distributions for differently polarized radiation. In the case of elliptically polarized radiation, the angular distribution in the plane perpendicular to the radiation can be cast into the form

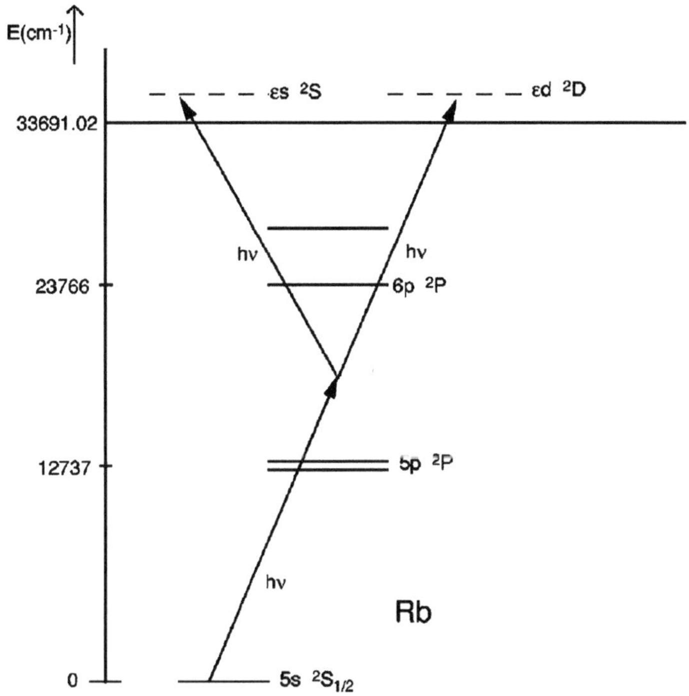

Fig. 4.38 Energy level diagram of atomic rubidium. The *arrows* indicate the transition pathways for non-resonant two-photon ionization. After Wang and Elliott (2000)

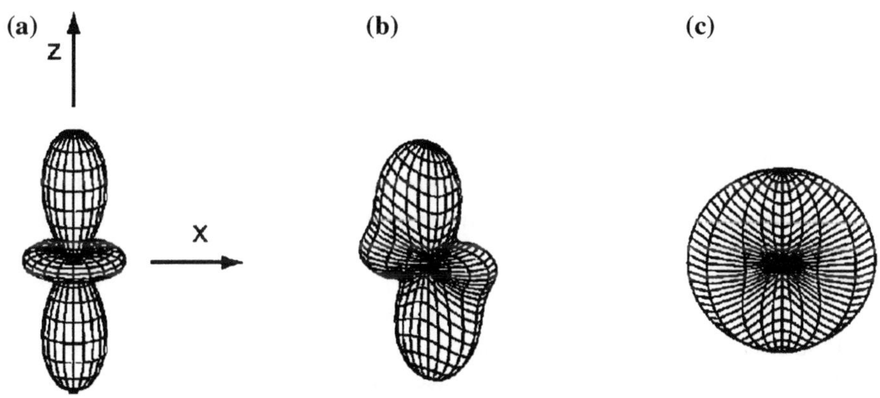

Fig. 4.39 Examples of calculated photoelectron angular distributions for (**a**) linearly polarized light (z axis along the polarization); **b** right elliptically polarized light; and **c** circularly polarized light. The distributions represent the probability density for detecting a photoelectron as a function of the electron emission angle. The direction of laser beam propagation is nearly into the plane of the page. After Wang and Elliott (2000)

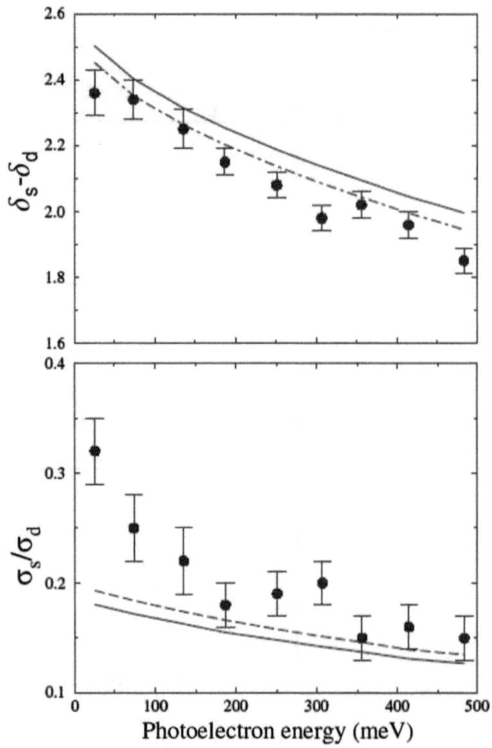

Fig. 4.40 The phase difference between continuum ε_s and ε_d wave functions for rubidium (*upper panel*), and the ratio of cross sections $\sigma_{s_{1/2}}/\sigma_d$, where $\sigma_d = \sigma_{d_{3/2}} + \sigma_{d_{5/2}}$ (*lower panel*), as functions of the photoelectron energy. (●): experimental data of Wang and Elliott (2000), (——): calculation using relativistic wave functions. (– – –): same calculation, except including only the 5p intermediate state in the relativistic dipole matrix element calculation. (– · –): calculation using a non-relativistic Hartree–Fock approach. Adapted from Colgan and Pindzola (2001)

$$\frac{d\sigma}{d\Omega} = \frac{\sigma_0}{4\pi}\Big(1 + a_2 \cos 2\theta + a_4 \cos 4\theta + b_2 \sin 2\theta\Big), \tag{4.93}$$

where the coefficients a_2, a_4, and b_4 depend on the Stokes parameters of the radiation and the two-photon amplitudes. Remind, that for one-photon ionization from unpolarized atom the elliptically polarized radiation, as follows from (4.23), leads to only one independent asymmetry parameter β. If the relativistic splitting of phases between the $\varepsilon d_{3/2}$ and $\varepsilon d_{5/2}$ channels is negligible (which was verified by Wang and Elliott 2000), the three above parameters determine the ratios of the partial cross sections σ_s/σ_d, $\sigma_{5/2}/\sigma_{3/2}$, and the phase difference $\delta_s - \delta_d$. Furthermore, the sign of the phase difference is accessible, because both sine and cosine enter the expressions for the coefficients a_2 and b_2. The extracted parameters σ_s/σ_d and $\delta_s - \delta_d$ are in rather good agreement with theory (Dodhy et al. 1985; Colgan and Pindzola 2001), as is exemplified in Fig. 4.40, whereas the ratio $\sigma_{5/2}/\sigma_{3/2}$ is in strong contradiction with these calculations and needs further experimental and theoretical study.

Complete experiments on two-colour two-photon non-resonant ionization have not yet been performed, as well as complete experiments for above threshold photoionization, when the energy of one of the photons is enough to ionize the atom. New opportunities in the field have been opened with the advent of free electron

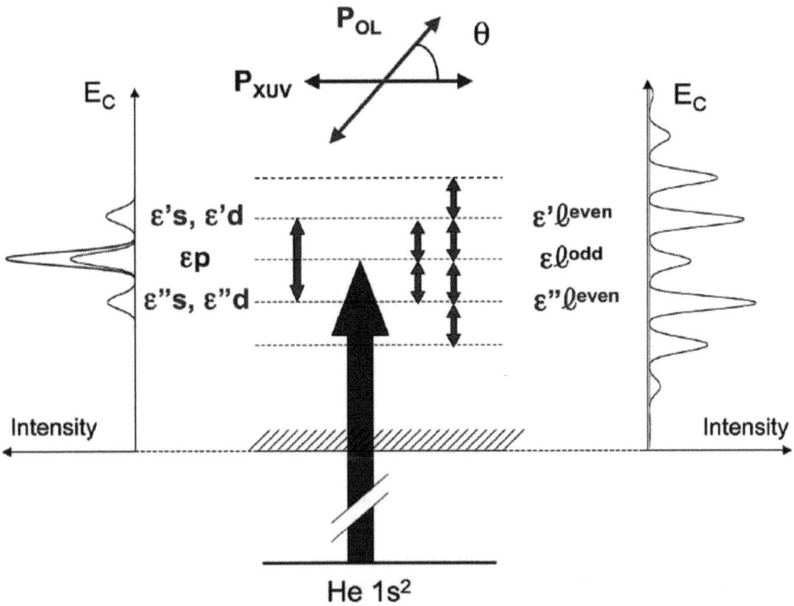

Fig. 4.41 Scheme of the two-colour above threshold ionization in He atoms by free electron (⬆) and infrared dressing laser. The process and the resulting photoelectron spectra are shown for, both, low (*left side*) and high (*right side*) dressing fields. The relative orientation of the electric field vectors of the ionizing XUV radiation ($\mathbf{P}_{XUV}$) and the optical dressing field ($\mathbf{P}_{OL}$) is shown at the *top*. After Meyer et al. (2008)

lasers in the self-amplifying regime with an unprecedent high intensity in the XUV. Combination of the XUV radiation ($h\nu_{XUV}$) with an optical laser beam in the visible or near infrared ($h\nu_{OL}$) permits the investigation of above threshold ionization by means of polarization (dichroic) methods (Meyer et al. 2008). Displayed in Fig. 4.41 is a scheme of photoionization of He by two linearly polarized radiation beams:

$$h\nu_{XUV} + h\nu_{OL} + \mathrm{He}(1s^2) \longrightarrow \mathrm{He}(1s) + e^-_{ph} \,. \tag{4.94}$$

The dominating contribution from two- and multiphoton ionization originates from processes in the course of which the emitted electron exchanges photons with a dressing laser field via stimulated emission (or absorption) resulting in a comb of sidebands on both sides of the main photoelectron line. Similar to two-photon resonance ionization, changing the polarization of either of the radiation beams gives rise to dichroic effects in the photoelectron spectrum and opens the possibility to control the relative contributions of multiphoton ionization channels with different angular momentum. At low optical field intensities, only one sideband is observed, dominated by εs and εd photoelectrons. I.e., for the He atom, with negligible relativistic effects and a final ionic S-state, only two two-photon ionization channels contribute to the first sidebands. The relative contribution of these channels can be determined

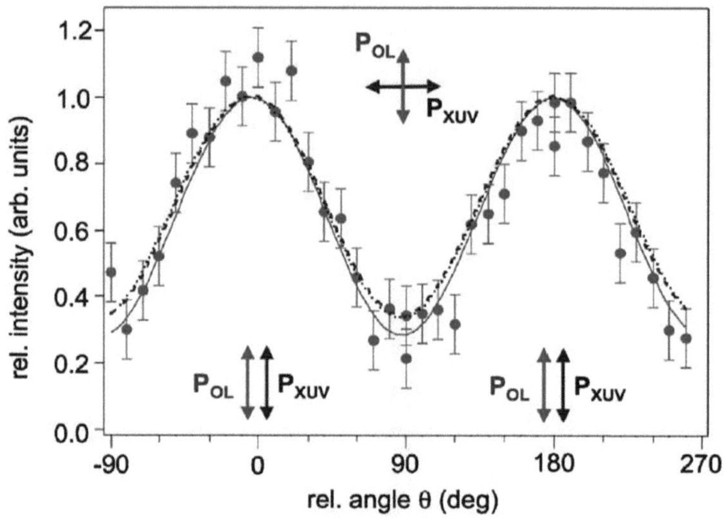

Fig. 4.42 Variation of the upper sideband yield in the low field regime (about 8×10^{10} W/cm^2) for the two-photon ionization in the 1s shell of He at 13.7 nm as a function of the relative angle between the linear polarization vectors of the free electron laser and the optical (800 nm) laser. (——): fit to the experimental data (●). The results of time-dependent second-order perturbation theory (— — —), and "soft-photon" approximation (······) are almost identical. After Meyer et al. (2008)

from the modulation of the ion yield as a function of the angle φ between the polarization directions. Applying second-order perturbation theory, the magnitude of the sidebands is modulated with the angle similar to (4.84), except that for He the p-wave is missing ($\sigma_1 = 0$). Figure 4.42 shows measured and calculated modulation, indicating that the contribution from the s and d channels are approximately equal. This is in contrast to the two-colour resonant ionization via discrete intermediate states, He*1snp ^{1}P ($n = 2, 3$) (Johansson et al. 2003), where a cross section more than 10 times higher for d- than for s-electrons has been determined, both, experimentally and theoretically.

For the resonant photoionization, the overlap between the discrete np and continuum εs or εd electron wave functions governs the branching ratio. Close to the ionization threshold for He, due to the centrifugal barrier, the first lobe of the εd wave overlaps considerably the lobe of the np ($n = 2, 3$) wave function. In contrast, the first oscillation of the εs wave is much closer to the nucleus, leading to compensation of positive and negative contributions to the transition matrix elements. Thus, near threshold, the εd wave dominates resonant two-colour ionization. For the above threshold ionization, the strengths of the p $\rightarrow$ s and p $\rightarrow$ d transitions are mainly determined by the overlap of the oscillatory asymptotics of the coupled continuum states, roughly by the squared cosine of their scattering phase difference (Mercouris et al. 1996). This quantity varies smoothly with the energy of the coupled continuum states and is accordingly in preference to the s channel for He at the experimental energies.

4.8 Photoionization in the Region of Resonance

Methods of complete photoionization experiments are common for photoionization into flat atomic continuum and in the region of autoionizing/Auger resonances. Nevertheless, there are features specific for photoionization in the resonance region, especially in the region of strong resonances with weak or negligible background of the direct photoionization, including a case of close and overlapping resonances. In this section we discuss some of these features, related to complete experiments.

4.8.1 Scanning Across Resonances

The emerging of new synchrotron radiation sources producing high-brilliance VUV and soft x-ray photon beams with high energy resolution provides the possibility to investigate polarization and correlation phenomena by scanning the excitation energy over an extended region around and across the resonances, even in cases, where the direct photoionization and therefore the off-resonance signal is extremely weak. As was shown by Grum-Grzhimailo et al. (2005), the Fano-like behavior in different correlation parameters for the same resonance possesses some universal features. Generally, in the region of an isolated resonance, the integral photoionization cross section to a particular ionic state $A^+(\alpha_f J_f)$ may be expressed (see the discussion in Sect. 2.3.6) by the parametric formula

$$\sigma_{\alpha_f J_f}(\epsilon) = \sigma^0_{\alpha_f J_f}\left(1 + \frac{2C_1\epsilon + C_2}{1+\epsilon^2}\right), \tag{4.95}$$

where $\sigma^0_{\alpha_f J_f}$ is the cross section without the resonance and C_1, C_2 are the profile parameters introduced by Starace (1977). The parameter C_1 describes the asymmetry of the profile, while the parameter C_2 determines the integral yield of the resonance into the cross section. Generally, any kind of angle-independent correlation parameter T, studied in complete photoionization experiments, like spin polarization and asymmetry parameters of photoelectrons, alignment and orientation of residual ions, and others, is expressed in terms of a bilinear combination of the dipole photoionization amplitudes of the form

$$T = \left(\sum_{\ell j J}|D(\ell j J)|^2\right)^{-1} \sum_{\substack{\ell j J \\ \ell' j' J'}} t(\ell j J; \ell' j' J') D(\ell j J) D^*(\ell' j' J'), \tag{4.96}$$

where $t(\ell j J; \ell' j' J')$ are the angular coupling coefficients specific for each parameter T. Substituting into (4.96) the partial dipole amplitudes in the region of an isolated resonance in the Fano form (Fano 1961; Fano and Cooper 1965; Kabachnik and Sazhina 1976; Starace 1977)

$$D_{\ell jJ} = D^0_{\ell jJ} + \frac{2\pi}{\Gamma} V_{\ell jJ} \sum_{\ell' j' J'} V^*_{\ell' j' J'} D^0_{\ell' j' J'} \frac{q+\mathrm{i}}{\epsilon - \mathrm{i}}, \tag{4.97}$$

where $D^0_{\ell jJ}$ is the photoionization amplitude in the absence of a resonance, $V_{\ell jJ}$ abbreviates the complex Coulombic (autoionization) decay amplitude, see (4.101), q is the Fano profile parameter of the photoabsorption cross section in the region of the resonance, Γ is the decay width of the resonance, and $\epsilon = (E - E_r)/(\Gamma/2)$, with E and E_r being the energies of the photon and the resonance, respectively, the parameter (4.96) can be reduced to the standard Fano form (Fano 1961; Fano and Cooper 1965)

$$T = \sigma_b^T + \sigma_a^T \frac{(q^T + \tilde{\epsilon})^2}{1 + \tilde{\epsilon}^2}, \tag{4.98}$$

where the parameters σ_b^T, σ_a^T and q^T are specific quantities for each individual T. In contrast, the scaled energy dependence $\tilde{\epsilon} = (E - \tilde{E}_r)/(\tilde{\Gamma}/2)$, which is given in terms of a scaled width

$$\tilde{\Gamma} = \left(1 + C_2 - C_1^2\right)^{\frac{1}{2}} \Gamma \equiv \chi \Gamma, \tag{4.99}$$

and the shifted resonance energy

$$\tilde{E}_r = E_r - \frac{C_1}{2} \Gamma \equiv E_r + \Delta, \tag{4.100}$$

are the same for all T parameters. Thus, the Fano profiles for all correlation parameters T of the form (4.96), and related to a particular final ionic state, show a kind of *scaling*; they possess similar width $\tilde{\Gamma}$ and similar shift Δ with respect to the energy of the resonance state. Particular cases of this general regularity have been found earlier by Grum-Grzhimailo and Zhadamba (1987) and Grum-Grzhimailo et al. (1991). Both quantities, $\tilde{\Gamma}$ and Δ, are proportional to the decay width of the resonance state, and the corresponding proportionality factors depend only on the two parameters C_1 and C_2 of the resonance profile in the integral cross section. The scaling factor χ depends on both parameters (see Fig. 4.43), while the shift Δ depends on C_1, only. The $3D$-plot of Fig. 4.43 shows regions of C_1 and C_2 corresponding to small and large values of χ. The large scaling factors χ, i.e. strong *broadening* of the resonance structures in the polarization and correlation parameters in comparison with the natural width, correspond to $C_2 \gg 1$, $C_2 \gg C_1^2$. The broadening is related to the relative strength of the resonance with respect to the direct ionization. For a strong resonance, the resonance channel dominates even in the regions outside the natural width, leading in these regions to the values of the parameters close to their resonant values. The values characteristic for the direct photoionization are reached only in the far wings of the resonance in the cross section.

The *broadening* effect is very well illustrated by the spin polarization data of Fig. 4.44 for the photoionization of thallium.

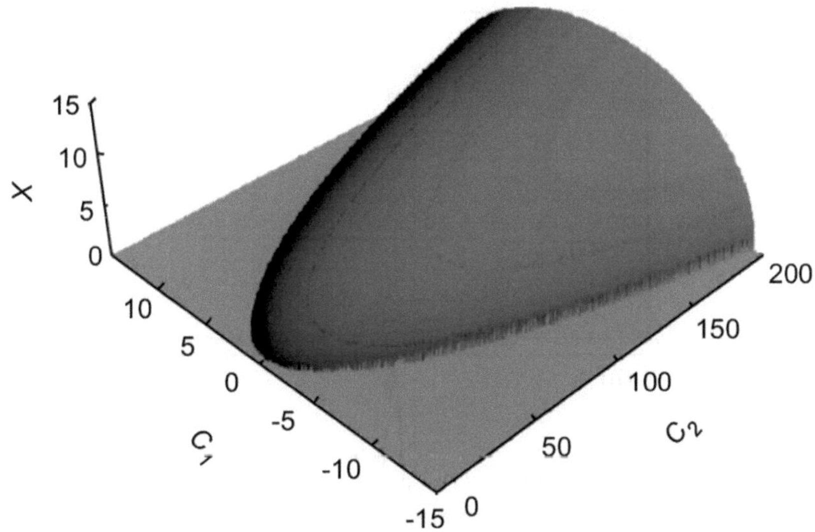

Fig. 4.43 Scaling factor χ as function of the parameters C_1 and C_2. The region with $\chi = 0$ corresponds to forbidden ($\sigma^0_{\alpha_f J_f} < 0$) values of C_1 and C_2. The parabolic boundary $C_1^2 = 1 + C_2$ between the regions with $\sigma^0_{\alpha_f J_f} > 0$ and $\sigma^0_{\alpha_f J_f} < 0$ corresponds to a single photoionization channel and, therefore, to a zero cross section (4.95) in the minimum "window" of the resonance profile. After Grum-Grzhimailo and Meyer (2006)

Fluorescence polarimetry provides further data for investigating the scaling phenomenon. We refer to the process (4.54) in the region of the strong $\mathrm{Xe}^*(4\mathrm{d}^{-1}_{5/2}6\mathrm{p})_{J=1}$ resonance and consider alignment and orientation of the ionic states $\mathrm{Xe}^{+*}\ (5\mathrm{p}^4(\mathrm{L}_c\mathrm{S}_c\mathrm{J}_c)6\mathrm{p}[K])_{J_f}$, as the T-parameters scanned through the resonance. Displayed in Fig. 4.45 are $\mathcal{A}_{20}$ and $\mathcal{A}_{10}$ as functions of the incident photon energy for two of the final ionic states. For comparison, the profile of the Xe^+ ion yield is shown in the upper panel of Fig. 4.45; its width and energy correspond to the integral cross section (4.95). As expected, the values $\tilde{\Gamma}$ and Δ extracted from the alignment and orientation of the ionic state are identical within the error bars, despite of completely different shapes of the corresponding resonance profiles. The resonance features in the parameters $\mathcal{A}_{20}$ and $\mathcal{A}_{10}$ are very broad pointing to the scaling factor χ of around 10, in accordance with the above discussion. The parameters C_1 and C_2 can be determined from the data (Grum-Grzhimailo et al. 2005), since the decay width of the Auger state, $\Gamma = 106.3(5)$ meV, is known (Masiu et al. 1995).

The scaling phenomenon has been also observed by Müller et al. (2006) in the photoelectron spin components for the 4d − 4f resonant photoemission from magnetized Gd and in the angular distribution of photoelectrons in the two-photon ionization of I (Tauro and Liu 2008).

The above described universal scaling phenomenon leads to important conclusions. For investigations of resonant states, both, experimentalists and theoreticians

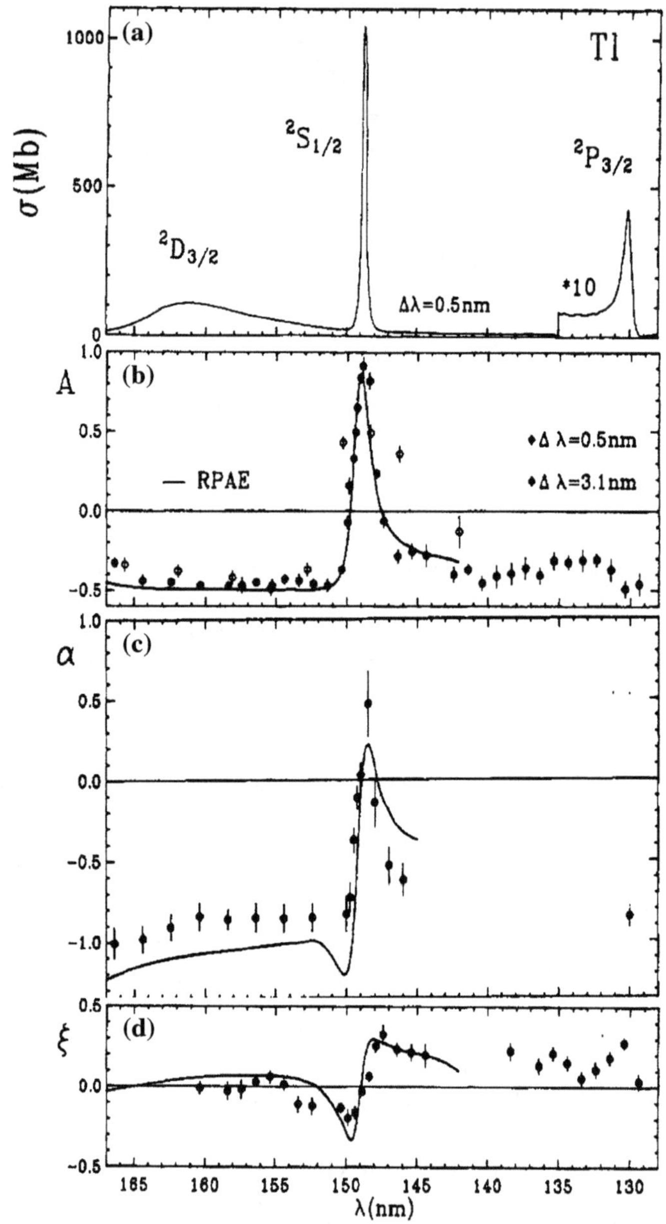

Fig. 4.44 Measured photoionization cross section σ of Tl ($6s^26p$) $^2P_{1/2}$ in the autionization region of the $6s6p^2$ configuration together with the experimental results (●) of the spin polarization parameters A, α, and ξ (Müller et al. 1990). (○): experimental data for the A parameter by Heinzmann et al. (1975); (——): RPAE calculation for A, ξ, and α by Cherepkov (1980), convoluted to the radiation bandwidth of the experiment ($\Delta\lambda = 0.5$ nm). After Müller et al. (1990)

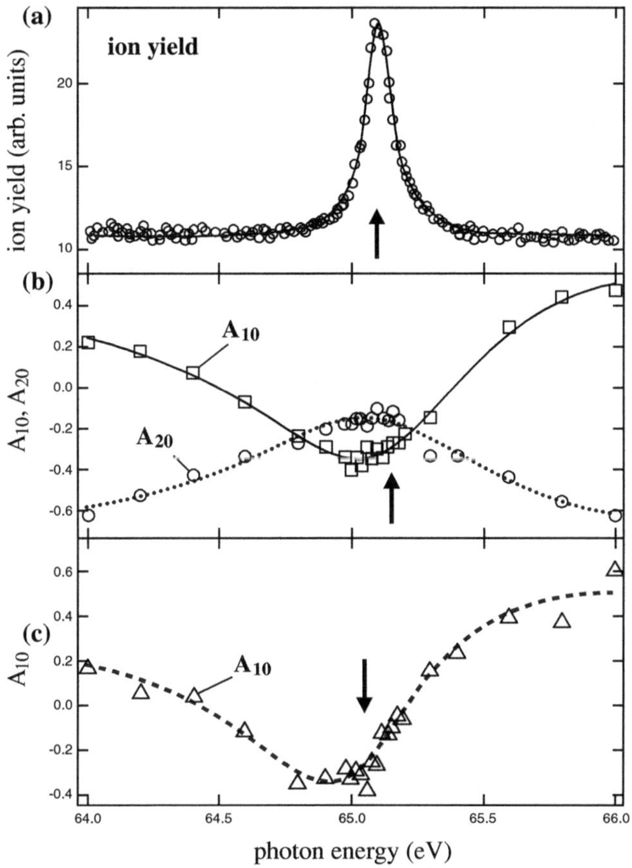

Fig. 4.45 **a** $Xe^*(4d_{5/2}^{-1}6p)_{J=1}$ resonance in the total ion yield; **b** alignment and orientation parameters of the photoion in the $(5p^4\,{}^3P_2\,6p[1])_{3/2}$ state; and **c** orientation parameter for the $(5p^4\,{}^1D_2\,6p[1])_{1/2}$ state. (States with $J = 1/2$ cannot be aligned). The energy dependence of the orientation and alignment parameters is fitted by (4.98). *Arrows* indicate the energy positions of the corresponding Fano profiles. After Grum-Grzhimailo et al. (2005)

can each study correlation parameters, see (4.96), which are most convenient for them, not necessarily the same. The comparison can, nevertheless, be made in terms of the scaled width $\tilde{\Gamma}$ (4.99) and the energy shift Δ (4.100). Moreover, measuring these two quantities yields the profile parameters C_1 and C_2 of the resonance profile in the integral cross section and vise versa, provided the decay width Γ is known. Thus, many independent studies turn out to be closely linked. Finally, a practical recipe for measuring the photoionization parameters for the complete experiments across the resonance is to choose the energy step of the incoming radiation in accordance with the width $\tilde{\Gamma}$, which can be much larger than the natural decay width Γ.

As follows from the above discussion, treating resonances with a weak direct photoionization background, separated even more than their natural widths, their resonance structures in the ionic alignment and orientation, as well as in other correlation parameters, can, nevertheless, still overlap strongly. Although the experimental observation of the effect is not straightforward, due to the small cross section between the resonances, the appearance of such phenomena can be revealed in combination with theoretical calculations. Such an example is provided by Lagutin et al. (2003) and Schartner et al. (2005) for alignment and orientation of the Kr II ($4p^4 5p$) states following Kr($3d^{-1}5p/6p$) resonance excitation.

Displayed in Fig. 4.46 are the alignment and orientation parameters of some of the Kr II ($4p^4 5p$) ionic states when the energy of the photon scans the region $h\nu = 91$–93 eV, where four resonance states are located. The strong resonance Kr$(3d_{5/2}^{-1}5p_{3/2})_{J=1}$ at $h\nu = 91.2$ eV is well isolated in the spectrum and the influence of the closest resonances on the cross section is negligible. The background of the direct ionization is very small in comparison with the resonant ionization. Similar to the case of the Xe$(4d_{5/2}^{-1}6p)_{J=1}$ resonance, the calculations show a large broadening of the resonance features in the alignment and orientation parameters in the region of the isolated Kr$(3d_{5/2}^{-1}5p_{3/2})_{J=1}$ Auger state. The influence of the interference with the direct photoionization is pronounced when the energy dependence is considered for the case of photoionization to the final ionic state with $J = 3/2$, see Fig. 4.46d, e, where the conjugate shake up is much stronger than in the case of the final ionic state with $J = 7/2$ (Fig. 4.46b, c). In the latter case, Fig. 4.46b, c, the interference between the resonances is more important. The interference effects are more pronounced for the weaker resonances at $h\nu = 92.2$–92.6 eV.

4.8.2 Photoinduced Resonant Auger Decay

We now turn to the case of a strong isolated resonance, where the direct photoionization can be completely neglected. This can happen when the energy of the photon is tuned to the resonance, or, when integral over the resonance quantities are measured. Then, the description of the photoionization can be further specified and reduced to the resonant Auger/autoionization decay of the photoexcited atomic state, lying in the continuum.

Consider ionization in the region of an isolated autoionizing resonance $|\alpha_r J_r\rangle$, characterized by the total angular momentum J_r and other quantum numbers α_r. Configuration interaction of the autoionizing state with the adjacent continuum is described by the matrix elements of the Coulomb interaction V

$$V(J_f \ell j, J_r) \equiv \langle (\alpha_f J_f, \ell j) J_r \| V \| \alpha_r J_r \rangle \,. \tag{4.101}$$

Using the theory of Fano (1961) and Fano and Cooper (1965), as applied to the angular distribution and spin polarization of photoelectrons in the region of the isolated resonance with vanishing probability of the direct photoionization, the amplitudes

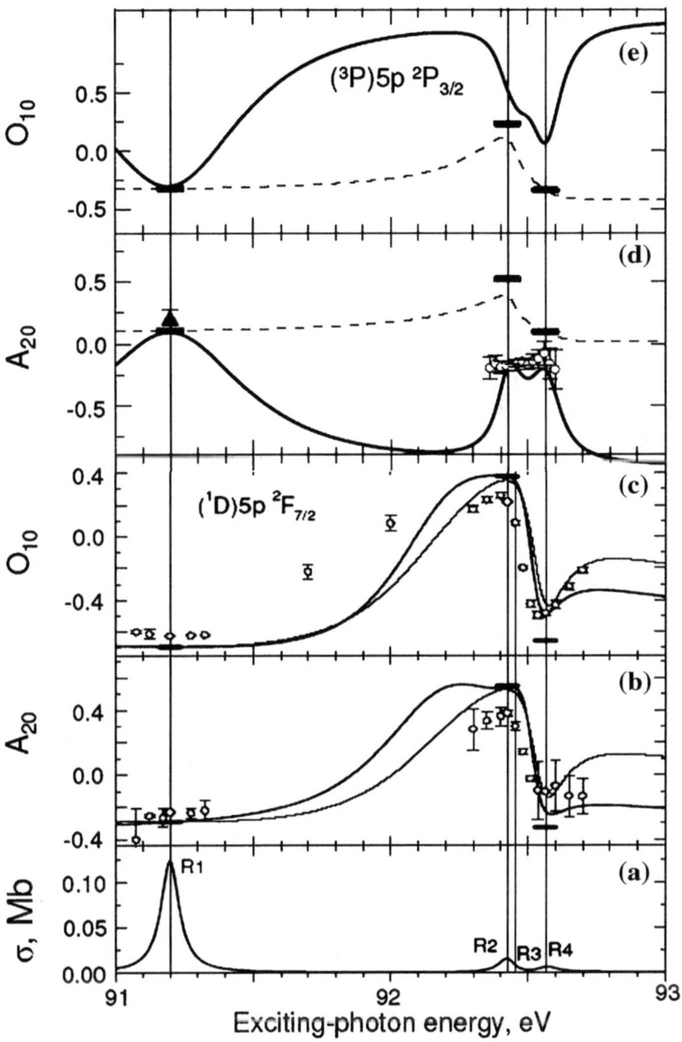

Fig. 4.46 Dependence of the alignment and orientation parameters on the excitation energy in the region of the Kr $J = 1$ Auger states: $4d_{5/2}^{-1}5p_{3/2}$ (R1), $4d_{5/2}^{-1}6p_{3/2}$ (R4), and perceptible mixtures of the $4d_{5/2}^{-1}5p_{1/2,3/2}$ states (R2, R3). (○): experimental data by Lagutin et al. (2003) and Schartner et al. (2005). Horizontal bars with lengths equal to the natural widths (the resonance R3 is negligible) indicate parameters computed within the two-step model for an isolated resonance. **a** calculated cross section; **b** and **c** thin lines, calculations neglecting interference between the resonances and between the resonant and direct amplitudes; thick lines, calculations accounting for the interference terms. **d** and **e** calculations with (——), and without (– – –) direct photoionization amplitude, respectively. Adapted from Schartner et al. (2005)

(4.15) in the channel with total angular momentum $J = J_r$ take the product form (Kabachnik and Sazhina 1976; Kabachnik et al. 1999)

$$D(J_f\ell j, J) = \langle \alpha_r J_r \| D \| \alpha_i J_i \rangle \frac{1}{E - E_r + i\Gamma_f/2} V(J_f\ell j, J_r)\, \delta_{J J_r}, \qquad (4.102)$$

where $\Gamma_f = 2\pi \sum_{\ell j} |V(J_f\ell j, J_r)|^2$ is the partial decay width of the resonance into the ionic state $|\alpha_f J_f\rangle$. Thus, the resonance selects only channels with sharp total angular momentum. Equation (4.102) symbolizes the two-step regime: the amplitude of the photoionization is factorized into a part describing the photoexcitation to the resonant state and a part describing the decay of this state. The Auger/autoionization decay amplitudes $V(J_f\ell j, J_r)$ are now becoming the principal objects of the complete experiment, because the first two factors are common for all non-zero partial amplitudes $D(J_f\ell j, J)$.

As an example, we discuss one of the first realizations of the complete experiment for the resonant Auger decay of the $\mathrm{Ar}^*(2\mathrm{p}^{-1}4\mathrm{s})_{J=1}$ photoexcited state (Ueda et al. 1999). In this experiment, the cascade of Ar, photoexcited to the $2\mathrm{p}_{3/2}^{-1}4\mathrm{s}$ state, by linearly polarized light has been studied

$$h\nu + \mathrm{Ar}(^1\mathrm{S}_0) \longrightarrow \mathrm{Ar}^*(2\mathrm{p}^5 3\mathrm{s}^2 3\mathrm{p}^6\, [^2\mathrm{P}_{3/2}]4\mathrm{s})_{J_r=1} \qquad (4.103\mathrm{a})$$

$$\hookrightarrow \mathrm{Ar}^{*+}(2\mathrm{p}^6 3\mathrm{s} 3\mathrm{p}^5\, [^1\mathrm{P}_1]4\mathrm{s})^2\mathrm{P}_{J_1} + e^-_{Auger_1}(l_1 j_1) \qquad (4.103\mathrm{b})$$

$$\hookrightarrow \mathrm{Ar}^{++}(2\mathrm{p}^6 3\mathrm{s}^2 3\mathrm{p}^4)^3\mathrm{P}_{J_2} + e^-_{Auger_2}(l_2 j_2)\ , \qquad (4.103\mathrm{c})$$

where $J_1 = 1/2, 3/2$ and $J_2 = 0, 1, 2$. The corresponding Auger lines at energy ~ 194 eV, transition (4.103b), and ~ 7 eV, transition (4.103c), have been observed earlier in an $e - e$ coincidence experiment by von Raven et al. (1990). The photon energy in (4.103a) has been tuned to the $\mathrm{Ar}(2\mathrm{p}_{3/2} \rightarrow 4\mathrm{s})$ excitation at 244.4 eV. The process (4.103a)–(4.103c), where the two Auger electrons are measured in coincidence, is an analogue of the photoelectron–Auger electron coincidences considered in Sect. 4.4. Different from the previous discussion is that the photoelectron is produced via the mechanisms of resonant Auger decay, i.e. during the *first-step* resonant Auger emission. For the closed-shell atom, only the value of the total angular momentum $J_R = 1$ is possible, which is valid also in the case of the direct photoionization. Therefore, all equations for the angular correlations derived for the direct photoionization remain valid for the Auger cascade (4.103a)–(4.103c) while taking into account as an additional equation (4.102). In practice, this leads to the substitution of $V(J_f\ell j, J_r = 1)$ instead of $D(J_f\ell j, J = 1)$ in the expressions for the photoionization parameters.

The angular distributions of, both, the first-step resonant Auger emission (4.103b) and the second-step Auger emission (4.103c), have been independently measured by

Ueda et al. (1999) together with the angular correlation between the two electrons. As in the case of direct photoionization (Sect. 4.4), the angular correlation was measured in the plane perpendicular to the photon beam. The first electron spectrometer has been mounted on a turntable and detected the resonant Auger electrons at kinetic energy of ~194 eV ejected in the first decay (4.103b). The second spectrometer was set in such a way that it detected Auger electrons at a kinetic energy of ~ 7 eV ejected in the second decay (4.103c) perpendicular to the linear polarization axis of the incident light. Figure 4.47 shows the non-coincident and coincident angular distributions of the first-step Auger electrons, which are given, correspondingly, by the expressions

$$I_1(\theta) = I_0^{(1)} \left(1 + \beta^{(1)} P_2(\cos\theta)\right), \tag{4.104}$$

and

$$I(\theta) = A_0 + A_2 \cos 2\theta + A_4 \cos 4\theta, \tag{4.105}$$

where θ is the angle between the light polarization vector and the emission direction of the resonant Auger electron $e^-_{Auger_1}$. The non-coincident angular distribution of the second-step Auger electrons $e^-_{Auger_2}$ are of the form similar to (4.104):

$$I_2(\theta) = I_0^{(2)} \left(1 + \beta^{(2)} P_2(\cos\theta)\right) . \tag{4.106}$$

By fitting expressions (4.104)–(4.106) to the experimental points the ratios of the coefficients A_2/A_0 and A_4/A_0 can be obtained together with the other two parameters $\beta^{(1)}$ and $\beta^{(2)}$. Thus, four parameters were experimentally determined. In the LSJ coupling approximation and neglecting spin-orbit interaction in the continuum, the four parameters are expressed in terms of only two independent quantities: the absolute ratio of the Auger decay amplitudes[2] V_d/V_s with the εd and εs electrons in the continuum, respectively, and the cosine of their phase difference $\Delta_{ds} = \delta_d - \delta_s$. This approximation is in close analogue with the Cooper–Zare model in direct photoionization.

To determine the two values from the experiment it is sufficient to measure only two independent parameters. For example, one can use the results of the coincidence experiment only: this variant is analogous to the complete experiment in direct photoionization made by Kämmerling and Schmidt (1991), which was discussed in Sect. 4.4. Alternatively, we can use the values $\beta^{(1)}$ and $\beta^{(2)}$ as obtained from the non-coincidence measurements. This variant is analogous to the complete experiment in direct photoionization made by Hausmann et al. (1988), which has been discussed in Sect. 4.3.2.

The most reliable values of the ratios V_d/V_s and of $\cos\Delta_{ds}$, however, can be obtained from the least squares method treating V_d/V_s and $\cos\Delta_{ds}$ as fitting parameters and four values $\beta^{(1)}$, $\beta^{(2)}$, A_2/A_0, and A_4/A_0 as data points to be fitted.

---

[2] A short notation for the decay amplitudes in analogy with (4.101) has been used.

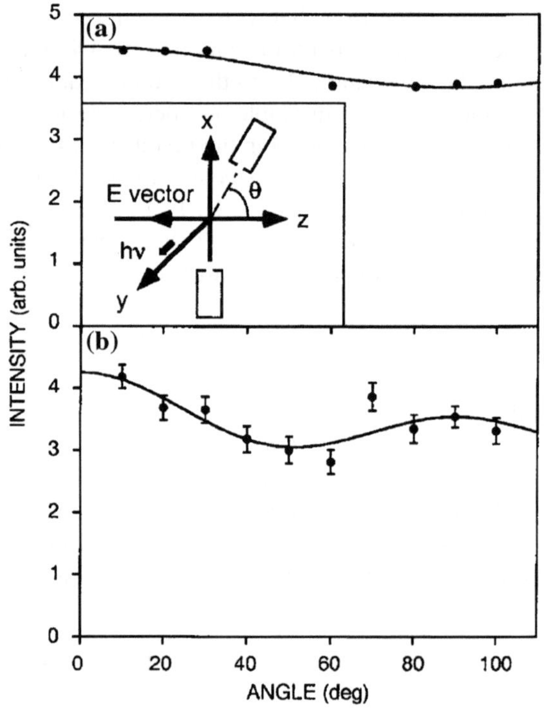

Fig. 4.47 Angular distributions of resonant Auger electrons ejected in the first-step decay of the Ar $2p_{3/2} \rightarrow 4s$ excitation; **a** without detecting the second-step Auger electrons, and **b** detecting the second-step Auger electrons in coincidence in the direction $\theta = 270°$. The *solid lines* in **a** and **b** correspond to the result of the fit using (4.104) and (4.105), respectively. In the *inset* the kinematics of the experiment is shown. After Ueda et al. (1999)

Some notes are appropriate to this general scheme of the complete experiment. First, the parameter $\beta^{(1)}$ is the average asymmetry parameter for the two decaying, experimentally unresolved, $^2P_{1/2,3/2}$ intermediate states (4.103b). In fact, they almost completely overlap, and in the analysis of the resonant Auger decay, (4.103a)–(4.103c), their energy splitting is ignored. Second, in the LSJ-coupling approximation, only one channel with emission of the $e^-_{Auger_2}(\varepsilon p)$ electron contributes to the decay (4.103c). In this case the amplitude of the second decay cancels out of any angular correlation coefficients, which allows to relate directly the asymmetry parameter $\beta^{(2)}$ to the dynamical parameters of the first-step resonant Auger decay (4.103a)–(4.103b). Furthermore, summing over the unresolved states of the final multiplet $^3P_{0,1,2}$ is necessary. Third, the analysis becomes more complicated due to the interference of the two overlapping intermediate states, which should be taken into account at the second-step Auger decay (4.103c). Doing that, equations derived by Kitajima et al. (2001) for the Auger decay of coherently excited states must be used.

The topic of the complete experiments for the resonant Auger decay within the two-step formulation is covered in reviews, for example Kabachnik (2004); Kabachnik et al. (2007).

4.9 Non-Dipole Effects

Over decades the dipole ($E1$) approximation was the basis for the theory of atomic and molecular photoionization in the VUV/soft-X-ray photon energy range. Though, early studies by Krause (1969), and Wuilleumier and Krause (1974), confirmed by many later measurements (see for example, Hemmers et al. 1996, 1997, 2003; Jung et al. 1996; Krässig et al. 1995, 2003; Lindle and Hemmers 1999; Martin et al. 1998; Ricz et al. 2003), showed that first order non-dipole contributions, originating from the interference $E1$-$E2$ and $E1$-$M1$ terms of (4.12), may substantially influence the angular distribution of the photoelectrons already at photon energies of some hundreds and even tens of eV. We also refer to reviews by Hemmers et al. (2004) and Guillemin et al. (2005) for the corresponding data. Since the end of the 1960s (Cooper and Manson 1969; Peshkin 1970), the non-dipole effects in atomic photoionization have been extensively studied theoretically. Besides the photoelectron angular distributions from isotropic targets, the non-dipole effects have been predicted for the macroscopic drug current (Amusia et al. 1974; Amusia 1984; Amusia et al. 2001), spin polarization of photoelectrons (Cherepkov and Semenov 2001a; Cherepkov et al. 2003; Amusia et al. 2005), and for the photoionization from polarized atoms in such phenomena as magnetic dichroism, and dichroism in the angular distribution of photoelectrons (Grum-Grzhimailo 2001). The influence of second-order non-dipole corrections on the angular distribution of photoelectrons has been also ascertained (Derevianko et al. 2000).

While, until today, complete photoionization experiments have been considered within the dipole approximation, non-dipole effects should be taken into account in future research within two respects. First, specific results may be modified by *unwanted* non-dipole parts of the photoionization amplitudes, and, second, in principle one can put a task to extract the non-dipole parts of the photoionization amplitudes.

The angular distribution of photoelectrons from polarized atoms expressing a symmetry axis (Grum-Grzhimailo 2001) may be described as

$$\frac{d\sigma}{d\Omega} = \frac{\sigma}{4\pi}\left(1 + \sum_{\substack{\pi L \pi' L' \\ k_0 k k_\gamma}} \mathcal{A}_{k_0 0}\, \beta^{\pi L,\pi' L'}_{k_0 k k_\gamma}\, F^{\pi L,\pi' L'}_{k_0 k k_\gamma}\right), \tag{4.107}$$

which is a generalization of (4.64) to arbitrary field multipoles. Here, the summation is performed over all contributing field multipoles πL, see (4.8). The geometrical factors $F^{\pi L,\pi' L'}_{k_0 k k_\gamma}$ are rotationally invariant, and the anisotropy parameters $\beta^{\pi L,\pi' L'}_{k_0 k k_\gamma}$ generalize the corresponding pure $E1$ dipole quantities of (4.64). In particular, we have the relations

$$F_{k_0 k k_\gamma} = F^{E1,E1}_{k_0 k k_\gamma}, \qquad \text{and} \qquad \beta_{k_0 k k_\gamma} = \beta^{E1,E1}_{k_0 k k_\gamma}. \tag{4.108}$$

In (4.107), σ is the total photoionization cross section with unpolarized light for randomly oriented atoms in the dipole $E1$ approximation, which is the same as in (4.64).

Restrictions on the summation indices in (4.107) follow from angular momentum and parity conservation:

$$\begin{aligned} k_\gamma &\leq L + L', \\ L + L' + \pi + \pi' + k &= \text{even}, \\ k_0 &\leq 2J_i, \\ k_0 &= \text{even for aligned atom}, \\ k_0 &= \text{odd and even for oriented atom}, \end{aligned}$$

where J_i is the total angular momentum of the initial atomic state. Some features of the angular distribution (4.107) can be established from these restrictions:

- In photoionization of aligned atoms, all odd-order non-dipole effects vanish when the photoelectron momentum and the initial atomic alignment lie in the plane perpendicular to the arbitrary polarized radiation beam.
- Only odd-order non-dipole corrections contribute into the magnetic dichroism in the angular distribution of photoelectrons when, both, the atomic orientation and the momentum of the photoelectrons lie in a plane perpendicular to the arbitrary polarized photon beam.
- The magnetic dichroism, observed with such geometrical arrangement, should be a pure non-dipole effect with the leading contribution from the first-order non-dipole corrections. In contrast, the odd-order non-dipole corrections into the magnetic dichroism vanish when the atomic orientation is parallel to the photon beam, while the momentum of the photoelectron is perpendicular to it.
- In ionization of polarized atoms, the azimuthal modulation of the photoelectron flux around the atomic polarization axis is a trigonometric polynomial, with exponents not higher than $L + L'$.

The above features can give a key to identification of the non-dipole effects. So far only unpolarized atoms were used in experimental studies of the non-dipole effects in atomic photoionization. Most important are the first-order non-dipole corrections originating from the interference of the electric dipole $E1$ amplitudes with electric quadrupole ($E2$) and magnetic dipole ($M1$) amplitudes. Within the first order non-dipole corrections, the angular distribution of photoelectrons (4.107) from unpolarized atoms, produced by linearly polarized light is usually parameterized by

$$\frac{d\sigma}{d\Omega} = \frac{\sigma}{4\pi}\Big(1 + \beta P_2(\cos\theta) + \big(\delta + \gamma\cos^2\theta\big)\sin\theta\cos\phi\Big), \tag{4.109}$$

(Cooper 1990, 1993), where the angle θ is counted relative to the polarization of the incoming radiation and ϕ is the azimuthal angle with respect to the direction of

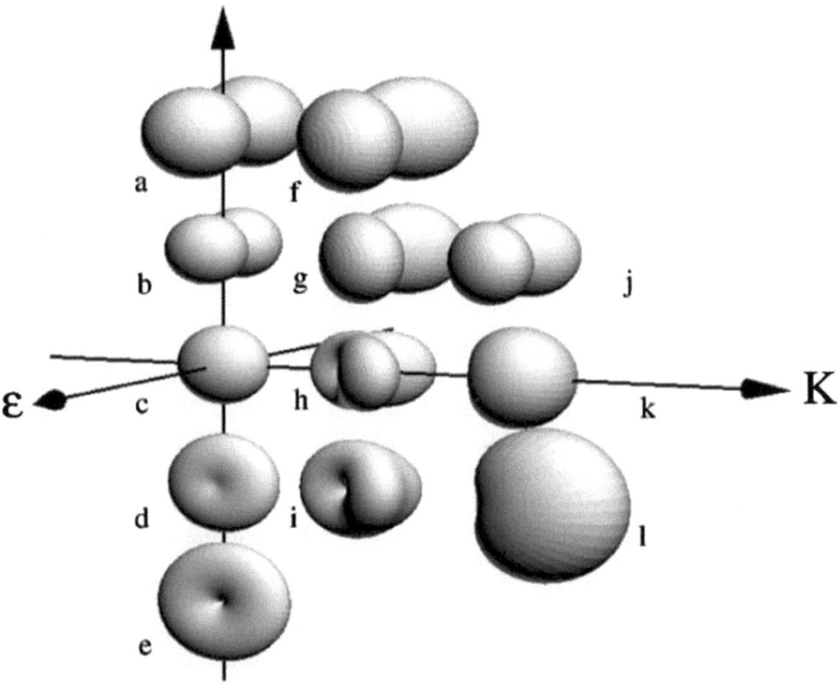

Fig. 4.48 Angular distribution patterns for different values of β, γ, and δ; see (4.109). K is the direction of the incoming photon beam, while ε denotes the axis of its linear polarization. The patterns show the photoelectron emission probability as a function of the emission angle. (*a*): $\beta = 2$ and $\gamma = \delta = 0$; (*b*): $\beta = 1$ and $\gamma = \delta = 0$; (*c*): $\beta = \gamma = \delta = 0$; (*d*): $\beta = -0.5$ and $\gamma = \delta = 0$; (*e*): $\beta = -1$ and $\gamma = \delta = 0$; (*f*): $\beta = 2$, $\gamma = 3$, and $\delta = 0$; (*g*): $\beta = 1$, $\gamma = 3.52$, and $\delta = 0$; (*h*): $\beta = 0$, $\gamma = \sqrt{27/4}$, and $\delta = 0$; (*i*): $\beta = -0.5$, $\gamma = 1.81$, and $\delta = 0$; (*j*): $\beta = 1$, $\gamma = 1.75$, and $\delta = 0.5$; (*k*): $\beta = 0$, $\gamma = 0.5$, and $\delta = 1$; (*l*): $\beta = -0.5$, $\gamma = 0$, and $\delta = \sqrt{27/4}$. After Hemmers et al. (2004)

the photon beam. In addition to the single asymmetry parameter β for the dipole approximation (4.21), two new non-dipole parameters, δ and γ, appear.

Figure 4.48 shows the variety of the angular distribution patterns (4.109). The pure dipole $E1$ patterns for different values of β are located on the vertical axis. The forward-backward symmetry of the angular distributions with respect to the direction of the photon beam is violated due to the non-dipole contributions leading to the drug current. The non-dipole term in (4.109) vanishes for photoelectrons emitted in the plane perpendicular to the incident photon beam. Therefore this plane is sometimes called the *dipole plane*. At the *magic angle*, $\theta = 54.7°$, the asymmetry in (4.109) remains only due to the non-dipole contribution. For the magic angle the combination

$$\zeta = 3\delta + \gamma \tag{4.110}$$

is measured.

The second-order corrections to the angular distribution of photoelectrons from unpolarized atoms arise from the interference terms $E1$-$E3$, $E1$-$M2$, $E2$-$E2$, $E2$-$M1$, $M1$-$M1$ and lead to the expression (Derevianko et al. 2000)

$$\frac{d\sigma}{d\Omega} = \frac{\sigma}{4\pi}\Big[1 + \big(\beta + \Delta\beta\big) P_2(\cos\theta) + \big(\delta + \gamma\cos^2\theta\big)\sin\theta\cos\phi + \eta P_2(\cos\theta)\cos 2\phi + \mu\cos 2\phi + \xi\big(1 + \cos 2\phi\big) P_4(\cos\theta)\Big]. \quad (4.111)$$

The second-order terms are associated mainly with the $E2$-$E2$ and $E1$-$E3$ contributions.

The non-dipole effects in the angular distribution (4.109) have been observed many times; they were investigated theoretically in the VUV and soft X-ray range using different atomic models by Bechler and Pratt (1989, 1990), Amusia et al. (1999), Nefedov et al. (2000), Dolmatov and Manson (2001) and others. Extensive calculations through the periodic table are available for the direct photoionization not only for the first-order non-dipole correction parameters γ and δ (Trzhaskovskaya et al. 2001, 2002), but also for the second-order parameters $\Delta\beta$, η, μ, ξ (Trzhaskovskaya et al. 2006) in (4.111).

We illustrate the non-dipole effects in the photoelectron angular distributions by a few examples. Figure 4.49 demonstrates a remarkable quadrupole autoionizing $E2$ resonance in the non-dipole asymmetry parameter γ. This resonance is completely unseen in the integral cross section, where the first-order corrections due to the $E1$-$E2$ interference vanish. Another resonance in the $E1$ dipole amplitude gives rise to the second structure in the non-dipole parameter. Note that the photon energies here are only a few tens of eV, which is already enough to produce strong non-dipole effects in the angular distribution of photoelectrons.

A pronounced variation of the non-dipole parameter γ for photoionization of the Xe 5s electron is shown in Fig. 4.50 in the region around the Cooper minimum for the ionization of the $4d^{10}$ subshell. Theoretically this variation can be reproduced only by accounting of the interchannel mixing, which includes 5s, 5p and 4d ionization channels.

Figure 4.51 illustrates the second-order non-dipole effects, becoming visible approximately from 1 keV in photoionization of the 2p subshell in Ne.

New observable quantities, such as the photoelectron spin polarization and magnetic dichroism in the angular distributions of photoelectrons can give additional independent photoionization non-dipole parameters. Preferential geometries for measurement of non-dipole effects in the magnetic dichroism in the angular distributions of photoelectrons have been analyzed by Grum-Grzhimailo (2001) and predictions have been given for the corresponding non-dipole parameters for ionization of laser excited sodium atoms (see Fig. 4.52). Thus, the dominating contribution in the CMDAD, (4.70), and the LMDAD, which are pure non-dipole effects in the geometry of the upper panels in Fig. 4.52, are given, respectively, by

$$\mathrm{CMDAD} = \frac{\sigma}{2\pi}\mathcal{A}_{10}(\lambda_s \sin\psi + \lambda_c \cos\psi), \quad (4.112)$$

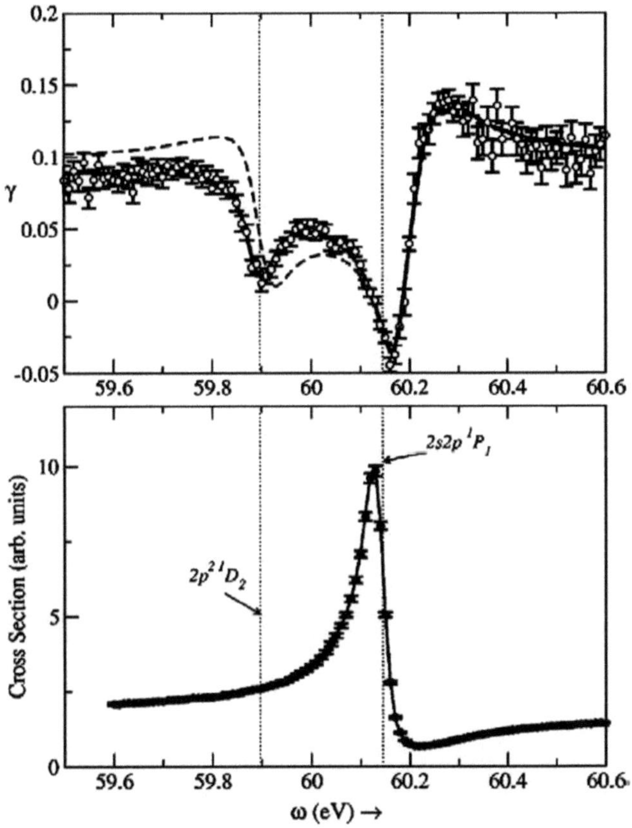

Fig. 4.49 Energy dependence of the total cross section (*bottom*), and asymmetry parameter γ (*top*) in the region of the helium (2s2p) $^1P_1$ and ($2p^2$) $^1D_2$ autoionizing levels. After Kanter et al. (2003)

and

$$\mathrm{LMDAD}_{\|(\perp)} = \frac{\sigma}{2\pi} \mathcal{A}_{10}(\nu_{\|(\perp)} \sin\psi + \nu'_{\|(\perp)} \sin 3\psi), \tag{4.113}$$

where $\mathcal{A}_{10}$ is the orientation parameter of the initial atomic state. The non-dipole parameters, including those for the angular distribution of photoelectrons from unpolarized atoms (δ and γ) are shown at the lower panels. They take rather large values for the photon energies as low as 10 eV (compare to the upper panel of Fig. 4.49). One of the reasons is an enhancement of the quadrupole amplitude of the 3p $\rightarrow$ εf transition close to the ionization threshold due to a strong overlap between the 3p and εf electron orbitals close to ionization threshold. Resonances in quadrupole channels corresponding to the np $\rightarrow$ εf transitions, revealed theoretically (Grum-Grzhimailo 2001; Cherepkov and Semenov 2001b; Cherepkov et al. 2003), lead to a prominent enhancement of all non-dipole effects in the angular distribution, magnetic dichroism, and spin polarization of photoelectrons at rather low photon energies.

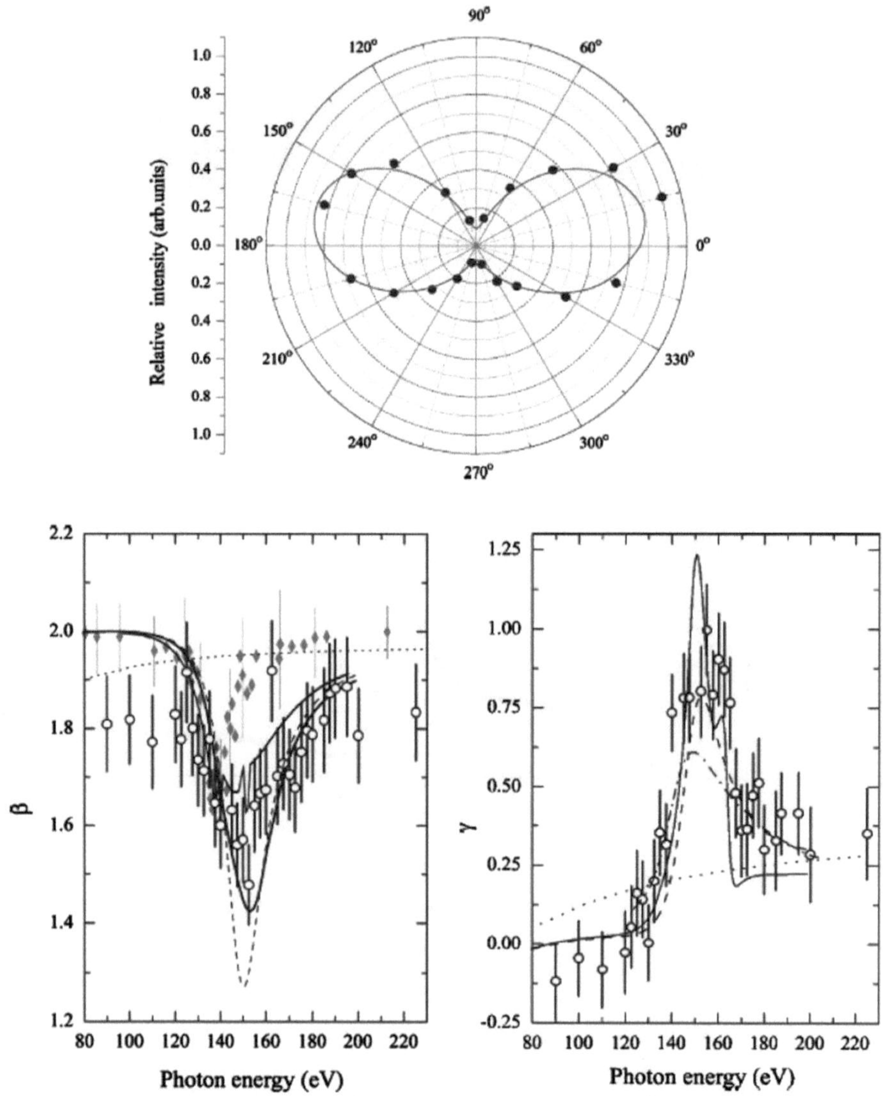

Fig. 4.50 Polar plot of the relative intensity (*upper panel*), and angular distribution parameter β of the Xe 5s photoelectron line (*left panel*), relative to the polarization vector measured by Ricz et al. (2003). (•): experimental data at 150 eV photon energy, (——): least-squares fit result. (*Right panel*): Comparison of the experimental non-dipole γ parameters, (○), with the corresponding theoretical values as a function of the photon energy. Theory (······): relativistic independent-particle model of Derevianko et al. (1999); (– · –): 13-channel random phase approximation with exchange of Amusia et al. (2001); (– – –): 13-channel and (——): 20-channel relativistic random phase approximation by Johnson and Cheng (2001). Adapted from Ricz et al. (2003)

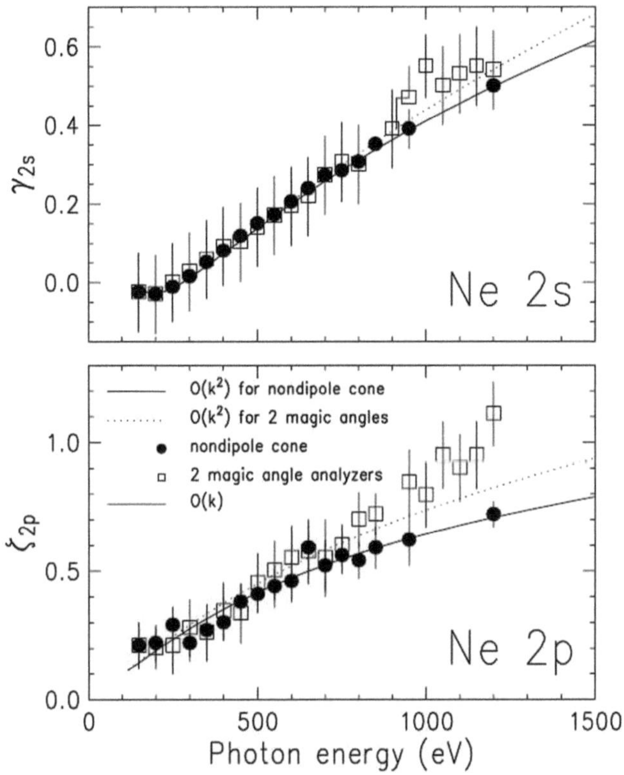

Fig. 4.51 Experimental and theoretical values of γ_{2s} (*top*), and ζ_{2p} (*bottom*), as a function of the photon energy for neon determined under different geometrical conditions. (□) and (······) relate to the magic angle geometry, where second-order effects are included; while (●) and (——) relate to the non-dipolar cone-geometry, where second-order effects are effectively excluded. Both, (······) and (——) curves include the second-order effects. (——) also represents first-order theory, independent of geometry. After Derevianko et al. (2000)

Large spin polarization of the photoelectrons is predicted in some cases, when it vanishes in the dipole approximation. For example, Fig. 4.53 shows a geometry with a linearly polarized radiation beam, where photoelectrons cannot be polarized in the dipole approximation, but according to theoretical predictions (Cherepkov et al. 2003) may be completely polarized when taking into account first-order non-dipole corrections. Unfortunately, the intensity of the photoelectron flux in the cases of large spin polarization are small and the effect is difficult to observe.

Eventually, we are drawing attention to non-dipole effects in molecular photoionization. In a pioneering study, Tully et al. (1968) obtained expressions for the angular distribution of photoelectrons from randomly oriented diatomic molecules. Few other papers went beyond the electric-dipole approximation, while studying molecular optical activity effects in the angular distribution and spin polarization produced

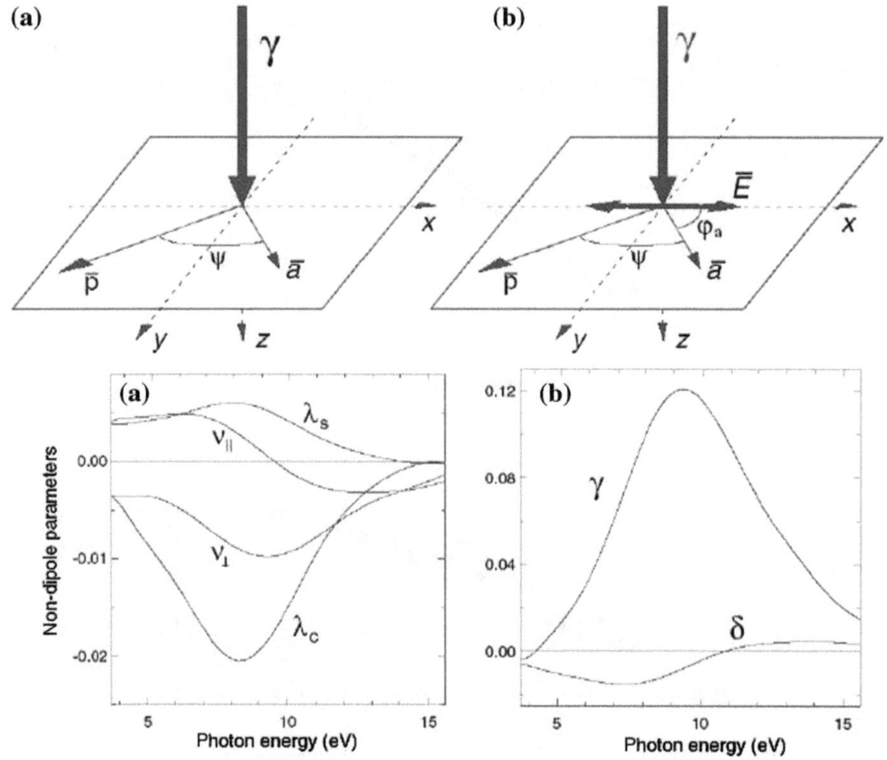

Fig. 4.52 *(Upper panels)* geometry for experiments to observe nondipole effects in magnetic dichroism in the angular distribution of photoelectrons for a circularly polarized photon beam (*CMDAD*) (**a**), and for a linearly polarized photon beam (*LMDAD*) (**b**). The momentum **p** of the photoelectron and the initial polarization of the target atom **a** are in the xy-plane; the photon beam γ is normal to the xy-plane incidenting along the z-axis. *(Lower panels)*: nondipole parameters for photoionization of the Na ($3p_{3/2}$) state as functions of the photon energy. **a** the non-dipole *CMDAD* and *LMDAD* parameters, see (4.112) and (4.113); **b** the non-dipole angular distribution parameters, see (4.109). Adapted from Grum-Grzhimailo et al. (2001)

by circularly polarized radiation (Ritchie 1976; Cherepkov 1982). A general description of photoelectron angular distribution from both, randomly oriented and fixed-in-space molecules, has been developed by Grum-Grzhimailo (2003). For randomly oriented, non-chiral molecules, the angular distribution of photoelectrons is described by the same equations as for unpolarized atoms. For example, for linearly polarized incoming radiation, the angular distribution within first-order non-dipole corrections takes the form (4.109). Numerical calculations and measurements of non-dipole parameters for particular molecules so far are rather limited (Seabra et al. 2005; Hemmers et al. 2005, 2006; Hosaka et al. 2006a, b; Southworth et al. 2006; Cherepkov and Semenov 2007; Grum- Grzhimailo et al. 2007a; Toffoli and Decleva 2006, 2008a, b; Bolognesi et al. 2008) and measurements are mostly related to the N_2 molecule.

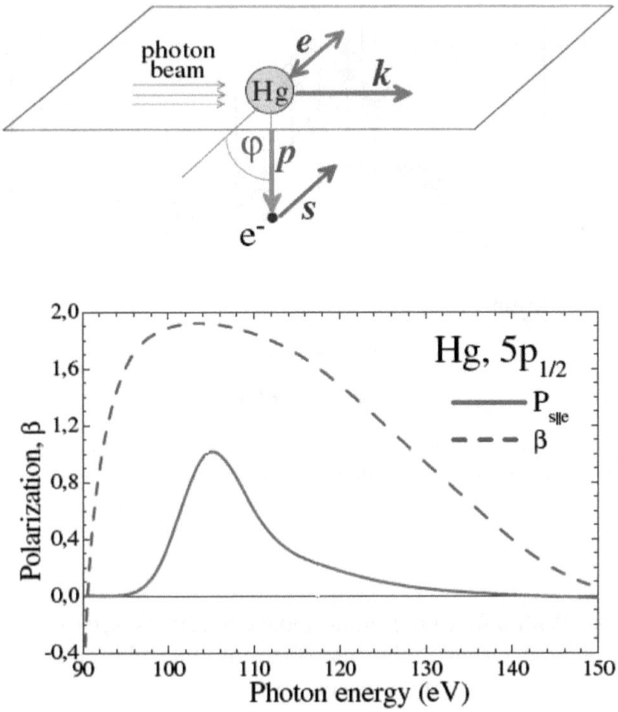

Fig. 4.53 (*Bottom*) the degree of transverse polarization of the photoelectrons (——) ejected from the $5p_{1/2}$ subshell of Hg by linearly polarized light for the displayed geometry (*top*), where **k**, **e**, **p** and **s** denote the directions of the photon beam, its linear polarization, photoelectron emission and spin polarization component, respectively. The asymmetry parameter β in the angular distribution of the photoelectrons is also shown (— — —). Adapted from Cherepkov et al. (2003)

Figure 4.54 demonstrates a good correspondence between theory and experiment for the non-dipole parameter ζ, see (4.110), in photoionization of the $3\sigma_g$ state of the N_2 molecule (Hemmers et al. 2005). Toffoli and Decleva (2006) made theoretical predictions also for higher photon energies. The pronounced oscillations in the value of the parameter ζ as a function of the photon energy take their origin from the bi-centric nature of the electron–ion interaction. According to Cohen and Fano (1966), the modulations superimpose on the slowly varying background due to the $(1 + \sin kR)/kR$ term, where R being the equilibrium bond length and k being the electron wave vector.

Especially intriguing is an interplay between non-dipole contributions and molecular chirality. The effect of circular dichroism (optical activity) in photoabsorption by randomly oriented chiral molecules, which is dominantly a result of the E1-M1 interference in the interaction of molecules with radiation, has long been known (Caldwell and Eyring 1971). Despite of this, very little has been done for a detailed understanding of the related class of phenomena in the differential characteristics of

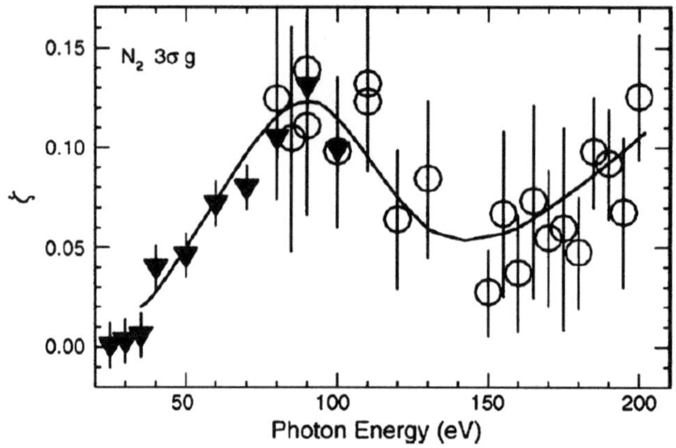

Fig. 4.54 Experimental data (▼): Wisconsin Synchrotron Radiation Center, and (○): Advanced Light Source of the Lawrence Berkeley National Laboratory, of the N_2 $3\sigma_g$ nondipole anisotropy parameter ζ in comparison to the "frozen-core" Hartree-Fock calculations.After Hemmers et al. (2005)

the photoprocess. Within first-order non-dipole corrections, see (4.109), the angular distribution of photoelectrons in the case of chiral molecules gets a new '*rotatory*' term, breaking the symmetry with respect to the plane perpendicular to the linear polarization of the incident radiation (Grum-Grzhimailo 2003)

$$\frac{d\sigma}{d\Omega} = \frac{\sigma}{4\pi}\Big(1 + \beta P_2(\cos\theta) + \big(\delta + \gamma\cos^2\theta\big)\sin\theta\cos\phi + \epsilon\sin 2\theta\sin\phi\Big)\ . \quad (4.114)$$

The contribution from the last term so far has not been observed and favorable targets and conditions to reveal the effect have not been analyzed. Other possible non-dipole effects in photoionization of chiral molecules have been discussed (Grum-Grzhimailo 2003; Grum-Grzhimailo et al. 2007b).

Eventually, we remark that for incoming circularly polarized radiation the angular distribution of photoelectrons from randomly oriented chiral molecules, already in the pure E1 dipole approximation, gets a new term proportional to $\cos\theta$, where θ is the photoelectron emission angle with respect to the incoming beam. This term produces an asymmetry in the foreward-backward photoelectron emission induced by circularly polarized light. Such an effect has been observed by Böwering et al. (2001) and Garcia et al. (2003).

Further extensions of complete photoionization experiments beyond the dipole approximation remain to be developed.

Chapter 5
Concluding Remarks

Starting from the end of the 1960s, when the concept of a perfect and-or complete experiment began to be fitted, the field have been growing vigorously, covering more and more atomic and molecular processes. It is not possible and it was not our task to cover the whole this rapidly progressing field in one book; we deeply apologize to those authors whose work we were not able to quote because of its limited scope. Selectively, only various parts of the physics indicated in the title of the book have illustratively been described and we regret to admit that we are not in the position to mention important sub-parts of the field.

As example we refer to impressive beginnings in approaches to complete experiments in molecular collision physics, based on vector correlations and imaging techniques. These developments lead to stereodynamics of molecular reactions and full information of the elementary process under study, as overviewed for example by Clark et al. (2006), Suits and Vasyutinskii (2008), Chichinin et al. (2009). We cannot ignore here complete experiments for molecular photoionization. Due to the violation of the spherical symmetry in molecules, the dipole selection rules do not restrict, in contrast to atomic photoionization discussed in this book, the number of partial-wave photoionization amplitudes. However, in practice the partial-wave expansion converges rather rapidly and this makes the complete experiment for molecular photoionization feasible (Allendorf et al. 1989; Leahy et al. 1991, 1992; Reid et al. 1992; Hikosaka and Eland 2000; Cherepkov et al. 2000; Cherepkov 2001; Adachi et al. 2002; Motoki et al. 2002; Geßner et al. 2002, and others), especially after introducing in the 1990s the technique of angle resolved photoelectron-photoion coincidence spectroscopy (AR-PEPICO) and the velocity imaging photoionization coincidence technique (Golovin 1991; Golovin et al. 1992; Shigemasa et al. 1995; Guyon et al. 1996; Heiser et al. 1997; Dörner et al. 1998; Hanold and Continetti 1998; Downie and Powis 1999; Lafosse et al. 2000; Takahashi et al. 2000). The AR-PEPICO, which in a few different options rapidly became a common experimental method, correlates the direction of the photoelectron emission with spacial orientation of the ionizing molecule, realizing photoionization of *fixed-in-space* molecules. The method works within the so-called axial recoil approximation, which means that the rotation of the

H. Kleinpoppen et al., *Perfect/Complete Scattering Experiments*,
Springer Series on Atomic, Optical, and Plasma Physics 75,
DOI: 10.1007/978-3-642-40514-3_5,

molecule has to be much slower than its fragmentation velocity. Complex angular distributions of photoelectrons from fixed-in-space photoelectron molecules give an opportunity to find the photoionization dipole amplitudes by fitting their absolute ratios and relative phases, as demonstrated by Shigemasa et al. (1998), Geßner et al. (2002), Lebech et al. (2003), Teramoto et al. (2007). Another method combining laser and electrostatic fields allows to fix the orientation of a broad class of molecules in the gaseous phase before photoionization (Auzinsh and Ferber 1995; Holmegaard et al. 2010).

Even those topics, which are more completely covered in the present book, omit some important points. We did not refer to Bederson's W-parameter (Andersen et al. 1997; Goldstein et al. 1972) by which, in early pioneering experiments, fractional depolarizations of polarized excited atomic beams, resulting from decaying back to the ground state, were measured. As a selected example we also did not refer to stepwise electron and laser excitation of atoms (MacGillivray and Standage 1988; Wang et al. 1995) by which a combination with the electron-photon coincidence techniques was established. Other areas we missed out, e.g. there are inner shell vacancy productions (Wille and Hippler 1986); symmetry properties and conservations laws for collisional excitation with planar symmetry (Andersen and Hertel 1986); first determination of rank 4 multipoles by a polarized photon–photon coincidence technique (Mikosza et al. 1993, 1997); analysis of individual contributions from channels with different symmetries in electron–electron coincidence in the single-photon double ionization (Malegat et al. 1997; Briggs and Schmidt 2000; Cvejanovic and Reddish 2000; Bolognesi et al. 2003; Avaldi and Huetz 2005) and in the direct double Auger decay Auger decay (Viefhaus et al. 2004a, b). Experimental and theoretical results showed that electron dichroism exists for chiral molecules containing heavy atoms (e.g. Hanne 2000; Kessler 1996). In this connection we draw attention to an electron spin effect detected in solid state physics, namely the effect observed by spin polarized electron energy loss spectroscopy with ultrathin Co films on copper (001)-crystals (Vollmer et al. 2003); the connection of this solid state effect to electron spin effects in collision physics has still to be worked out.

We also draw attention to complete experiments in intense laser fields, chemical effects in $(e, 2e)$ processes, as well as $(e, 2e)$ experiments from atomic solid state physics, determination of 16-pole moments, spin exchange in spin polarized atoms by ion impact, spin asymmetry measurements, relativistic effects in polarized electron-atom collisions, quantum chaos, dissociative recombination, spin polarized metastable atomic hydrogen, and Paul trap resonances in He^+-He collisions. This list goes on and on.

Complete experiments described in this book are based on stationary approach to atomic and molecular processes. In particular, the concept of complete experiment is formulated in terms of time-independent amplitudes. A relevance of a new language is connected with the revolutionary breakthrough into femtosecond and attosecond time domain with new radiation sources, such as high harmonic generation (HHG) and free-electron lasers (FELs) generating short wavelength pulses. Combining ultrafast extreme ultraviolet/X-ray lasers and/or optical lasers in the pump-probe arrangement provides an opportunity to follow atomic and molecular evolution with femtosecond time resolution which brings the need for description of the processes in terms of evolving density matrices and wave packets.

We may sum up the field by mentioning regular international conferences presenting many new developments in the field of complete experiments and related fields, such as the bi-annual International Symposium on Polarization and Correlation in Electronic and Atomic Collisions held together with the $(e, 2e)$, Double Photoionization and Related Topics Symposium; the International Conference on Photonic, Electronic and Atomic Collisions (ICPEAC); the International Conference on Multiphoton Processes; the International Conference on Many Particle Spectroscopy of Atoms, Molecules, Clusters and Surfaces; the International Conference on X-Ray and Inner-Shell Processes held recently together with the International Conference on Vacuum Ultraviolet Radiation Physics; the European Conference on Atomic and Molecular Physics, and many others. This indicates that a kind of renaissance of research on atomic physics had taken place during about the second part of the twentieth century and the beginning of the new millennium, providing most detailed knowledge of elementary processes, which are the core of our understanding of many phenomena in nature and technologies. Let us go farther and endeavour.

References

Åberg, T., & Howat, G. (1982). Theory of the Auger effect. In S. Flügge & W. Mehlhorn (Eds.), *Handbuch der Physik, 31* (p. 469). Berlin: Springer.

Adachi, J., Motoki, S., Cherepkov, N. A., & Yagishita, A. (2002). *Journal of Physics B: Atomic, Molecular and Optical Physics, 35*, 5023.

Aksela, S., Aksela, H., Levasalmi, M., Tan, K. H., & Bancroft, G. M. (1987). *Physical Review A, 36*, 3449.

Aksela, H., Sairanen, O.-P., Aksela, S., Kivimäki, A., Nõmmiste, E., Tulkki, J., et al. (1995). *Physical Review A, 51*, 1291.

Aksela, H., Jauhiainen, J., Nõmmiste, E., Sairanen, O.-P., Karvonen, J., Kukk, E., et al. (1996). *Physical Review A, 54*, 2874.

Aksela, H., & Mursu, J. (1996). *Physical Review A, 54*, 2882.

Aksela, H., Mursu, J., Jauhiainen, J., Nõmmiste, E., Karvonen, J., & Aksela, S. (1997). *Physical Review A, 55*, 3532.

Alguard, M. J., Hughes, V. W., Lubell, M. S., & Wainwright, P. F. (1977). *Physical Review Letters, 39*, 334.

Allendorf, S. W., Leahy, D. J., Jacobs, D. C., & Zare, R. N. (1989). *Journal of Chemical Physics, 91*, 2216.

Aloïse, S., O'Keeffe, P., Cubaynes, D., Meyer, M., & Grum-Grzhimailo, A. N. (2005). *Physical Review Letters, 94*, 223002.

Al-Shamma, S. H., & Kleinpoppen, H. (1978). *10th International Conference on Physics of Electronic and Atomic Collisions, Paris* (p. 518).

Amusia, M. Ya. (1996). In U. Becker & D. A. Shirley (Eds.), *VUV and soft X-ray photoionization* (p. 1). Plenum Press: New York.

Amusia, M. Ya., Arifov, P. U., Baltenkov, A. S., Grinberg, A. A., & Shapiro, S. G. (1974). *Physics Letters A, 47*, 66.

Amusia, M. Ya. (1984). *Comments on Atomic and Molecular Physics, 14*, 103.

Amusia, M. Ya. (1990). *Atomic photoeffect*. New York: Plenum Press.

Amusia, M. Ya., & Chernysheva, L. V. (1997). *Computation of atomic processes—A handbook for the ATOM programs*. Bristol: IOP Publishing.

Amusia, M. Ya., Baltenkov, A. S., Felfli, Z., & Msezane, A. Z. (1999). *Physical Review A, 59*, R2544.

Amusia, M. Ya., Baltenkov, A. S., Chernysheva, L. V., Felfli, Z., Msezane, A. Z., & Nordgren, J. (2001). *Physical Review A, 63*, 052512.

Amusia, M. Ya., Cherepkov, N. A., Chernysheva, L. V., Felfli, Z., & Msezane, A. Z. (2005). *Journal of Physics B: Atomic, Molecular and Optical Physics, 38*, 1133.

Andersen, N., & Hertel, I. V. (1986). *Comments on Atomic and Molecular Physics, 19*, 1.

H. Kleinpoppen et al., *Perfect/Complete Scattering Experiments*, Springer Series on Atomic, Optical, and Plasma Physics 75, DOI: 10.1007/978-3-642-40514-3, © Springer-Verlag Berlin Heidelberg 2013

Andersen, N., Gallagher, J. W., & Hertel, I. V. (1988). *Physics Reports*, *165*, 1.
Andersen, N., Bartschat, K., Broad, J. T., & Hertel, I. V. (1997). *Physics Reports*, *278*, 107.
Andersen, N., & Bartschat, K. (2000). *Polarization, alignment, and orientation in atomic collisions*. Heidelberg: Springer.
Archard, G. D. (1963). Private communication.
Asaad, W. N. (1959). *Proceedings of the Royal Society A*, *249*, 555.
Asaad, W. N. (1963a). *Nuclear Physics*, *44*, 399.
Asaad, W. N. (1963b). *Nuclear Physics*, *44*, 415.
Asaad, W. N. (1965). *Nuclear Physics*, *66*, 494.
Auzinsh, M., & Ferber, R. (1995). *Optical polarization of molecules*. Cambridge: Cambridge University Press.
Auzinsh, M., Budker, D., & Rochester, S. M. (2010). *Optically polarized atoms*. Oxford: Oxford University Press.
Avaldi, L., Dawber, G., Camilloni, R., King, G. C., Roper, M., Siggel, M. R. F., et al. (1994). *Journal of Physics B: Atomic, Molecular and Optical Physics*, *27*, 3953.
Avaldi, L., & Huetz, A. (2005). *Journal of Physics B: Atomic, Molecular and Optical Physics*, *38*, S861.
Awe, B., Kemper, F., Rosicky, F., & Feder, F. (1983). *Journal of Physics B: Atomic, Molecular Physics*, *16*, 603.
Aymar, M., Greene, C. H., & Luc-Koenig, E. (1996). *Reviews of Modern Physics*, *68*, 1015.
Baier, S., Grum-Grzhimailo, A. N., & Kabachnik, N. M. (1994a). *Journal of Physics B: Atomic, Molecular and Optical Physics*, *27*, 3363.
Baier, S., Schulze, M., Staiger, H., Zimmermann, P., Lorenz, C., Pahler, M., et al. (1994b). *Journal of Physics B: Atomic, Molecular and Optical Physics*, *27*, 1341.
Balashov, V. V., Grum-Grzhimailo, A. N., Dolinov, V. K., Krementsova, Yu. N., Smirnov, Yu. F., & Yudin, N. P. (1984). *Theoretical practicum in nuclear and atomic physics*. Moscow: Energoatomizdat.
Balashov, V. V., Grum-Grzhimailo, A. N., & Zhadamba, B. (1988). *Optika i Spektroskopiya*, *65*, 529.
Balashov, V. V., Grum-Grzhimailo, A. N., & Kabachnik, N. M. (2000). *Polarization and correlation phenomena in atomic collisions. A practical theory course*. New York: Kluwer Academic/Plenum Publishers.
Bambynek, W., Crasemann, B., Freund, H. U., Mark, H., Price, R. E., & Swift, C. D. (1972). *Reviews of Modern Physics*, *44*, 716.
Band, I. M., Kharitonov, Yu. I., & Trzhaskovskaya, M. B. (1979). *Atomic Data and Nuclear Data Tables*, *23*, 443.
Bartschat, K., Hudson, E. T., Scott, M. P., Burke, P. G., & Burke, V. M. (1996). *Journal of Physics B: Atomic, Molecular and Optical Physics*, *29*, 2875.
Bates, D. R., & Williams, J. F. (1965). *Proceedings of the Physical Society*, *83*, 425.
Baum, G., Freienstein, P., Frost, L., Granitza, B., Raith, W., & Steidl, H. (1990). In P. A. Neill, K. Becker, & M. H. Kelley (Eds.), National Institute of Standards and Technology Special Publication, 789, p. 121. Gaithersburg.
Baum, G., Moede, M., Raith, W., & Schröder, W. (1985). *Physical Review B*, *18*, 531.
Baum, G., Moede, M., Raith, W., & Sillmen, U. (1986). *Physical Review Letters*, *57*, 1885.
Baum, G., Raith, W., & Steidl, H. (1988). *Zeitschrift für Physik D*, *10*, 171.
Baum, G., Fink, M., Raith, W., Steidel, H., & Taborski, J. (1989). *Physical Review A*, *40*, 6734.
Baum, G., Fink, M., Raith, W., Steidel, H., & Taborski, W. (1993). *Journal of Physics B: Atomic, Molecular and Optical Physics*, *26*, 331.
Baum, G., Raith, W., Roth, B., Tondera, M., Bartschat, K., Bray, I., et al. (1999). *Physical Review Letters*, *82*, 1128.
Baum, G., Pavlovic, N., Roth, B., Bartschat, K., Fang, Y., & Bray, I. (2002). *Physical Review A*, *66*, 022705.

Baum, G., Förster, S., Pavlovic, N., Roth, B., Bartschat, K., & Bray, I. (2004). *Physical Review A, 70*, 012707.
Bechler, A., & Pratt, R. H. (1989). *Physical Review A, 39*, 1774.
Bechler, A., & Pratt, R. H. (1990). *Physical Review A, 42*, 6400.
Beck, D., & Loesch, H. J. (1966). *Zeitschrift für Physik, 195*, 444.
Becker, U. (1990). In A. Dalgarno, R. S. Freud, P. M. Koch, M. S. Lubell & T. B. Lucatorto (Eds.), *The physics of electronic and atomic collisions* (p. 162). *AIP Conference Proceedings 205*, New York.
Becker, U., & Shirley, D. A. (1996). In U. Becker & D. A. Shirley (Eds.), *VUV and soft X-ray photoionization* (p. 135). Plenum Press: New York.
Becker, U., Szostak, D., Kerkhoff, H. G., Kupsch, M., Langer, B., Wehlitz, R., et al. (1989). *Physical Review A, 39*, 3902.
Becker, U. (1998). *Journal of Electron Spectroscopy and Related Phenomena, 96*, 105.
Becker, U., & Langer, B. (1998). *Physica Scripta, T78*, 13.
Becker, U., & Crowe, A. (Eds.) (2001). *Complete scattering experiments*. New York: Kluwer Academic/Plenum Publishers.
Beckord, K. (1989). *Diplom thesis*. Germany: University of Bielefeld.
Bederson, B. (1969a). *Comments on Atomic and Molecular Physics, 1*, 41.
Bederson, B. (1969b). *Comments on Atomic and Molecular Physics, 1*, 65.
Bederson, B. (1970). *Comments on Atomic and Molecular Physics, 2*, 160.
Bellicard, J. B., Moussa, A., & Haynes, S. K. (1956). *Nuclear Physics, 3*, 307.
Bellicard, J. B., Moussa, A., & Haynes, S. K. (1957). *Journal de Physique et Le Radium, 18*, 115.
Berezhko, E. G., & Kabachnik, N. M. (1977). *Journal of Physics B: Atomic and Molecular Physics, 10*, 2467.
Berezhko, E. G., Kabachnik, N. M., & Rostovsky, V. S. (1978a). *Journal of Physics B: Atomic and Molecular Physics, 11*, 1749.
Berezhko, E. G., Ivanov, V. K., & Kabachnik, N. M. (1978b). *Physics Letters A, 66*, 474.
Berezhko, E. G., & Kabachnik, N. M. (1979). *Journal of Physics B: Atomic and Molecular Physics, 12*, 2993.
Berger, O., & Kessler, J. (1986). *Journal of Physics B: Atomic, Molecular and Optical Physics, 19*, 3539.
Berkowitz, J. (1978). *Photoabsorption, photoionization, and photoelectron spectroscopy*. New York: Academic Press.
Berrah, N. et al. (1993). 18th International Conference on Photonic, Electronic and Atomic Collisions (p. 17). Arhus, Denmark.
Berrington, K. A., Eissner, W. B., & Norrigton, P. H. (1995). *Computer Physics Communications, 92*, 290.
Bethe, H. A. (1933). *Handbuch der Physik. XXVI/1* (2nd ed.).
Bethke, C., Kisker, E., Weber, N. B., & Hillebrecht, F. U. (2005). *Physical Review A, 71*, 024413.
Beutler, H. Z. (1934). *Zeitschrift für Physik, 91*, 202.
Beutler, H. Z. (1935). *Zeitschrift für Physik, 93*, 177.
Beyer, H. J., Kleinpoppen, H., McGregor, I., & McIntgre Jr., L. C. (1982). Journal of Physics B: Atomic and Molecular Physics 15, L545.
Beyer, H. J., West, J. B., Ross, K. J., Ueda, K., Kabachnik, N. M., Hamdy, H., et al. (1995). *Journal of Physics B: Atomic, Molecular and Optical Physics, 28*, L47.
Beyer, H. J., West, J. B., Ross, K. J., Ueda, K., Kabachnik, N. M., Hamdy, H., et al. (1996). *Journal of Electron Spectroscopy and Related Phenomena, 79*, 339.
Beynon, J. D. E., & Cairns, R. B. (1965). *Proceedings of the Physical Society, 86*, 1343.
Beynon, J. D. E. (1966). *Proceedings of the Physical Society, 89*, 59.
Biedenharn, L. C., & Rose, M. E. (1953). *Reviews of Modern Physics, 25*, 729.
Bizau, J. M., Cubaynes, D., & Wuilleumier, F. J. (1993). In F. J. Wuilleumier, W. Petroff, & I. Nenner (Eds.), *Proceedings of the 10th VUV Conference* (p. 165). World Scientific: Singapore.

Bizau, J. M., Wuilleumier, F. J., Ederer, D. L., Keller, J. C., LeGouet, J. L., Picque, J. L., et al. (1985). *Physical Review Letters*, *55*, 1281.
Blatt, J. M., & Biedenharn, L. C. (1952). *Reviews of Modern Physics*, *24*, 258.
Blum, K. (1996). *Density matrix theory and applications* (2nd ed.). Plenum Press: New York.
Blum, K., & Kleinpoppen, H. (1974). *Physical Review A*, *9*, 1902.
Blum, K., & Kleinpoppen, H. (1979). *Physics Reports*, *52*, 203.
Blum, K., Lohmann, B., & Taute, E. (1986). *Journal of Physics B: Atomic and Molecular Physics*, *19*, 3815.
Bobashev, S. V. (1970). *Pis'ma Zh. Eksp. Teor. Fiz. 11*, 389 (Trans.) (1970). *JETP Letters*, *11*, 260.
Bobashev, S. V., Elizarov, A. Yu., Korshunov, V. V., Prilipko, V. K., & Cherepkov, N. A. (1994). *Zh. Eksp. Teor. Fiz., 106*, 90, (Trans.) *Soviet Physics—JETP*, *79*, 47.
Bolognesi, P., Kheifets, A. S., Bray, I., Malegat, L., Selles, P., Kazansky, A. K., et al. (2003). *Journal of Physics B: Atomic, Molecular and Optical Physics*, *36*, L241.
Bolognesi, P., de Fanis, A., Coreno, M., & Avaldi, L. (2004). *Physical Review A*, *70*, 022701.
Bolognesi, P., Toffoli, D., Decleva, P., Feyer, V., Pravica, L., & Avaldi, L. (2008). *Journal of Physics B: Atomic, Molecular and Optical Physics*, *41*, 221002.
Bonhoff, S. (1998). PhD Thesis, University of Münster, Germany.
Bonhoff, K., Nahrup, S., Lohmann, B., & Blum, K. (1996). *Journal of Chemical Physics*, *104*, 7921.
Bonhoff, S., Bonhoff, K., Schimmelpfennig, B., & Nestmann, B. (1997). *Journal of Physics B: Atomic, Molecular and Optical Physics*, *30*, 2821.
Bonhoff, K., Bonhoff, S., Schimmelpfennig, B., Nestmann, B., & Blum, K. (1998). *Journal of Physics B: Atomic, Molecular and Optical Physics*, *31*, 1511.
Bonhoff, S., Bonhoff, K., & Blum, K. (1999). *Journal of Physics B: Atomic, Molecular and Optical Physics*, *32*, 1139.
Born, M., & Wolf, E. (1970). *Principles of optics*. New York: Pergamon Press.
von dem Borne, A., Dohrmann, T., Verweyen, A., Sonntag, B., Godehusen, K., & Zimmermann, P. (1997). *Physical Review Letters*, *78*, 4019.
Böwering, N., Lischke, T., Schmidtke, B., Müller, N., Khalil, T., & Heinzmann, U. (2001). *Physical Review Letters*, *86*, 1187.
Braune, M., Reinköster, A., Viefhaus, J., Lohmann, B., & Becker, U. (2007). *25th International Conference on Photonic, Electronic and Atomic Collisions* (p. Fr034). Freiburg, Germany.
Bray, I., & Stelbovics, A. T. (1992). *Physical Review A*, *46*, 6995.
Briggs, J. S., & Schmidt, V. (2000). *Journal of Physics B: Atomic, Molecular and Optical Physics*, *33*, R1.
Brink, D. M., & Satchler, G. R. (1962). *Angular momentum*. Oxford: Oxford University Press.
Buck, U., & Pauly, H. (1968). *Zeitschrift für Naturforschung*, *23A*, 475.
Buck, U., & Pauly, H. J. (1971). *Journal of Chemical Physics*, *54*, 1929.
Buck, U., Köhler, K. A., & Pauly, H. (1971). *Zeitschrift für Physik*, *244*, 180.
Bullard, E. C., & Massey, H. S. W. (1931). *Proceedings of the Royal Society A*, *130*, 579.
Burhop, E. H. S. (1935). *Proceedings of the Royal Society A*, *148*, 272.
Burke, P. G., Ormonds, S., & Whitaker, W. (1967). *Proceedings of the Physical Society*, *92*, 313.
Burke, P. G., & Mitchell, J. F. B. (1974). *Journal of Physics B: Atomic and Molecular Physics*, *7*, 214.
Burnett, G. C., Monroe, T. J., & Dunning, F. B. (1994). *Review of Scientific Instruments*, *65*, 1893.
Busalla, A., & Blum, K. (1997). *The Journal of Physical Chemistry A*, *101*, 7476.
von Busch, F., Strunck, H. J., & Schlier, C. (1967). *Zeitschrift für Physik*, *199*, 518.
Bußert, W., & Klar, H. (1983). *Zeitschrift für Physik A*, *312*, 315.
Cairns, R. B., Harrison, H., & Schoen, R. I. (1970). *Journal of Chemical Physics*, *53*, 96.
Caldwell, D. J., & Eyring, H. (1971). *The theory of optical activity*. New York: Wiley.
Caldwell, C. D., & Zare, R. N. (1977). *Physical Review A*, *16*, 255.
Callan, E. J. (1961). *Physical Review*, *124*, 793.
Campbell, E. B., Schmidt, H., & Hertel, I. V. (1988). *Advances in Chemistry*, *72*, 37.
Camus, P., Dieulin, M., El Hindy, A., & Aymar, M. (1983). *Physica Scripta*, *T27*, 125.

Carlson, T. A., Krause, M. O., Grimm, F. A., Keller, P. R., & Taylor, J. W. (1982). *Chemical Physics Letters*, *87*, 552.

Carre, B., d'Oliveira, P., Ferray, M., Fournier, P., Gounand, F., Cubayanes, D., et al. (1990). *Zeitschrift für Physik D*, *15*, 117.

Casey, W. R., & Albridge, R. G. (1969). *Zeitschrift für Physik*, *219*, 216.

Chadwick, J. (1929). In H. S. W. Massey, *Atomic and molecular collisions*, Chap. 10.6. Taylor & Francis: London.

Chang, T. N. (1972). *Journal of Physics B: Atomic and Molecular Physics*, *8*, 743.

Chang, T. N. (1982). *Journal of Physics B: Atomic and Molecular Physics*, *15*, L81.

Charlton, M., Andersen, L. H., Brun-Nielsen, L., Deutsch, B. I., Hvelpund, P., Jacobsen, F. M., et al. (1988). *Journal of Physics B: Atomic, Molecular and Optical Physics*, *21*, L545.

Chattarji, D. (1976). *The theory of Auger transitions*. London, New York, San Francisco: Academic Press.

Chatwell, M., Clendenin, J., Maruyama, T., & Schultz, D. (1993). *Proceedings of the Workshop on Photocathodes for Polarized Electron Sources, SLAC, Conference 9309288* (p. 397). National Technical Information Service U.S. Department of Commerce.

Chaudry, M. A., Duncan, A. J., Hippler, R., & Kleinpoppen, H. (1987). *Physical Review Letters*, *59*, 2036.

Chen, M. H. (1993). *Physical Review A*, *47*, 3733.

Cherepkov, N. A. (1974). *Zh. Eksp. Teor. Fiz. 65*, 933 (Trans.) *Soviet Physics—JETP*, *38*, 463.

Cherepkov, N. A. (2001). In U. Becker & A. Crowe (Eds.), *Complete scattering experiments* (p. 167). Kluwer Academic/Plenum Publishers: New York.

Cherepkov, N. A. (1972). *Physics Letters A*, *40*, 119.

Cherepkov, N. A. (1979). *Journal of Physics B: Atomic and Molecular Physics*, *12*, 1279.

Cherepkov, N. A. (1980). *Optics and Spectroscopy*, *49*, 582.

Cherepkov, N. A. (1982). *Chemical Physics Letters*, *87*, 344.

Cherepkov, N. A. (1983). *Advances in Atomic and Molecular Physics*, *19*, 395.

Cherepkov, N. A., & Kuznetsov, V. V. (1989). *Journal of Physics B: Atomic, Molecular and Optical Physics*, *22*, L405.

Cherepkov, N. A., Kuznetsov, V. V., & Verbitskii, V. A. (1995). *Journal of Physics B: Atomic, Molecular and Optical Physics*, *28*, 1221.

Cherepkov, N. A., Raseev, G., Adachi, J., Hikosaka, Y., Ito, K., Motoki, S., et al. (2000). *Journal of Physics B: Atomic, Molecular and Optical Physics*, *33*, 4213.

Cherepkov, N. A., & Semenov, S. K. (2001a). *Journal of Physics B: Atomic, Molecular and Optical Physics*, *34*, L211.

Cherepkov, N. A., & Semenov, S. K. (2001b). *Journal of Physics B: Atomic, Molecular and Optical Physics*, *34*, L495.

Cherepkov, N. A., Semenov, S. K., Drescher, M., & Heinzmann, U. (2003). *Journal of Physics B: Atomic, Molecular and Optical Physics*, *36*, 3063.

Cherepkov, N. A., & Semenov, S. K. (2004). *Journal of Physics B: Atomic, Molecular and Optical Physics*, *37*, 1267.

Cherepkov, N. A., & Semenov, S. K. (2007). *International Journal of Quantum Chemistry*, *107*, 2889.

Chichinin, A. I., Gericke, K.-H., Kauczok, S., & Maul, C. (2009). *International Reviews in Physical Chemistry*, *28*, 607.

Clark, A. P., Brouard, M., Quadrini, F., & Vallance, C. (2006). *Physical Chemistry Chemical Physics*, *8*, 5591.

Cleff, B., & Mehlhorn, W. (1969). *Zeitschrift für Physik*, *219*, 311.

Cocke, L. C. (2004). *Physica Scripta*, *T110*, 9.

Cohen, H. D., & Fano, U. (1966). *Physical Review*, *150*, 30.

Colgan, J., & Pindzola, M. S. (2001). *Physical Review Letters*, *86*, 1998.

Collins, R. E., Goldstein, M., Bederson, B., & Rubin, K. (1967). *Physical Review Letters*, *19*, 1366.

Connerade, J. P. (2005). *Highly excited atoms*. Cambridge: Cambridge University Press.

Continetti, R. E., & Hayden, C. C. (2004). In X. Yang & K. Liu (Eds.), *Modern trends in reaction dynamics* (p. 475). World Scientific: Singapore.
Cooper, J. W., & Zare, R. N. (1969). In S. Geltman, K. T. Mahanthappa, & W. E. Britten (Eds.), *Lectures in theorerical physics* (Vol. XI-C, p. 317). Gordon and Breach: New York.
Cooper, J. W. (1962). *Physical Review*, *128*, 681.
Cooper, J. W. (1964). *Physical Review Letters*, *13*, 762.
Cooper, J. W., & Zare, R. N. (1968). *Journal of Chemical Physics*, *48*, 942.
Cooper, J. W., & Manson, S. T. (1969). *Physical Review*, *177*, 157.
Cooper, J. W. (1990). *Physical Review A*, *42*, 6942.
Cooper, J. W. (1993). *Physical Review A*, *47*, 1841.
Cowan, R. D. (1981). *The theory of atomic structure and spectra*. Berkeley: UCLA Press.
Crowe, A., Nogueira, J. C., & Liew, Y. C. (1983). *Journal of Physics B: Atomic and Molecular Physics*, *16*, 481.
Cubaynes, D., Bizau, J. M., Wuilleumier, F. J., Carré, B., & Gounand, F. (1989). *Physical Review Letters*, *63*, 2460.
Cubaynes, D., Meyer, M., Grum-Grzhimailo, A. N., Bizau, J. M., Kennedy, E. T., Bozek, J., et al. (2004). *Physical Review Letters*, *92*, 233002.
Cubaynes, D., Guilbaud, S., Wuilleumier, F. J., Meyer, M., Heinecke, E., Riek, K., et al. (2009). *Physical Review A*, *80*, 023410.
Cvejanovic, S., & Reddish, T. J. (2000). *Journal of Physics B: Atomic, Molecular and Optical Physics*, *33*, 4691.
Damascelli, A. (2004). *Physica Scripta*, *T109*, 61.
Dehmer, J. L., & Berkowitz, J. (1974). *Physical Review A*, *10*, 484.
Dehmer, J. L., Chupka, W. A., Berkowitz, J., & Jivery, W. T. (1975). *Physical Review A*, *12*, 1966.
Deichsel, H. Z. (1961). *Zeitschrift für Physik*, *164*, 156.
Derenbach, H., & Schmidt, V. (1984). *Journal of Physics B: Atomic and Molecular Physics*, *17*, 83.
Derevianko, A., Johnson, W. R., & Cheng, K. T. (1999). *Atomic Data and Nuclear Data Tables*, *73*, 153.
Derevianko, A., Hemmers, O., Oblad, S., Glans, P., Wang, H., Whitfield, S. B., et al. (2000). *Physical Review Letters*, *84*, 2116.
Deshmukh, P. C., & Manson, S. T. (1983). *Physical Review A*, *28*, 209.
Deslattes, R. D., & Kessler, E. G., Jr. (2003). *Reviews of Modern Physics*, *75*, 35.
Devons, S., & Goldfarb, L. J. B. (1957). In S. Flügge (Ed.), *Handbuch der Physik* (Vol. 42, p. 362). Springer: Berlin.
Dill, D., & Fano, U. (1972). *Physical Review Letters*, *29*, 1203.
Dill, D. (1973). *Physical Review A*, *7*, 1976.
Dill, D., Starace, A. F., & Manson, S. (1975). *Physical Review A*, *11*, 1596.
Dill, D., Swanson, J. R., Wallace, S., & Dehmer, J. L. (1980). *Physical Review Letters*, *45*, 1393.
Dodhy, A., Compton, R. N., & Stockdale, J. A. D. (1985). *Physical Review Letters*, *54*, 422.
Dohrmann, Th., von dem Borne, A., Verweyen, A., Sonntag, B., Wedowski, M., Godehusen, K., et al. (1996). *Journal of Physics B: Atomic, Molecular and Optical Physics*, *29*, 5699.
Dolmatov, V. K., & Manson, S. T. (2001). *Physical Review A*, *63*, 022704.
Dorn, A., Nienhaus, J., Wetzstein, M., Winnewisser, C., Eichmann, U., Sandner, W., et al. (1995). *Journal of Physics B: Atomic, Molecular and Optical Physics*, *28*, L225.
Dörner, R., Bräuning, H., Jagutzki, O., Mergel, V., Achler, M., Moshammer, R., et al. (1998). *Physical Review Letters*, *81*, 5776.
Dörner, R., Glessen, H., Moshammer, R., & Rottke, H. (2001). *Physikalische Blätter*, *57*(4), 49.
Downie, P., & Powis, I. (1999). *Physical Review Letters*, *82*, 2864.
Duncanson, J. A., Jr., Strand, M. P., Lindgard, A., & Berry, R. S. (1976). *Physical Review Letters*, *37*, 987.
Duong, H. T., Pinard, J., & Vialle, J. L. (1978). *Journal of Physics B: Atomic and Molecular Physics*, *11*, 797.

D'yakov, B. B., & Rogachev, I. M. (1962). *Bulletin of the Academy of Sciences of the USSR, Physical Series, 26*, 191.
Dyall, K. G., Grant, I. P., Johnson, C. T., Parpia, F. A., & Plummer, E. P. (1989). *Computer Physics Communications, 55*, 425.
Ederer, D. L., Rubensson, J. E., Mueller, D. R., Shuker, R., O'Brien, W. L., Jai, J., et al. (1992). *Nuclear Instruments and Methods A, 319*, 250.
Ehresmann, A., Vollweiler, F., Schmoranzer, H., Sukhorukov, V. L., Lagutin, B. M., Petrov, I. D., et al. (1994). *Journal of Physics B: Atomic, Molecular and Optical Physics, 27*, 1489.
Ehrhardt, H., & Morgan, L. A. (Eds.). (1994). *Electron collisions with molecules, clusters and surfaces*. Plenum Press: New York.
Ehrhardt, H., Jung, K., & Schubert, E. E. (1980). In H. Kleinpoppen & J. F. Williams (Eds.), *Coherence and correlations in atomic collisions* (p. 41). Plenum Press: New York.
Ehrhardt, H., Schulz, M., Tekaat, T., & Willmann, K. (1969). *Physical Review Letters, 22*, 89.
Elizarov, A. Yu., & Cherepkov, N. A. (1986). Pis'ma Zh. Eksp. Teor. Fiz. 44, 3, Trans. (1986). *JETP Letters, 44*, 1.
Elsener, K. (1989). *Comments on Atomic and Molecular Physics, XXII*(263), 263.
Eminyan, M., McAdam, K. B., Slevin, J., & Kleinpoppen, H. (1973). *Physical Review Letters, 31*, 576.
Eminyan, M., McAdam, K. B., Slevin, J., & Kleinpoppen, H. (1974). *Journal of Physics B: Atomic and Molecular Physics, 7*, 1519.
Eminyan, M., McAdam, K. B., Slevin, J., Standage, M. C., & Kleinpoppen, H. (1975). *Journal of Physics B: Atomic and Molecular Physics, 8*, 2058.
Erman, P., Rossi, J., Bonacalza, E. C. O., & Miskel, J. (1964). *Arkiv för fysik, 26*, 135.
de Fanis, A., West, J. B., Ross, K. J., Ueda, K., Beyer, H. J., & Chernysheva, L. V. (1999). *Journal of Physics B: Atomic, Molecular and Optical Physics, 32*, 5739.
Fano, U. (1957). *Reviews of Modern Physics, 29*, 74.
Fano, U., & Racah, G. (1959). *Irreducible tensorial sets*. New York: Academic Press.
Fano, U. (1961). *Physical Review, 124*, 1866.
Fano, U., & Cooper, J. W. (1965). *Physical Review, 137*, A1364.
Fano, U. (1969a). *Physical Review, 178*, 131.
Fano, U. (1969b). *Physical Review, 184*, 250.
Fano, U., & Dill, D. (1972). *Physical Review A, 6*, 185.
Fano, U., & Macek, J. (1973). *Reviews of Modern Physics, 45*, 553.
Farhat, A., Humphrey, M., Langer, B., Berrah, N., Bozek, J. D., & Cubaynes, D. (1997). *Physical Review A, 56*, 501.
Faxen, H., & Holtsmark, J. (1927). *Zeitschrift für Physik, 45*, 307.
Ferch, J., Simon, F., & Stakeljahn, G. (1987). *15th International Conference on Photonic, Electronic and Atomic Collisions* (p. 132). Brighton, UK.
Fergusson, A. T. (1965). *Angular correlation methods in gamma-ray spectroscopy*. Amsterdam: North-Holland.
Fink, M. G. J., & Johnson, W. R. (1986). *Physical Review A, 34*, 3754.
Fink, R. F., & Sorensen, S. L. (2000). *Journal of Chemical Physics, 112*, 6666.
Fink, R. F., Piancastelli, M. N., Grum-Grzhimailo, A. N., & Ueda, K. (2009). *Journal of Chemical Physics, 130*, 014306.
Flechter, G. D., Alguard, M. J., Gay, T. J., Hughes, V. W., Wainwright, R. F., Lubell, M. S., et al. (1985). *Physical Review A, 31*, 2854.
Flügge, S., Mehlhorn, W., & Schmidt, V. (1972). *Physical Review Letters, 29*, 7.
Franck, J., & Hertz, G. (1914). *Verh. der Phys. Ges., 16*, 457.
Frauenfelder, H., & Henley, E. M. (1979). *Teilchen und Kerne*. München, Wien: Oldenbourg.
Fritsch, A. R. (1986). private communication.
Fritzsche, S. (2001). *Journal of Electron Spectroscopy and Related Phenomena, 114–116*, 1155.
Fritzsche, S., Nikkinen, J., Huttula, S. M., Aksela, H., Huttula, M., & Aksela, S. (2007). *Physical Review A, 75*, 012501.

Fritzsche, S., Grum-Grzhimailo, A. N., Gryzlova, E. V., & Kabachnik, N. M. (2008). *Journal of Physics B: Atomic, Molecular and Optical Physics*, *41*, 165601.

Fritzsche, S., Grum-Grzhimailo, A. N., Gryzlova, E. V., & Kabachnik, N. M. (2009). *Journal of Physics B: Atomic, Molecular and Optical Physics*, *42*, 145602.

Froese-Fischer, C., Brage, T., & Jönsson, P. (1997). *Computational atomic structure. An MCHF approach*. Bristol: IOP Publishing.

Froese-Fischer, C. (2000). *Computer Physics Communications*, *128*, 635.

Fukuzawa, H., Gryzlova, E. V., Motomura, K., Yamada, A., Ueda, K., Grum-Grzhimailo, A. N., et al. (2010). *Journal of Physics B: Atomic, Molecular and Optical Physics*, *43*, 111001.

Fursa, D. V., & Bray, I. (1995). *Physical Review A*, *52*, 1279.

Garcia, G. A., Nahon, L., Lebech, M., Houver, J.-C., Dowek, D., & Powis, I. (2003). *Journal of Chemical Physics*, *119*, 8781.

Garton, W. R. S. (1959). *Journal of Scientific Instruments*, *36*, 11.

Geffrion, C., & Nadeau, G. (1959). U.S. Air Force Office of Scientific Research Report TR 59–145, unpublished.

Gehenn, W., & Reichert, E. (1977). *Journal of Physics B: Atomic and Molecular Physics*, *10*, 3105.

Geltman, S., & Burke, P. G. (1970). *Journal of Physics B: Atomic and Molecular Physics*, *3*, 1062.

Geßner, O., Hikosaka, Y., Zimmermann, B., Hempelmann, A., Lucchese, R. R., Eland, J. H. D., et al. (2002). *Physical Review Letters*, *88*, 193002.

Gisselbrecht, M., Marquette, A., & Meyer, M. (1998). *Journal of Physics B: Atomic, Molecular and Optical Physics*, *31*, L977.

Godehusen, K., Zimmermann, P., Verweyen, A., von dem Borne, A., Wernet, Ph., & Sonntag, B. (1998). *Physical Review A*, *58*, R3371.

Goldstein, M., Kasdan, A., & Bederson, B. (1972). *Physical Review A*, *5*, 660.

Golovin, A. V. (1991). *Optics and Spectroscopy (USSR)*, *71*, 537.

Golovin, A. V., Cherepkov, N. A., & Kuznetsov, V. V. (1992). *Zeitschrift für Physik D*, *24*, 371.

de Gouw, J. A., van Eck, J., Peters, A. C., van der Weg, J., & Heideman, H. G. M. (1995). *Journal of Physics B: Atomic, Molecular and Optical Physics*, *28*, 2127.

Graham, R. L., Brown, F., Ewan, G. T., & Uhler, J. (1961). *Canadian Journal of Physics*, *39*, 1086.

Grant, I. P., & Parpia, F. A. (1992). Private communication.

Grant, I. P. (1970). *Advances in Physics*, *19*, 747.

Grant, I. P., McKenzie, B., Norrington, P., Mayers, D., & Pyper, N. (1980). *Computer Physics Communications*, *21*, 207.

Greene, C. H., & Zare, R. N. (1982). *Physical Review A*, *25*, 2031.

Greene, C. H., & Zare, R. N. (1983). *Journal of Chemical Physics*, *78*, 6741.

Greene, C. H., & Aymar, M. (1991). *Physical Review A*, *44*, 1773.

Greiner, W., & Reinhardt, J. (1995). *Physica Scripta*, *T56*, 203.

Grum-Grzhimailo, A. N., & Meyer, M. (2005). In B. Piraux (Ed.), *Electron and photon impact ionization and related topics* (p. 103). IOP Conference Series, Vol. 183. IOP Publishing: Bristol.

Grum-Grzhimailo, A. N., & Meyer, M. (2006). In A. Lahmam-Bennani & B. Lohmann (Eds.), *Ionization, correlation and polarization in atomic collisions* (p. 191). AIP Conference Proceedings (Vol. 811). Melville, New York.

Grum-Grzhimailo, A. N., and Zhadamba, B. (1987). Vestnik MGU Ser.3, Fiz. Astron. 28 (N1), 19.

Grum-Grzhimailo, A. N., Dorn, A., & Mehlhorn, W. (2001). In U. Becker & A. Crowe (Eds.), *Complete scattering experiments* (p. 111). Kluwer Academic/Plenum Publishers: New York.

Grum-Grzhimailo, A. N., Danzan, S., Lhagva, O., & Strakhova, S. I. (1991). *Zeitschrift fur Physik D*, *18*, 147.

Grum-Grzhimailo, A. N., & Dorn, A. (1995). *Journal of Physics B: Atomic, Molecular and Optical Physics*, *28*, 3197.

Grum-Grzhimailo, A. N., Dorn, A., & Mehlhorn, W. (1999). Comments on Atomic and Molecular Physics. *Comments on Modern Physics D*, *1*, 29.

Grum-Grzhimailo, A. N. (2001). *Journal of Physics B: Atomic, Molecular and Optical Physics*, *34*, L359.

Grum-Grzhimailo, A. N. (2003). *Journal of Physics B: Atomic, Molecular and Optical Physics, 36*, 2385.

Grum-Grzhimailo, A. N., Fritzsche, S., O'Keeffe, P., & Meyer, M. (2005). *Journal of Physics B: Atomic, Molecular and Optical Physics, 38*, 2545.

Grum-Grzhimailo, A. N., Lucchese, R. R., Liu, X.-J., Prümper, G., Morishita, Y., Saito, N., et al. (2007a). *Journal of Electron Spectroscopy and Related Phenomena, 155*, 100.

Grum-Grzhimailo, A. N., Lucchese, R. R., Prümper, G., & Ueda, K. (2007b). *Journal of Electron Spectroscopy and Related Phenomena, 155*, 104.

Grum-Grzhimailo, A. N., & Meyer, M. (2009). *European Physical Journal, 169*, 43.

Grum-Grzhimailo, A. N., Gryzlova, E. V., Strakhova, S. I., Kabachnik, N. M., & Fritzsche, S. (2009). *Journal of Physics: Conference Series, 194*, 012004.

Gryzlova, E. V., Grum-Grzhimailo, A. N., Fritzsche, S., & Kabachnik, N. M. (2010). *Journal of Physics B: Atomic, Molecular and Optical Physics, 43*, 225602.

Guillemin, R., Hemmers, O., Lindle, D. W., & Manson, S. T. (2005). *Radiation Physics and Chemistry, 73*, 311.

Guyon, P. M., Golovin, A. V., Quayle, C. J. K., Vervloet, M., & Richard-Viard, M. (1996). *Physical Review Letters, 76*, 600.

Haberland, R., Fritsche, L., & Noffke, J. (1986). *Physical Review A, 33*, 2305.

Haffner, H., Kleinpoppen, H., & Krüger, H. (1965). *Physics Letters, 18*, 270.

Hamdy, H., Beyer, H. J., West, J., & Kleinpoppen, H. (1991). *Journal of Physics B: Atomic, Molecular and Optical Physics, 24*, 4957.

Hanle, W. (1924). *Zeitschrift für Physik, 30*, 93.

Hanne, G. F., Malegat, L., & Schmidt-Böcking, H. (Eds.) (2003). AIP Conference Proceedings (Vol. 697). Melville, New York.

Hanne, G. F. (1988). *American Journal of Physics, 56*(8), 696.

Hanne, G. F. (2000). *TSINGUA Science and Technology, 5*, 343.

Hanold, K. A., & Continetti, R. E. (1998). *Chemical Physics, 239*, 493.

Hansen, J. C., Duncanson, J. A., Jr., Chien, R. L., & Berry, R. S. (1980). *Physical Review A, 21*, 222.

Happer, W., Jau, Y.-Y., & Walker, T. (2010). *Optically pumped atoms*. KGaA, Weinheim: WILEY-VCH GmbH & Co.

Hausmann, A., Kämmerling, B., Kossmann, H., & Schmidt, V. (1988). *Physical Review Letters, 61*, 2669.

He, L. W., Burkhardt, C. E., Ciocca, M., Leventhal, J. J., Zhou, H. L., & Manson, S. T. (1995). *Physical Review A, 51*, 2085.

Heckenkamp, H. C., Schäfers, F., Schönhense, G., & Heinzmann, U. (1986). *Zeitschrift für Physik D, 2*, 257.

Heinzmann, U., & Cherepkov, N. A. (1996). In U. Becker & D. A. Shirley (Eds.), *VUV and soft X-ray photoionization* (p. 521). Plenum Press: New York.

Heinzmann, U., Kessler, J., & Lorenz, J. (1970). *Zeitschrift für Physik, 240*, 42.

Heinzmann, U., Heuer, H., & Kessler, J. (1975). *Physical Review Letters, 34*, 441.

Heinzmann, U. (1980a). *Journal of Physics B: Atomic and Molecular Physics, 13*, 4353.

Heinzmann, U. (1980b). *Journal of Physics B: Atomic and Molecular Physics, 13*, 4367.

Heiser, F., Geßner, O., Viefhaus, J., Wieliczek, K., Hentges, R., & Becker, U. (1997). *Physical Review Letters, 79*, 2435.

Helm, H., Bjerre, N., Dyer, M. J., Huestis, D. L., & Saeed, M. (1993). *Physical Review Letters, 70*, 3221.

Hemmers, O., Heiser, F., Eiben, J., Wehlitz, R., & Becker, U. (1993). *Physical Review Letters, 71*, 987.

Hemmers, O., Glans, P., Hansen, D. L., Wang, H., Whitfield, S. B., Lindle, D. W., et al. (1996). *Synchrotron Radiation News, 9*, 40.

Hemmers, O., Fisher, G., Glans, P., Hansen, D. L., Wang, H., Whitfield, S. B., et al. (1997). *Journal of Physics B: Atomic, Molecular and Optical Physics, 30*, L727.

Hemmers, O., Guillemin, R., Kanter, E. P., Krässig, B., Lindle, D. W., Southworth, S. H., et al. (2003). *Physical Review Letters, 91*, 053002.
Hemmers, O., Guillemin, R., & Lindle, D. W. (2004). *Radiation Physics and Chemistry, 70*, 123.
Hemmers, O., Guillemin, R., Rolles, D., Wolska, A., Lindle, D. W., Kanter, E. P., et al. (2005). *Journal of Electron Spectroscopy and Related Phenomena, 144–147*, 155.
Hemmers, O., Guillemin, R., Rolles, D., Wolska, A., Lindle, D. W., Kanter, E. P., et al. (2006). *Physical Review Letters, 97*, 103006.
Henk, J., & Ernst, A. (2002). *Journal of Electron Spectroscopy and Related Phenomena, 125*, 107.
Hergenhahn, U., Snell, G., Drescher, M., Schmidtke, B., Müller, N., Heinzmann, U., Wiedenhöft, M., & Becker, U. (1999). *Physical Review Letters, 82*, 5020, and private communication.
Hergenhahn, U., Kabachnik, N. M., & Lohmann, B. (1991). *Journal of Physics B: Atomic, Molecular and Optical Physics, 24*, 4759.
Hergenhahn, U., & Becker, U. (1995). *Journal of Electron Spectroscopy and Related Phenomena, 76*, 225.
Hermann, J., Menner, B., Reisacher, E., Zehnle, E., & Kempter, V. J. (1980). *Journal of Physics B: Atomic and Molecular Physics, 13*, L165.
Hertel, I. V. (1976). In H. Kleinpoppen, & M. R. C. McDowell (Eds.), *Electron and photon interactions with atoms* (p. 375). Plenum Press: New York.
Hertel, I. V., & Stoll, W. (1974a). *Journal of Physics B: Atomic and Molecular Physics, 7*, 570.
Hertel, I. V., & Stoll, W. (1974b). *Journal of Physics B: Atomic and Molecular Physics, 7*, 583.
Hertel, I. V., & Stoll, W. (1977). *Advances in Atomic and Molecular Physics, 13*, 113.
Herting, C., & Hanne, G. F. (2003). In G. F. Hanne, L. Malegat, & H. Schmidt-Böcking (Eds.), *AIP Conference Proceedings, 697*, 181. Melville, New York.
Herting, C., Hanne, G. F., Bartschat, K., Grum-Grzhimailo, A. N., Srivastava, R., & Stauffer, A. D. (2002). *Journal of Physics B: Atomic, Molecular and Optical Physics, 35*, 4439.
Hertz, H. (1887). *Wiener Annalen, 31*, 983.
Herzberg, G. (1965). *Molecular spectra and molecular structure, I. spectra of diatomic molecules*. Princeton: Van Nostrand.
Hikosaka, Y., & Eland, J. H. D. (2000). *Journal of Physics B: Atomic, Molecular and Optical Physics, 33*, 3137.
Hikosaka, Y., Fushitani, M., Matsuda, A., Tseng, C.-M., Hishikawa, A., Shigemasa, E., et al. (2010). *Physical Review Letters, 105*, 133001.
Hillig, H., Cleff, B., Mehlhorn, W., & Schmitz, W. (1974). *Zeitschrift für Physik, 268*, 225.
Hils, D., & Kleinpoppen, H. (1978). *Journal of Physics B: Atomic and Molecular Physics, 11*, L283.
Hils, D., McCusker, V., Kleinpoppen, H., & Smith, S. J. (1972). *Physical Review Letters, 29*, 398.
Hils, D., Jitschin, W., & Kleinpoppen, H. (1981). *Applied Physics, 25*, 39.
Hils, D., Jitschin, W., & Kleinpoppen, H. (1982). *Journal of Physics B: Atomic and Molecular Physics, 15*, 3347.
Hippler, R., Faust, M., Wolf, R., Kleinpoppen, H., & Lutz, H. O. (1987). *Physical Review A, 36*, 4644.
Hollywood, M. T., Crowe, A., & Williams, J. F. (1979). *Journal of Physics B: Atomic and Molecular Physics, 12*, 819.
Holmegaard, L., Hansen, J. L., Kalh⊘j, L., Kragh, S.L., Stapelfeldt, H., Filsinger, F., et al. (2010). *Nature Physics, 6*, 428.
Holtsmark, J. (1930). *Zeitschrift für Physik, 66*, 49.
Hood, S. T., Weigold, E., & Dixon, A. J. (1979). *Journal of Physics B: Atomic and Molecular Physics, 12*, 631.
Hosaka, K., Adachi, J., Golovin, A. V., Takahashi, M., Teramoto, T., Watanabe, Jahnke, T., et al. (2006b). *Physical Review A, 73*, 022716.
Hosaka, K., Adachi, J., Golovin, A. V., Takahashi, M., Teramoto, T., Watanabe, N., et al. (2006a). *Journal of Physics B: Atomic, Molecular and Optical Physics, 39*, L25.
Huang, K. N. (1980). *Physical Review A, 22*, 223.
Huang, K. N., Johnson, W. R., & Cheng, K. T. (1981). *Atomic Data & Nuclear Data Tables, 26*, 33.

Hudson, R. D. (1964). *Physical Review, 135*, A1212.
Hudson, R. D., & Carter, V. L. (1967). *The Journal of the Optical Society of America, 57*, 651.
Humpert, H. J., Schwier, H., Hippler, R., & Lutz, H. O. (1985). In H. Kleinpoppen, J. S. Briggs & H. O. Lutz (Eds.), *Fundamental Processes in Atomic Collision Physics* (Vol. 134, p. 649). Plenum Press: New York.
Huzinaga, S. (1965). *The Journal of Chemical Physics, 42*, 1293.
Ivanov, V. K., Medvedev, S. Yu., & Sosnivker, V. A. (1979). Physical-Technical Institute. Report No: Leningrad. 615.
Jacobs, V. L. (1972). *Journal of Physics B: Atomic and Molecular Physics, 5*, 2257.
Jetzke, S., Zarmba, J., & Faisal, F. H. M. (1989). *Zeitschrift für Physik D, 11*, 63.
Jiménez-Mier, J., Caldwell, C. D., & Ederer, D. L. (1986). *Physical Review Letters, 57*, 2260.
Jiménez-Mier, J., Caldwell, C. D., Flemming, M. G., Whitfield, S. B., & van der Meulen, P. (1993). *Physical Review A, 48*, 442.
Johansson, A., Raarup, M. K., Li, Z. S., Lokhnygin, V., Descamps, D., Lyngå, C., et al. (2003). *European Physical Journal D: Atomic, Molecular, Optical and Plasma Physics, 22*, 3.
Johnson, R. E. (1982). *Introduction to atomic and molecular collisions*. New York: Plenum Press.
Johnson, W. R., & Cheng, K. T. (1992a). *Physical Review Letters, 69*, 1144.
Johnson, W. R., & Cheng, K. T. (1992b). *Physical Review A, 46*, 2952.
Johnson, W. R., & Cheng, K. T. (2001). *Physical Review A, 63*, 022504.
Jost, K., Bissling, P. G. F., Eschen, F., Felsmann, M., & Walter, L. (1983). *13th International Conference on Photonic, Electronic and Atomic Collisions* (p. 11). Berlin, Germany.
Jung, M., Krässig, B., Gemmel, D. S., Kanter, E. P., LeBrun, T., Southworth, S. H., et al. (1996). *Physical Review A, 54*, 2127.
Kabachnik, N. M., & Sazhina, I. P. (1976). *Journal of Physics B: Atomic and Molecular Physics, 9*, 1681.
Kabachnik, N. M., & Sazhina, I. P. (1984). *Journal of Physics B: Atomic and Molecular Physics, 17*, 1335.
Kabachnik, N. M., Sazhina, I. P., Lee, I. S., & Lee, O. V. (1988). *Journal of Physics B: Atomic, Molecular and Optical Physics, 21*, 3695.
Kabachnik, N. M., & Lee, O. V. (1989). *Journal of Physics B: Atomic, Molecular and Optical Physics, 22*, 2705.
Kabachnik, N. M., & Sazhina, I. P. (1990). *Journal of Physics B: Atomic, Molecular and Optical Physics, 23*, L353.
Kabachnik, N. M., Lohmann, B., & Mehlhorn, W. (1991). *Journal of Physics B: Atomic, Molecular and Optical Physics, 24*, 2249.
Kabachnik, N. M. (1992). *Journal of Physics B: Atomic, Molecular and Optical Physics, 25*, L389.
Kabachnik, N. M., Tulkki, J., Aksela, H., & Ricz, S. (1994). *Physical Review A, 49*, 4653.
Kabachnik, N. M., & Ueda, K. (1995). *Journal of Physics B: Atomic, Molecular and Optical Physics, 28*, 5013.
Kabachnik, N. M., Sazhina, I. P., & Ueda, K. (1999). *Journal of Physics B: Atomic, Molecular and Optical Physics, 32*, 1769.
Kabachnik, N. M., & Grum-Grzhimailo, A. N. (2001). *Journal of Physics B: Atomic, Molecular and Optical Physics, 34*, L63.
Kabachnik, N. M., & Sazhina, I. P. (2002). *Journal of Physics B: Atomic, Molecular and Optical Physics, 35*, 3591.
Kabachnik, N. M. (2004). *Journal of Electron Spectroscopy and Related Phenomena, 137–140*, 305.
Kabachnik, N. M., & Ueda, K. (2004). *Physica Scripta, T110*, 73.
Kabachnik, N. M. (2005). *Journal of Physics B: Atomic, Molecular and Optical Physics, 38*, L19.
Kabachnik, N. M., Fritzsche, S., Grum-Grzhimailo, A. N., Meyer, M., & Ueda, K. (2007). *Physics Reports, 451*, 155.
Kaminski, H., Kessler, J., & Kollath, K. J. (1980). *Physical Review Letters, 45*, 1161.

Kämmerling, B., Schmidt, V., Mehlhorn, W., Peatman, W. B., Schaefers, F., & Schroeter, J. (1989). *Journal of Physics B: Atomic, Molecular and Optical Physics*, *22*, L597.
Kämmerling, B., Krässig, B., & Schmidt, V. (1990). *Journal of Physics B: Atomic, Molecular and Optical Physics*, *23*, 4487.
Kämmerling, B., & Schmidt, V. (1991). *Physical Review Letters*, *67*, 1848.
Kämmerling, B., Krässig, B., Schwarzkopf, O., Ribeiro, J. P., & Schmidt, V. (1992a). *Journal of Physics B: Atomic, Molecular and Optical Physics*, *25*, L5.
Kämmerling, B., Hausmann, A., Läuger, J., & Schmidt, V. (1992b). *Journal of Physics B: Atomic, Molecular and Optical Physics*, *25*, 4773.
Kämmerling, B., Krässig, B., & Schmidt, V. (1993). *Journal of Physics B: Atomic, Molecular and Optical Physics*, *26*, 261.
Kämmerling, B., & Schmidt, V. (1993). *Journal of Physics B: Atomic, Molecular and Optical Physics*, *26*, 1141.
Kämmerling, B., Läuger, J., & Schmidt, V. (1994). *Journal of Electron Spectroscopy and Related Phenomena*, *67*, 363.
Kanter, E. P., Krässig, B., Southworth, S. H., Guillemin, R., Hemmers, O., Lindle, D. W., et al. (2003). *Physical Review A*, *68*, 012714.
Karaganov, V., Teubner, P. J. O., & Brunger, M. J. (2001). In D. H. Madison, & M. Schulz (Eds.), *AIP Conference Proceedings* (Vol. 604, p. 196). Melville, New York.
Karule, E., & Peterkop, R. (1973). Private communication; but also see, (1972). Akademya Nauk Latviiskoy, SSR., Izvestiya N3, 4.
Karvonen, J., Kivimäki, A., Aksela, H., Aksela, S., Camilloni, R., Avaldi, L., et al. (1999). *Physical Review A*, *59*, 315.
Kauppila, W. E., & Stein, T. S. (1982). *Canadian Journal of Physics*, *60*, 461.
Kazantsev, S. A., & Hénoux, J. C. (1995). *Polarization spectroscopy of ionized gases*. Dordrecht: Kluwer Academic Publishers.
Keller, F., & Combet-Farnoux, F. (1979). *Journal of Physics B: Atomic and Molecular Physics*, *12*, 2821.
Keller, F., & Combet-Farnoux, F. (1982). *Journal of Physics B: Atomic and Molecular Physics*, *15*, 2657.
Kennedy, D. J., & Manson, S. T. (1972). *Physical Review A*, *5*, 227.
Kerling, C., Böwering, N., & Heinzmann, U. (1990). *Journal of Physics B: Atomic, Molecular and Optical Physics*, *23*, L629.
Kessler, J. (1985). *Polarized electrons* (2nd ed.). Springer: Berlin.
Kessler, J. (1996). In D. M. Campbell, & H. Kleinpoppen (Eds.), *Selected topics in electron physics* (p. 151). Plenum Press: New York.
Kessler, J., & Lorenz, J. (1970). *Physical Review Letters*, *24*, 87.
Kessler, J. (1981). *Comments on Atomic and Molecular Physics*, *10*, 47.
Khalid, S. M., & Kleinpoppen, H. (1983). *Physical Review A*, *27*, 236.
Khalil, T., Schmidtke, B., Drescher, M., & Heinzmann, U. (2002). *Physical Review Letters*, *89*, 053001.
Kheifets, A. S. (2007). *Journal of Physics B: Atomic, Molecular and Optical Physics*, *40*, F313.
Kheifets, A. S. (2009). *Journal of Physics B: Atomic, Molecular and Optical Physics*, *42*, 134016.
Kim, Y. S., Pratt, R. H., Ron, A., & Tseng, H. K. (1980). *Physical Review A*, *22*, 567.
Kitajima, M., Okamoto, M., Shimizu, Y., Chiba, H., Fritzsche, S., Kabachnik, N. M., et al. (2001). *Journal of Physics B: Atomic, Molecular and Optical Physics*, *34*, 3829.
Klar, H. (1979). *Journal of Physics B: Atomic and Molecular Physics*, *12*, L409.
Klar, H. (1980a). *Journal of Physics B: Atomic and Molecular Physics*, *13*, 2037.
Klar, H. (1980b). *Journal of Physics B: Atomic and Molecular Physics*, *13*, 4741.
Klar, H., & Kleinpoppen, H. (1982). *Journal of Physics B: Atomic and Molecular Physics*, *15*, 933.
Klar, H., Roy, A. C., Schlemmer, P., Jung, K., & Ehrhardt, H. J. (1987). *Journal of Physics B: Atomic and Molecular Physics*, *20*, 821.
Kleiman, U. (2001). PhD Thesis, University of Münster.

Kleiman, U., Lohmann, B., & Blum, K. (1999a). *Journal of Physics B: Atomic, Molecular and Optical Physics*, *32*, 309.
Kleiman, U., Lohmann, B., & Blum, K. (1999b). *Journal of Physics B: Atomic, Molecular and Optical Physics*, *32*, L219.
Kleiman, U. (2002). *Journal of Physics B: Atomic, Molecular and Optical Physics*, *35*, 947.
Kleiman, U., & Lohmann, B. (2003). *Journal of Electron Spectroscopy and Related Phenomena*, *131–132*, 29.
Kleiman, U., & Becker, U. (2005). *Journal of Electron Spectroscopy and Related Phenomena*, *142*, 45.
Kleinpoppen, H. (1967). Columbia University, New York, unpublished report.
Kleinpoppen, H. (1988). In H. O. Lutz, J. S. Briggs, & H. Kleinpoppen (Eds.), *Fundamental processes in energetic atomic collisions* (p. 393). Plenum Press: New York.
Kleinpoppen, H., & McGregor, I. (1980). In H. Kleinpoppen & J. F. Williams (Eds.), *Coherence and correlations in atomic collisions* (p. 109). Plenum Press: New York.
Kleinpoppen, H., Lohmann, B., Grum-Grzhimailo, A., & Becker, U. (2005). In H. H. Stroke (Ed.), *Advances in atomic, molecular, and optical physics* (Vol. 51, p. 471). Elsevier Academic Press: New York.
Kleinpoppen, H., & Raible, V. (1965). *Physics Letters*, *18*, 24.
Kleinpoppen, H. (1971). *Physical Review A*, *3*, 2015.
Kleinpoppen, H. (1976). *Comments on Modern Physics D*, *2*, 35.
Kleinpoppen, H. (1980). *Comments on Atomic and Molecular Physics*, *9*, 201.
Kleinpoppen, H., & Williams, J. F. (Eds.). (1980). *Coherence and correlations in atomic collisions*. New York: Plenum Press.
Kohmoto, M., & Fano, U. (1981). *Journal of Physics B: Atomic and Molecular Physics*, *14*, L447.
Kollath, R. (1949). *Physikalische Blätter*. V. Jhrg., *66*.
Koschmieder, H., & Raible, V. (1975). *Review of Scientific Instruments*, *46*, 536.
Kostroun, V. O., Chen, M. H., & Crasemann, B. (1971). *Physical Review A*, *3*, 533.
Krässig, B., Jung, M., Gemmel, D. S., Kanter, E. P., LeBrun, T., Southworth, S. H., et al. (1995). *Physical Review Letters*, *75*, 4736.
Krässig, B., Bilheux, J. C., Dunford, R. W., Gemmel, D. S., Hasegawa, S., Kanter, E. P., et al. (2003). *Physical Review A*, *67*, 022707.
Krause, M. O., Vestal, M. L., Johnston, W. H., & Carlson, T. A. (1964). *Physical Review A*, *133*, 385.
Krause, M. O., & Carlson, T. A. (1966). *Physical Review*, *149*, 52.
Krause, M. O. (1969). *Physical Review*, *177*, 151.
Krause, M. O., Carlson, T. A., & Woodruff, P. R. (1981). *Physical Review A*, *24*, 1374.
Kronast, W., Huster, R., & Mehlhorn, W. (1984). *Journal of Physics B: Atomic and Molecular Physics*, *17*, L51.
Kronast, W., Huster, R., & Mehlhorn, W. (1986). *Zeitschrift für Physik D*, *2*, 285.
Kuchiev, M. Yu., & Sheinerman, S. A. (1989). *Soviet Physics Uspekhi*, *32*, 569.
Kupliauskiene, A., Rakštikas, N., & Tutlys, V. (2001). *Journal of Physics B: Atomic, Molecular and Optical Physics*, *34*, 1783.
Kupliauskiene, A., & Tutlys, V. (2003). *Physica Scripta*, *67*, 290.
Kurka, M., Rudenko, A., Foucar, L., Kühnel, K. U., Jiang, Y. H., Ergler, Th., et al. (2009). *Journal of Physics B: Atomic, Molecular and Optical Physics*, *42*, 141002.
Kutzner, M. (1996). In U. Becker & D. A. Shirley (Eds.), *VUV and soft X-ray photoionization* (p. 47). Plenum Press: New York.
Kutzner, M., Shamblin, Q., Vance, S. E., & Winn, D. (1997). *Physical Review A*, *55*, 248.
Labzowsky, L., Plunien, G., & Shabaev, V. (2006). *Physica Scripta*, *73*, C119.
Lacoursière, J., Meyer, M., Nahon, L., Morin, P., & Larzillière, M. (1994). *Nuclear Instruments and Methods A*, *351*, 545.
Lafosse, A., Lebech, M., Brenot, J. C., Guyon, P. M., Jagutzki, O., Spielberger, L., et al. (2000). *Physical Review Letters*, *84*, 5987.

Lagutin, B. M., Petrov, I. D., Demekhin, Ph. V., Sukhorukov, V. L., Vollweiler, F., Liebel, H., et al. (2000). *Journal of Physics B: Atomic, Molecular and Optical Physics*, *33*, 1337.

Lagutin, B. M., Petrov, I. D., Sukhorukov, V. L., Demekhin, P. V., Zimmermann, B., Mickat, S., et al. (2003). *Journal of Physics B: Atomic, Molecular and Optical Physics*, *36*, 3251.

Lahmam-Bennani, A., & Lohmann, B. (Eds.). (2005). *AIP Conference Proceedings* (Vol. 811). Melville, New York.

Langer, B., Berrah, N., Farhat, A., Hemmers, O., & Bozek, J. D. (1996). *Physical Review A*, *53*, R1946.

Langer, B., Berrah, N., Farhat, A., Humphrey, M., Cubayanes, D., Menzel, A., et al. (1997). *Journal of Physics B: Atomic, Molecular and Optical Physics*, *30*, 4255.

Larkins, F. P. (1972). In R. W. Fink (Ed.), *Inner shell ionization phenomena* (p. 1543). Oak Ridge.

Leahy, D. J., Reid, K. L., & Zare, R. N. (1991). *The Journal of Chemical Physics*, *95*, 1757.

Leahy, D. J., Reid, K. L., Park, H., & Zare, R. N. (1992). *The Journal of Chemical Physics*, *97*, 4948.

Lebech, M., Houver, J. C., Lafosse, A., Dowek, D., Alcaraz, C., Nahon, L., et al. (2003). *The Journal of Chemical Physics*, *118*, 9653.

Lee, C. M. (1974). *Physical Review A*, *10*, 1598.

Lehmann, J., Bonhoff, K., Bonhoff, S., Lohmann, B., & Blum, K. (1997). In J. L. Duggan, & I. L. Morgan (Eds.), *Application of accelerators in research and industry. AIP Conference Proceedings* (Vol. 392, p. 63). New York.

Lehmann, J., & Blum, K. (1997). *Journal of Physics B: Atomic, Molecular and Optical Physics*, *30*, 633.

Lenard, P. (1903). *Annals of Physics*, *12*, 714.

L'Huiller, A., Lampe, L. A., Mainfroy, G., & Manus, G. (1983). *Physical Review A*, *27*, 2503.

Lindle, D. W., Ferrett, T. A., Becker, U., Kobrin, P. H., Truesdale, C. M., Kerkhoff, H. G., et al. (1985). *Physical Review A*, *31*, 714.

Lindle, D. W., Heimann, P. A., Ferrett, T. A., Kobrin, P. H., Truesdale, C. M., Becker, U., et al. (1986). *Physical Review A*, *33*, 319.

Lindle, D. W., Medhurst, L. J., Ferrett, T. A., Heimann, M. N., Piancastelli, M. N., Liu, S. H., et al. (1988). *Physical Review A*, *38*, 2371.

Lindle, D. W., & Hemmers, O. (1999). *Journal of Electron Spectroscopy and Related Phenomena*, *100*, 297.

Listengarten, M. A. (1962). Izvest. Akad. Nauk. S.S.S.R., Ser. Fiz. *24*, 182.

Lockwood, G. J., & Everhart, E. (1962). *Physical Review*, *125*, 567.

Lohmann, B. (1988). PhD Thesis, University of Münster, Germany.

Lohmann, B. (1998). *Angle and spin resolved Auger processes on free atoms and molecules. Habilitation thesis*. University of Münster: Germany.

Lohmann, B., & Kleiman, U. (2001). In J. Berakdar, & J. Kirschner (Eds.), *Many particle spectroscopy of atoms, molecules, clusters, and surfaces* (p. 173). Kluwer/Plenum: New York.

Lohmann, B., Langer, B., Snell, G., Kleiman, U., Canton, S., Martins, M., et al. (2003b). In G. F. Hanne, L. Malegat, & H. Schmidt-Böcking (Eds.), *AIP Conference Proceedings* (Vol. 697, p. 133). Melville, New York.

Lohmann, B., Langer, B., Snell, G., Kleiman, U., Canton, S., Martins, M., et al. (2005). *Physical Review A, 71*, 020701(R).

Lohmann, B., Srivastava, R., Kleiman, U., & Blum, K. (2002). In D. H. Madison, & M. Schulz (Eds.), *AIP Conference Proceedings* (Vol. 604, p. 229). Melville, New York.

Lohmann, B., Zimmermann, B., Kleinpoppen, H., & Becker, U. (2003a). In B. Bederson & H. Walther (Eds.), *Advances in atomic, molecular, and optical physics* (Vol. 49, p. 217). Elsevier Academic Press: New York.

Lohmann, Birgit, & Weigold, E. (1981). *Physics Letters A*, *86*, 139.

Lohmann, B. (1984). *Diplom thesis*. Germany: University of Münster.

Lohmann, B. (1990). *Journal of Physics B: Atomic, Molecular and Optical Physics*, *23*, 3147.

Lohmann, B., & Fritzsche, S. (1994). *Journal of Physics B: Atomic, Molecular and Optical Physics, 27*, 2919.

Lohmann, B., Fritzsche, S., & Larkins, F. P. (1996). *Journal of Physics B: Atomic, Molecular and Optical Physics, 29*, 1191.

Lohmann, B. (1999a). *Journal of Physics B: Atomic, Molecular and Optical Physics, 32*, L643.

Lohmann, B. (1999b). *Australian Journal of Physics, 52*, 397.

Lohmann, B., & Kleiman, U. (2006). *Journal of Physics B: Atomic, Molecular and Optical Physics, 39*, 271.

Lohmann, B. (2008). *Angle and spin resolved Auger emission: Theory and applications to atoms and molecules*. Berlin: Springer.

Lörch, H., Bizau, J. M., Scherer, N., Diehl, S., Cubaynes, D., Zerouni, O., et al. (1999). *Journal of Physics B: Atomic, Molecular and Optical Physics, 32*, 2215.

Lower, J., Panajotovic, R., & Weigold, E. (2004). *Physica Scripta, T110*, 216.

Lowry, T. F., Tobouliam, D. H., & Ederer, D. L. (1965). *Physical Review A, 137*, 1054.

Luke, T. M. (1982). *Journal of Physics B: Atomic and Molecular Physics, 15*, L1217.

Lutz, H. O. (1992). In P. J. O. Teubner & E. Weigold (Eds.), *Proceedings International Symposium Correlations and Polarization in Electronic and Atomic Collisions and (e, 2e) reactions. IOP Conference Series* (Vol. 122, p. 335).

Macek, J., & Wang, C. (1986). *Physical Review A, 34*, 1787.

MacGillivray, W. R., & Standage, M. C. (1988). *Physics Reports, 168*, 1.

Madden, R. P., & Codling, K. (1965). *The Astrophysical Journal, 141*, 364.

Madison, D. H., & Schulz, M. (Eds.). (2001). *AIP Conference Proceedings* (Vol. 604). Melville, New York.

Madison, D. H., & Shelton, W. N. (1973). *Physical Review A, 7*, 449.

Madison, D. H., & Winters, K. H. (1981). *Physical Review Letters, 47*, 1885.

Madison, D. H., Bray, I., & McCarthy, I. E. (1991). *Journal of Physics B: Atomic, Molecular and Optical Physics, 24*, 3861.

Malegat, L., Selles, P., Lablanquie, P., Mazeau, J., & Huetz, A. (1997). *Journal of Physics B: Atomic, Molecular and Optical Physics, 30*, 263.

Manakov, N. L., Ovsiannikov, V. D., & Rapoport, L. P. (1986). *Physics Reports, 141*, 319.

Manakov, N. L., Marmo, S. I., & Meremianin, A. V. (1996). *Journal of Physics B: Atomic, Molecular and Optical Physics, 29*, 2711.

Manakov, N. L., Maquet, A., Marmo, S. I., Veniard, V., & Ferrante, G. (1999). *Journal of Physics B: Atomic, Molecular and Optical Physics, 32*, 3747.

Manson, S. T., & Dill D. (1978). In C. R. Brundle & A. D. Baker (Eds.), *Electron spectroscopy. II. Theory, techniques and applications* (p. 157). Academic Press: New York.

Manson, S. T., & Cooper, J. W. (1968). *Physical Review, 165*, 126.

Manson, S. T., & Starace, A. F. (1982). *Reviews of Modern Physics, 54*, 389.

Marr, G. V. (1963). *Proceedings of the Physical Society of London, 81*, 9.

Martin, N. L. S., & de Harak, B. A. (Eds.). (2010). *Journal of Physics: Conference Series, 212*, IOP Publishing: Bristol.

Martin, N. L. S., Thompson, D. B., Bauman, R. P., Caldwell, C. D., Krause, M. O., Frigo, S. P., et al. (1998). *Physical Review Letters, 81*, 1199.

Masiu, S., Shigemasa, E., Yagishita, A., & Sellin, I. A. (1995). *Journal of Physics B: Atomic, Molecular and Optical Physics, 28*, 4529.

Massey, H. S. W., & Mohr, C. B. O. (1941). *Proceedings of the Royal Society, A177*, 341.

Massey, H. S. W., & Burhop, E. H. S. (1969). *Electronic and ionic impact phenomena* (Vol. 1). Oxford: Clarendon Press.

Mayer-Kuckuk, T. (1977). *Atomphysik*. Stuttgart: Teubner.

McClelland, J. J., Kelley, M. H., & Cellotta, R. J. (1987). *Physical Review Letters, 58*, 2198.

McEachran, R. P., & Stauffer, A. D. (1984). *Journal of Physics B: Atomic and Molecular Physics, 17*, 2507.

McEachran, R. P., & Stauffer, A. D. (1986). *Journal of Physics B: Atomic and Molecular Physics, 19*, 3523.
McEachran, R. P., & Stauffer, A. D. (1987). *Journal of Physics B: Atomic and Molecular Physics, 20*, 3483.
McFarlane, C. S. (1975). *Journal of Physics B: Atomic and Molecular Physics, 8*, 895.
McGuire, E. J. (1970). *Physical Review A, 2*, 273.
McLaughlin, K. W., Yenen, O., Jaecks, D. H., Gay, T. J., Calabrese, D., & Thaden-Jordan, B. (2002). *Physical Review Letters, 88*, 123003.
Mehlhorn, W. (1983). In H. O. Lutz, J. Briggs, & H. Kleinpoppen (Eds.), *Fundamental processes in energetic atomic collisions* (p. 253). Plenum Press: New York.
Mehlhorn, W. (1990). In T. A. Carlson, M. O. Krause, & S. T. Manson (Eds.), *X-ray and inner shell processes, AIP Conference Proceedings* (Vol. 215, p. 465). New York.
Mehlhorn, W. (2000). In R. W. Dunford, D. S. Gemmell, E. P. Kanter, & B. Krässig (Eds.), *X-ray and inner shell processes, AIP Conference Proceedings* (Vol. 506, p. 33). Melville, New York.
Mehlhorn, W., & Asaad, W. N. (1966). *Zeitschrift für Physik, 191*, 231.
Mehlhorn, W., & Taulbjerg, K. (1980). *Journal of Physics B: Atomic and Molecular Physics, 13*, 445.
Menner, B., Hall, T., Zehnle, E., & Kempter, V. J. (1981). *Journal of Physics B: Atomic and Molecular Physics, 14*, 3693.
Mercouris, T., Komninos, Y., Dionissopoulou, S., & Nicolaides, C. A. (1996). *Journal of Physics B: Atomic, Molecular and Optical Physics, 29*, L13.
Meyer, M., Müller, B., Nunnemann, A., Prescher, Th., v. Raven, E., Richter, M., et al. (1987). *Physical Review Letters, 59*, 2963.
Meyer, M., Pahler, M., Drescher, T., v. Raven, E., Richter, M., Sonntag, B., et al. (1990). *Physica Scripta, T 31*, 28.
Meyer, M., Nahon, L., Lacoursiére, J., Gisselbrecht, M., Morin, P., & Larzilliére, M. (1996). *Nuclear Instruments and Methods A, 351*, 545.
Meyer, M., Marquette, A., Grum-Grzhimailo, A. N., Kleiman, U., & Lohmann, B. (2001). *Physical Review A, 64*, 022703.
Meyer, M., Cubaynes, D., Wuilleumier, F. J., Heinecke, E., Richter, T., Zimmermann, P., et al. (2006). *Journal of Physics B: Atomic, Molecular and Optical Physics, 39*, L153.
Meyer, M., Cubaynes, D., Glijer, D., Dardis, J., Hayden, P., Hough, P., et al. (2008). *Physical Review Letters, 101*, 193002.
Mies, H. (1968). *Physical Review, 175*, 164.
Mikosza, A. G., Hippler, R., Wang, J. B., & Williams, J. F. (1993). *Physical Review Letters, 71*, 235.
Mikosza, A. G., Williams, J. F., & Wang, J. B. (1997). *Physical Review Letters, 79*, 3375.
Miller, D. L., Dow, J. D., Houlgate, R. G., Marr, G. V., & West, J. B. (1977). *Journal of Physics B: Atomic and Molecular Physics, 10*, 3205.
Mitani, T., Okamoto, H., Yoshihiro, T., Watanabe, M., & Fukui, K. (1989). *Review of Scientific Instruments, 60*, 1569.
Mitsuke, K., Hikosaka, Y., & Iwasaki, K. (2000). *Journal of Physics B: Atomic, Molecular and Optical Physics, 33*, 391.
Miyauchi, N., Adachi, J., Yagishita, A., Sako, T., Koike, F., Sato, T., et al. (2011). *Journal of Physics B: Atomic, Molecular and Optical Physics, 44*, 071001.
Morgan, L. A., & McDowell, M. R. C. (1975). *Journal of Physics B: Atomic and Molecular Physics, 8*, 1073.
Morin, P., Adam, M. Y., Nenner, I., Delwiche, J., Hubin-Franskin, M. J., & Lablanquie, P. (1983). *Nuclear Instruments and Methods in Physics Review, 208*, 761.
Moshammer, R., Jiang, Y. H., Foucar, L., Rudenko, A., Ergler, Th., Schröter, C. D., et al. (2007). *Physical Review Letters, 98*, 203001.
Motoki, S., Adachi, J., Ito, K., Ishii, K., Soejima, K., Yagishita, A., et al. (2002). *Physical Review Letters, 88*, 063003.

Mott, N. F. (1929). *Proceedings of the Royal Society, 124*, 425.
Mott, N. F. (1932). *Proceedings of the Royal Society, 135*, 429.
Müller, R., Schulz, J., Wernet, Ph., Godehusen, K., Martins, M., Sonntag, B., et al. (2007). *Physical Review A, 75*, 012718.
Müller, N., Böwering, N., Svensson, A., & Heinzmann, U. (1990). *Journal of Physics B: Atomic, Molecular and Optical Physics, 23*, 2267S.
Müller, N., Khalil, T., Pohl, M., Uphues, T., Polcik, M., Rader, O., et al. (2006). *Physical Review B, 74*, 161401.
Mursu, J., Aksela, H., Sairanen, O.-P., Kivimäki, A., Nõmmiste, E., Ausmees, A., et al. (1996). *Journal of Physics B: Atomic, Molecular and Optical Physics, 29*, 4387.
Nefedov, V. I., Yarzhemsky, V. G., Nefedova, I. S., & Trzhaskovskaya, M. B. (2000). *Journal of Electron Spectroscopy and Related Phenomena, 113*, 91.
Neudatchin, V. G., Popov, Yu. V., & Smirnov, Yu., F. (1999). Uspekhi Fiz. Nauk *169*, 1111; and English version. *Physics-Uspekhi, 42*, 1017.
Newsom, G. H. (1971). *Astrophysical Journal, 166*, 243.
Oed, A. (1971). *Physics Letters, 34A*, 435.
O'Keeffe, P., Aloïse, S., Meyer, M., & Grum-Grzhimailo, A. N. (2003). *Physical Review Letters, 90*, 023002.
O'Keeffe, P., Aloïse, S., Fritzsche, S., Lohmann, B., Kleiman, U., Meyer, M., et al. (2004). *Physical Review A, 70*, 012705.
O'Keeffe, P., & Bolognesi, P. (2010). *Physical Review A, 82*, 052522.
O'Neil, R. W., van der Burgt, P. J. M., Dziczek, D., Bowe, P., Chwirot, S., & Slevin, J. A. (1998). *Physical Review Letters, 80*, 1630.
Oppenheimer, J. R. (1927a). *Proceedings of the National Academy of Sciences, 13*, 800.
Oppenheimer, J. R. (1927b). *Zeitschrift für Physik, 43*, 27.
Oppenheimer, J. R. (1928). *Physical Review, 32*, 361.
Osimitsch, S. (1989). PhD Thesis, University of Bielefeld, Germany.
Osimitsch, S., Jitschin, W., Reihl, H., Kleinpoppen, H., Lutz, H. O., Mo, O., et al. (1989). *Physical Review A, 40*, 2958.
Ottley, T. W., & Kleinpoppen, H. (1975). *Journal of Physics B: Atomic and Molecular Physics, 8*, 621.
Pahler, M., Lorenz, C., von Raven, E., Rüder, P., Sonntag, B., Baier, S., et al. (1992). *Physical Review Letters, 68*, 2285.
Palenius, H. P., Kohl, J. L., & Larkinsen, W. H. (1976). *Physical Review A, 13*, 1805.
Parpia, F. A., & Froese Fischer, C. (1996). *Computer Physics Communications, 94*, 249.
Pauly, H. J. (1988). In G. Scoles (Ed.), *Atomic and molecular beam methods* (Vol. 1, p. 124). Oxford University Press: Oxford.
Pavlovic, Z. M., Boness, J. W., Herzenberg, A., & Schulz, G. J. (1972). *Physical Review A, 6*, 676.
Penney, W. G. (1932a). *Proceedings of the National Academy of Sciences, 18*, 231.
Penney, W. G. (1932b). *Physical Review, 39*, 467.
Percival, I. C., & Seaton, M. J. (1958). *Philosophical Transactions of the Royal Society A, 251*, 113.
Peshkin, M. (1970). *Advances in Chemical Physics, 18*, 1.
Pieper, W., & Greiner, W. (1969). *Zeitschrift für Physik, 218*, 327.
da Pieve, F., di Matteo, S., Sébilleau, D., Gunnella, R., Stefani, G., & Natoli, C. R. (2007a). *Physical Review A, 75*, 052704.
da Pieve, F., Fritzsche, S., Stefani, G., & Kabachnik, N. M. (2007b). *Journal of Physics B: Atomic, Molecular and Optical Physics, 40*, 329.
Pizzoferrato, R., Casalboni, M., Francini, R., Grassano, U. M., Antonageli, F., Piacentini, M., et al. (1986). *Europhysics Letters, 2*, 571.
Plotzke, O., Prümper, G., Zimmermann, B., Becker, U., & Kleinpoppen, H. (1996). *Physical Review Letters, 77*, 2642.
Prümper, G. (1998). PhD Thesis, Technical University Berlin, Germany.

Prümper, G., Zimmermann, B., Becker, U., & Kleinpoppen, H. (2001a). In U. Becker & A. Crowe (Eds.), *Complete scattering experiments* (p. 141). Kluwer Academic/Plenum Publishers: New York.

Prümper, G., Zimmermann, B., Plotzke, O., Becker, U., & Kleinpoppen, H. (1997). *Europhysics Letters*, *38*, 19.

Prümper, G., Zimmermann, B., Langer, B., Viefhaus, J., Hentges, R., Cherepkov, N. A., et al. (2000). *Physical Review Letters*, *85*, 5074.

Prümper, G., Geßner, O., Zimmermann, B., Viefhaus, J., Hentges, R., Kleinpoppen, H., et al. (2001b). *Journal of Physics B: Atomic, Molecular and Optical Physics*, *34*, 2707.

Raith, W. (1988). In J. S. Briggs, H. Kleinpoppen, & H. O. Lutz (Eds.), *Fundamental processes of atomic dynamics* (p. 229). Plenum Press: New York.

Ramsauer, C. (1921). *Annalen der Physik*, *64*, 513.

von Raven, E., Meyer, M., Pahler, M., & Sonntag, B. (1990). *Journal of Electron Spectroscopy and Related Phenomena*, *52*, 677.

Reich, H. (1987). *Diplom thesis*. Heidelberg: Max-Planck-Institut für Kernphysik.

Reid, K. L., Leahy, D. J., & Zare, R. N. (1992). *Physical Review Letters*, *68*, 3527.

Reid, K. L. (2004). *Annual Review of Physical Chemistry*, *54*, 397.

Reihl, H. (1993). PhD Thesis, University of Bielefeld, Germany.

Ricz, S., Sankari, R., Kövér, Á., Jurvansuu, M., Varga, D., Nikkinen, J., et al. (2003). *Physical Review A*, *67*, 012712.

Rioual, S., Rouvellou, B., Avaldi, L., Battera, G., Camilloni, R., Stefani, G., et al. (2001). *Physical Review Letters*, *86*, 1470.

Ritchie, B. (1976). *Physical Review A*, *14*, 359.

Rolles, D., Prümper, G., Fukuzawa, H., & Liu, X. J. (2008). *Physical Review Letters*, *101*, 263002.

Rosenthal, H., & Foley, H. M. (1969). *Physical Review Letters*, *23*, 1480.

Rotenberg, M., Bivins, R., & Metropolis, N. (1959). *The 3-j and 6-j symbols*. Cambridge: Technology Press.

Rouvellou, B., Rioual, S., Avaldi, L., Camilloni, R., Stefani, G., & Turri, G. (2003). *Physical Review A*, *67*, 012706.

Rouzée, A., Johnsson, P., Gryzlova, E. V., Fukuzawa, H., Yamada, A., Siu, W., et al. (2011). *Physical Review A*, *83*, 031401(R).

Rubenstein, R. A. (1955). PhD Thesis, University of Illinois.

Sackmann, S., Lutz, H. O., & Briggs, J. (1974). *Physical Review Letters*, *12*, 805.

Samson, J. A. R. (1982). In W. Mehlhorn (Ed.), *Handbuch der Physik* (Vol. 31, p. 12). Springer: Berlin.

Samson, J. A. R. (1964). *The Journal of the Optical Society of America*, *54*, 876.

Samson, J. A. R., & Gardner, J. L. (1974). *Physical Review Letters*, *33*, 671.

Samson, J. A. R., & Starace, A. F. (1975). *Journal of Physics B: Atomic and Molecular Physics*, *8*, 1806.

Sanche, L., & Burrows, P. D. (1972). *Physical Review Letters*, *29*, 1639.

Santra, S., Mitra, D., Sarkar, M., & Bhattacharaya, D. (2007). *Physical Review A*, *75*, 022901.

Sarkadi, L., Vajnai, T., Palinkas, J., Kover, A., Vegh, J., & Mukoyama, T. (1990). *Journal of Physics B: Atomic, Molecular and Optical Physics*, *23*, 3643.

Schaphorst, S. J., Krause, M. O., Caldwell, C. D., Saha, H. P., Pahler, M., & Jiménez-Mier, J. (1995). *Physical Review A*, *52*, 4657.

Schaphorst, S. J., Qian, Q., Krässig, B., van Kampen, P., Scherer, N., & Schmidt, V. (1997). *Journal of Physics B: Atomic, Molecular and Optical Physics*, *30*, 4003.

Schartner, K. H., Schill, R. H., Hassekamp, D., Mickat, S., Kammer, S., Werner, L., et al. (2005). *Journal of Physics B: Atomic, Molecular and Optical Physics*, *38*, 4155.

Schattke, W., & Van Hove, M. A. (Eds.). (2003). *Solid-state photoemission and related methods: Theory and experiment*. Weinheim: WILEY-VCH GmbH & Co. KGaA.

Scherer, N., Lörch, H., Kerkau, T., & Schmidt, V. (2004). *Journal of Physics B: Atomic, Molecular and Optical Physics*, *37*, L121.

Schimmelpfennig, B., Nestmann, B., & Peyerimhoff, S. D. (1995). *Journal of Electron Spectroscopy and Related Phenomena, 74*, 173.

Schimmelpfennig, B., & Peyerimhoff, S. D. (1996). *Chemical Physics Letters, 253*, 377.

Schmidt, V. (1990). In T. A. Carlson, M. O. Krause, & S. T. Manson (Eds.), *X-Ray and inner-shell processes. AIP Conference Proceedings* (Vol. 215, p. 559). AIP: New York.

Schmidt, V. (1973). *Physics Letters A, 45*, 63.

Schmidt, V. (1992). *Reports on Progress in Physics, 55*, 1483.

Schmidt, V. (1997). *Electron spectrometry of atoms using synchrotron radiation.* Cambridge: Cambridge University Press.

Schmidt-Böcking, H., Dörner, R., & Ullrich, J. (2002). *Europhysics News, 33*(6), 210.

Schmidtke, B., Drescher, M., Cherepkov, N. A., & Heinzmann, U. (2000a). *Journal of Physics B: Atomic, Molecular and Optical Physics, 33*, 2451.

Schmidtke, B., Khalil, T., Drescher, M., Müller, N., Kabachnik, N. M., & Heinzmann, U. (2000b). *Journal of Physics B: Atomic, Molecular and Optical Physics, 33*, 5225.

Schmidtke, B., Khalil, T., Drescher, M., Müller, N., Kabachnik, N. M., & Heinzmann, U. (2001). *Journal of Physics B: Atomic, Molecular and Optical Physics, 34*, 4293.

Schohl, S., Klar, D., Cherepkov, N. A., Petrov, I. D., Ueda, K., Baier, S., et al. (1997). *Journal of Physics B: Atomic, Molecular and Optical Physics, 30*, 609.

Scholz, T. T., Walters, H. R. J., Burke, P. G., & Scott, M. P. (1991). *Journal of Physics B: Atomic, Molecular and Optical Physics, 24*, 2097.

Schönhense, G., Schäfers, F., Heinzmann, U., & Kessler, J. (1982). *Zeitschrift für Physik A, 304*, 31.

Schönhense, G. (1981). *Journal of Physics B: Atomic and Molecular Physics, 14*, L187.

Schönhense, G., & Heinzmann, U. (1984). *Physical Review A, 29*, 987.

Schulz, G. J. (1963). *Physical Review Letters, 10*, 103.

Schulz, J., Tchaplyguine, M., Rander, T., Bergersen, H., Lindblad, A., Öhrwall, G., et al. (2005). *Physical Review A, 72*, 032718.

Schulz, J., Määttä, M., Heinäsmäki, S., Huttula, M., Sankari, R., Kukk, E., et al. (2006). *Physical Review A, 73*, 062721.

Scofield, J. H. (1976). *Physical Review A, 14*, 1418.

Seabra, G. M., Kaplan, I. G., & Ortiz, J. V. (2005). *The Journal of Chemical Physics, 123*, 114105.

Senftleben, A., Al-Hagan, O., Pflüger, T., Ren, X., Madison, D., Dorn, A., et al. (2010). *The Journal of Chemical Physics, 133*, 044302.

Shakeshaft, A. (1978). *Physical Review, 18*, 1930.

Shanker, R., Wille, U., Bilau, R., Hippler, R., McMurray, W. R., & Lutz, H. O. (1984). *Journal of Physics B: Atomic and Molecular Physics, 17*, 1353.

Shannon, S. P., & Codling, K. (1978). *Journal of Physics B: Atomic and Molecular Physics, 11*, 1193.

Shanti, N., Deshmukh, P. C., & Manson, S. T. (1988). *Physical Review A, 37*, 4720.

Sheinerman, S. A., & Schmidt, V. (1997). *Journal of Physics B: Atomic, Molecular and Optical Physics, 30*, 1677.

Shigemasa, E., Adachi, J., Oura, M., & Yagishita, A. (1995). *Physical Review Letters, 74*, 359.

Shigemasa, E., Adachi, J., Soejima, K., Watanabe, N., Yagishita, A., & Cherepkov, N. A. (1998). *Physical Review Letters, 80*, 1622.

Shull, C. G., Chase, C. T., & Myers, F. E. (1943). *Physical Review, 63*, 29.

Siegbahn, K., Nordling, C., Fahlman, A., Nordberg, R., Hamrin, K., Hedman, J., et al. (1967). *ESCA—atomic, molecular and solid state structure studied by means of electron spectroscopy. Nova acta Regiae Societatis Scientiarum Upsaliensis IV* (Vol. 20).

Siegel, A., Ganz, J., Bussert, W., & Hotop, H. (1983). *Journal of Physics B: Atomic and Molecular Physics, 16*, 2945.

Silim, H. A. (1985). PhD Thesis, Mansoure University, Egypt.

Silim, H. A., Beyer, H. J., El-Sheikh, A., & Kleinpoppen, H. (1987). *Physical Review, 35*, 4454.

Siska, P. E., Parson, J. M., Schafer, T. P., & Lee, Y. T. (1979). In H. S. W. Massey, *Atomic and molecular collisions*. Taylor and Francis: New York.
Snell, Drescher, M., Müller, N., Heinzmann, U., Hergenhahn, U., & Becker, U., (1999b). *Journal of Physics B: Atomic, Molecular and Optical Physics*, *32*, 2361.
Snell, G., Drescher, M., Müller, N., Heinzmann, U., Hergenhahn, U., & Viefhaus, J., et al. (1996). *Physical Review Letters*, *76*, 3923.
Snell, G., Langer, B., Drescher, M., Müller, N., Zimmermann, B., Hergenhahn, U., et al. (1999a). *Physical Review Letters*, *82*, 2480.
Snell, G., Viefhaus, J., Dunning, F. B., & Berrah, N. (2000a). *Review of Scientific Instruments*, *71*, 2608.
Snell, G., Kukk, E., Langer, B., & Berrah, N. (2000b). *Physical Review A*, *61*, 042709.
Snell, G., Hergenhahn, U., Müller, N., Drescher, M., Viefhaus, J., Becker, U., et al. (2001). *Physical Review A*, *63*, 032712.
Snell, G., Langer, B., Young, A. T., & Berrah, N. (2002). *Physical Review A*, *66*, 022701.
Sonntag, B., & Zimmermann, P. (1992). *Reports on Progress in Physics*, *55*, 911.
Sonntag, B., & Zimmermann, P. (1995). *Physikalische Blätter*, *51*, 279.
Sorokin, A. A., Bobashev, S. V., Feigl, T., Tiedtke, K., Wabnitz, H., & Richter, M. (2007). *Physical Review Letters*, *99*, 213002.
Southworth, S. H., Kobrin, P. H., Truesdale, C. M., Lindle, D. W., Owaki, S., & Shirley, D. A. (1981). *Physical Review A*, *24*, 2257.
Southworth, S. H., Becker, U., Truesdale, C. M., Kobrin, P. H., Lindle, D. W., Owaki, S., et al. (1983). *Physical Review A*, *28*, 261.
Southworth, S. H., Parr, A. C., Hardis, J. E., Dehmer, J. L., & Holland, D. M. P. (1986). *Nuclear Instruments and Methods in Physics Research A*, *246*, 782.
Southworth, S., Dunford, R., Kanter, E., Krassig, B., Young, L., LaJohn, L., et al. (2006). *Radiation Physics and Chemistry*, *75*, 1574.
Spicher, G., Olsson, B., Raith, W., Sinepius, G., & Sperber, W. (1990). *Physical Review Letters*, *64*, 1019.
Standage, M. C., & Kleinpoppen, H. (1975). *Physical Review Letters*, *36*, 577.
Starace, A. F. (1982). In W. Mehlhorn (Ed.), *Handbuch der Physik* (Vol. 31, p. 1). Springer: Berlin.
Starace, A. F. (1977). *Physical Review A*, *16*, 231.
Starke, K. (2000). *Magnetic dichroism in core-lavel photoemission*. Berlin: Springer.
Steidl, H., Reichert, E., & Deichsel, H. (1965). *Physics Letters*, *17*, 31.
Stewart, A. L., & Webb, T. G. (1963). *Proceedings of the Physical Society*, *86*, 532.
Suits, A. G., & Vasyutinskii, O. S. (2008). *Chemical Reviews*, *108*, 3706.
Takahashi, M., Cave, J. P., & Eland, J. H. D. (2000). *Review of Scientific Instruments*, *71*, 1337.
Taouil, L., Duguet, A., Lahman-Bennani, A., Lohmann, B., Rasch, J., Whelan, C. T., et al. (1999). *Journal of Physics B: Atomic, Molecular and Optical Physics*, *32*, L5.
Tauro, S., & Liu, K. (2008). *Journal of Physics B: Atomic, Molecular and Optical Physics*, *41*, 225001.
Taylor, A. J., & Burke, P. G. (1967). *Proceedings of the Physical Society*, *92*, 336.
Teramoto, T., Adachi, J., Hosaka, K., Yamazaki, M., Yamanouchi, K., Cherepkov, N. A., et al. (2007). *Journal of Physics B: Atomic, Molecular and Optical Physics*, *40*, F241.
Toffoli, O., & Decleva, P. (2006). *Journal of Physics B: Atomic, Molecular and Optical Physics*, *39*, 2681.
Toffoli, O., & Decleva, P. (2008a). *The Journal of Chemical Physics*, *128*, 234101.
Toffoli, O., & Decleva, P. (2008b). *Physical Review A*, *78*, 063402.
Tolk, N. H., Tulley, J. C., White, C. W., Krauss, A. A., Simms, D. L., & Robbins, M. F. (1967). *Physical Review A*, *13*, 969.
Tonuma, T., Yagishita, A., Shibita, H., Koizumi, T., Matsuo, T., Shima, K., et al. (1987). *Journal of Physics B: Atomic and Molecular Physics*, *20*, L 31.
Torop, L., Morton, J., & West, J. B. (1976). *Journal of Physics B: Atomic and Molecular Physics*, *9*, 2035.

Townsend, J. S., & Bailey, V. A. (1922). *Philosophical Magazine, 43*, 593.
Trzhaskovskaya, M. B., Nefedov, V. I., & Yarzhemsky, V. G. (2001). *Atomic and Nuclear Data Tables, 77*, 97.
Trzhaskovskaya, M. B., Nefedov, V. I., & Yarzhemsky, V. G. (2002). *Atomic and Nuclear Data Tables, 82*, 257.
Trzhaskovskaya, M. B., Nikulin, V. K., Nefedov, V. I., & Yarzhemsky, V. G. (2006). *Atomic and Nuclear Data Tables, 92*, 245.
Tulkki, J., Aksela, S., Aksela, H., Shigemasa, A., & Furusawa, Y. (1992). *Physical Review A, 45*, 4640.
Tulkki, J., Aksela, H., & Kabachnik, N. M. (1993). *Physical Review A, 48*, 2957.
Tully, J. C., Berry, R. S., & Dalton, B. J. (1968). *Physical Review, 176*, 95.
Turri, G., Lohmann, B., Langer, B., Snell, G., Becker, U., & Berrah, N. (2007). *Journal of Physics B: Atomic, Molecular and Optical Physics, 40*, 3453.
Ueda, K., West, J., Ross, K. J., Beyer, H. J., & Kabachnik, N. M. (1998). *Journal of Physics B: Atomic, Molecular and Optical Physics, 31*, 4801.
Ueda, K., Shimizu, Y., Chiba, H., Sato, Y., Kitajima, M., Tanaka, H., et al. (1999). *Physical Review Letters, 83*, 5463.
Ullrich, J., Moshammer, R., Dorn, A., Dörner, R., Schmidt, L. Ph. H., & Schmidt-Böcking, H., (2003). *Reports on Progress in Physics, 66*, 1463.
Ullrich, J., Moshammer, R., & Dörner, R. (1998). *Physikalische Blätter, 54*(2), 140.
Varshalovich, D. A., Moskalev, A. N., & Khersonskii, V. K. (1988). *Quantum theory of angular momentum*. Singapore: World Scientific.
Vegh, L., & Macek, J. H. (1994). *Physical Review A, 50*, 4031.
Verweyen, A., Grum-Grzhimailo, A. N., & Kabachnik, N. M. (1999). *Physical Review A, 60*, 2076.
Viefhaus, J., Cvejanović, S., Langer, B., Lischke, T., Prümper, G., Rolles, D., et al. (2004a). *Physical Review Letters, 92*, 083001.
Viefhaus, J., Grum-Grzhimailo, A. N., Kabachnik, N. M., & Becker, U. (2004b). *Journal of Electron Spectroscopy and Related Phenomena, 141*, 121.
Vollmer, R., Etzkorn, M., Anilkumar, P. S., Ibach, H., & Kirschner, J. (2003). *Physical Review Letters, 91*, 147201.
Vredenborg, A., Lehmann, C. S., Irimia, D., Roeterdink, W. G., & Janssen, M. H. M. (2011). *ChemPhysChem, 12*, 1393.
Wang, Y. D., Callaway, J., & Unnikrishnan, K. (1994). *Physical Review A, 49*, 1854.
Wang, J. B., Mikosza, A. G., Hippler, R., & Williams, J. F. (1995). *Journal of Electron Spectroscopy and Related Phenomena, 74*, 121.
Wang, Z. M., & Elliott, D. S. (2000). *Physical Review A, 62*, 053404.
Watson, W. S., & Stuart, D. S. (1974). *Journal of Physics B: Atomic and Molecular Physics, 7*, L466.
Wedowski, M., Godehusen, K., Weisbarth, F., Zimmermann, P., Dohrmann, Th., von dem Borne, A., et al. (1995). *Journal of Electron Spectroscopy and Related Phenomena, 75*, 61.
Wedowski, M., Godehusen, K., Weisbarth, F., Zimmermann, P., Martins, M., Dohrmann, Th., et al. (1997). *Physical Review A, 55*, 1922.
Weigold, E., & McCarthy, I. E. (1999). *Electron momentum spectroscopy*. New York: Kluwer Academic/Plenum Publishers.
Wentzel, G. (1927). *Zeitschrift fur Physik, 43*, 524.
Wernet, Ph., Schulz, J., Sonntag, B., Godehusen, K., Zimmermann, P., Grum-Grzhimailo, A. N., et al. (2001). *Physical Review A, 64*, 042707.
Wernet, Ph., Schulz, J., Sonntag, B., Godehusen, K., Zimmermann, P., Martins, M., et al. (2000). *Physical Review B, 62*, 14331.
West, J. B., Ueda, K., Kabachnik, N. M., Ross, K. J., Beyer, H. J., & Kleinpoppen, H. (1996). *Physical Review A, 53*, R9.
West, J. B., Ross, K. J., Ueda, K., & Beyer, H. J. (1998). *Journal of Physics B: Atomic, Molecular and Optical Physics, 31*, L647.

Whitfield, S. B., Caldwell, C. D., Huang, D. X., & Krause, M. O. (1992). *Journal of Physics B: Atomic, Molecular and Optical Physics, 25*, 4755.

Wiedenhoeft, M., Canton, S. E., Wills, A. A., Gorczyca, T., Viefhaus, J., Becker, U., et al. (2008). *Journal of Physics B: Atomic, Molecular and Optical Physics, 41*, 095202.

Wille, U., & Hippler, R. (1986). *Physics Reports, 132*, 129.

Williams, J. F. (1975). *Journal of Physics B: Atomic and Molecular Physics, 8*, 1683.

Williams, J. F. (2000). *Comments on Atomic and Molecular Physics, 2*, 165.

Wolcke, A., Bartschat, K., Blum, K., Borgmann, H., Hanne, G. F., & Kessler, J. (1983). *Journal of Physics B: Atomic and Molecular Physics, 16*, 639.

Wolff, B., Rottke, H., Feldmann, D., & Welge, K. H. (1988). *Zeitschrift fur Physik, 10*, 35.

Wood, R. P., Greene, C. H., & Armstrong, D. (1993). *Physical Review A, 47*, 229.

Wuilleumier, F. J., & Krause, M. O. (1974). *Physical Review A, 10*, 242.

Wuilleumier, F. J., & Meyer, M. (2006). *Journal of Physics B: Atomic, Molecular and Optical Physics, 39*, R425.

van Wyngaarden, W. L., & Walters, H. R. J. (1986). *Journal of Physics B: Atomic and Molecular Physics, 19*, 929.

Yalim, H. A., Cvejanovic, D., & Crowe, A. (1999). *Journal of Physics B: Atomic, Molecular and Optical Physics, 32*, 3437.

Yamaoka, H., Oura, M., Takahiro, K., Takeshima, N., Kawatsura, K., Mizumaki, M., et al. (2002). *Physical Review A, 65*, 062713.

Yamaoka, H., Oura, M., Takahiro, K., Morikawa, T., Ito, S., Mizumaki, M., et al. (2003). *Journal of Physics B: Atomic, Molecular and Optical Physics, 36*, 3889.

Yeh, J. J., & Lindau, I. (1985). *Atomic and Nuclear Data Tables, 32*, 1.

Yeh, J. J. (1993). *Atomic calculation of photoionization cross sections and asymmetry parameters*. New York: Gordon and Breach.

Yenen, O., McLaughlin, K. W., Jaecks, D. H., Sant'Anna, M. M., & Seddon, E. A. (2001). *Physical Review Letters, 86*, 979.

Young, L., Kanter, E. P., Krässig, B., Li, Y., March, A. M., Pratt, S. T., et al. (2010). *Nature, 456*, 56.

Zähringer, K., Meyer, H. D., & Cederbaum, L. S. (1992). *Physical Review A, 46*, 5643.

Zare, R. N. (1988). *Angular momentum. Understanding spatial aspects in chemistry and physics*. New York: Wiley-Interscience.

Zatsarinny, O. (1995). *Journal of Physics B: Atomic, Molecular and Optical Physics, 28*, 4759, and private communication.

Zatsarinny, O. (2006). *Computer Physics Communications, 174*, 273.

Zeldovich, Ya., B., & Popov, V. S. (1971). *Uspekhi Fizicheskikh Nauk, 105*, 403 (Trans.) (1972). *Soviet Physics Uspekhi, 14*, 673.

Zheng, Q., Edwards, A. K., Wood, R. M., & Mangan, M. A. (1995). *Physical Review A, 52*, 3940.

Zimmermann, B., Snell, G., Schmidtke, B., Viefhaus, J., Cherepkov, N. A., Langer, B., et al. (2001). *Physical Review A, 64*, 062501.

Index

H. Kleinpoppen et al., *Perfect/Complete Scattering Experiments*,
Springer Series on Atomic, Optical, and Plasma Physics 75,
DOI: 10.1007/978-3-642-40514-3, © Springer-Verlag Berlin Heidelberg 2013

F

L

M

T

MIX
Papier aus verantwortungsvollen Quellen
Paper from responsible sources
FSC® C105338

If you have any concerns about our products,
you can contact us on
ProductSafety@springernature.com

In case Publisher is established outside the EU,
the EU authorized representative is:
Springer Nature Customer Service Center GmbH
Europaplatz 3, 69115 Heidelberg, Germany

Printed by Libri Plureos GmbH
in Hamburg, Germany